Protobiology: Physical Basis of Biology

Author

Koichiro Matsuno, Ph.D.
Professor
Systems Laboratory
Nagaoka University of Technology
Nagaoka, Japan

CRC Press
Taylor & Francis Group
Boca Raton London New York

CRC Press is an imprint of the
Taylor & Francis Group, an **informa** business

First published 1989 by CRC Press
Taylor & Francis Group
6000 Broken Sound Parkway NW, Suite 300
Boca Raton, FL 33487-2742

Reissued 2018 by CRC Press

Library of Congress Cataloging-in-Publication Data

Matsuno, Kōichirō.
Protobiology : physical basis of biology / author, Koichiro Matsuno.
p. cm.
Includes bibliographies and index.
ISBN 0-8493-6403-5
1. Biophysics. 2. Biology--Miscellanea. I. Title.
QH505 .M27 1989
574.19'1--dc19 88-38428

A Library of Congress record exists under LC control number: 88038428

Publisher's Note
The publisher has gone to great lengths to ensure the quality of this reprint but points out that some imperfections in the original co may be apparent.

Disclaimer
The publisher has made every effort to trace copyright holders and welcomes correspondence from those they have been unabl contact.

ISBN 13: 978-1-315-89699-1 (hbk)
ISBN 13: 978-1-351-07609-8 (ebk)

Visit the Taylor & Francis Web site at http://www.taylorandfrancis.com and the
CRC Press Web site at http://www.crcpress.com

PREFACE

Biology, as a distinctive discipline of physical sciences, concerns itself with a very unique, strange, and complex mode of matter. At the same time, biology is no exception in being subject to physics as the most fundamental basis of the material sciences. What makes biology stand out among the material sciences is the way physics is practiced. My principal objective in the present monograph is to seek a way of practicing physics in a manner that can give birth to biology.

It goes without saying that contemporary biology provides a rich field to which physics is successfully applied. A partial catalog of successes includes spectroscopy, molecular dynamics, molecular kinetics, statistical mechanics, energetics, and nonequilibrium thermodynamics, to name only a few. Physics is quite competent in deciphering what is "being" in biology. Nevertheless, these successes do not tell us the whole story about how physics is and should be practiced in biology. The presence of the domain of physics which must be further cultivated becomes evident when one raises the question of how a physical object that one can observe and measure in biology came into existence in the first place. The physics of becoming, in contrast to the physics of being, is necessary in biology. It is this physics of becoming that should be responsible for giving birth to biology in the matrix of physics.

A unique characteristic of physics, or, more specifically, of the physics of being that has been practiced, is its claim for an exhaustive specification of boundary conditions for every degree of freedom under examination, irrespective of whether in experiment or in theory. Uncontrollable boundary conditions are not allowed to intervene according to the rules of the game. For instance, any experiment that goes beyond the control of an experimenter fails to yield testable and dependable measurements. No theoretician allows any physical variables or parameters that defy one's control over results in the attempted theories. This line of thought naturally leads to a tentative conclusion, at least in theory, that the controllability over every degree of freedom would yield a complete fixedness or identifiability of boundary conditions by extending the capability of control to its extreme. A complete fixedness of initial and boundary values for every degree of freedom in motion, when combined together with the equations of motion, unambiguously specifies in complete detail how later development would proceed. There is no room left for anything unexpected to emerge. So far, so good with the physics of being. However, the price we have to pay for the idea of completely fixed boundary conditions is a denial of the physics of becoming.

Consequently, the question arises: are completely fixed boundary conditions acceptable in physics without any dispute, even in theory? The answer I shall try to defend is a "no."

Given, for instance, an interacting many-particle system, a theoretician can set such a time-independent boundary condition that the total energy is conserved at a certain level at any moment. Likewise, every particle in the system recognizes that the total energy is conserved at the level that the theoretician sets. But, there is an important difference between the theoretician looking down and those particles looking up in their ways of recognizing the conservation of energy. Those particles in the system recognize and realize the macroscopic conservation of energy only through their microscopic interaction, and it takes time, due to the undeniable finiteness of the propagation velocity of interaction change in the medium. Every single particle is involved in the synthetic process of realizing the conservation of energy *a posteriori*. The conservation of energy, when viewed from particles as if bottom up, serves as a final cause to drive microscopic interactions. In contrast, the boundary condition which the theoretician sets in a completely identifiable manner imposes a global actualization of the conservation of energy in the medium right at the moment it is set. The theoretician always recognizes the conservation *a priori* as long as the boundary condition is completely identifiable. The conservation of energy, when viewed by the physicist as if top down, is simply a matter of imposing a constraint that serves as a mechanistic cause to drive microscopic interactions.

The conservation of energy, under the completely fixed boundary conditions which the theoretician sees, deprives those particles in the system of the synthetic capability of achieving the very same conservation internally, because the conservation of energy has already been imposed without leaving any room for internal determination. A remedy that saves the endogenous capacity of particles for conservation is an appraisal of those boundary conditions that are only partially fixed as a matter of principle. Completely fixed boundary conditions conceived in theory in a time-independent manner could not survive in reality unless the propagation velocity of the change in the interaction would diverge. Instead, partially fixed boundary conditions can gain currency in physics once one recognizes that any action in physics is propagated at a velocity not exceeding light velocity.

Every electron in an isolated molecule knows or detects that the total energy is conserved, and similarly a theoretician knows the same. But the way of knowing differs. Those electrons recognize it through microscopic interaction communicating the conservation of energy as a process, while the theoretician knows it simply by macroscopically imposing the boundary condition in an unambiguous manner. How can the theoretician appreciate the difference? An appraisal of partially fixed boundary conditions leaving electrons room for internal determination is one way out, though the theoretician asks completely fixed ones for their own sake. Completely fixed boundary conditions upon those electrons are too much to ask for in the physics of becoming.

Physics of becoming is about how and to what extent partially fixed boundary conditions vary with time, in contrast to the physics of being which is under completely fixed boundary conditions. One may expect that changes in partially fixed boundary conditions can be an emergent property.

Protobiology, as a physics of becoming, is an attempt to bridge the gap between the physics of being and chemistry on the one hand, and biology on the other. The physics of being has been competent in reductionistic analysis of the material hierarchy from cells downward to molecules, atoms, nuclei, nucleons, and even further down to quarks. In contrast, a physics of becoming makes its *raison d'être* evident when one moves up the material hierarchy from below. What is unique to the physics of becoming, that has not explicitly been recognized in the physics of being, is the nexus connecting different levels of the material hierarchy. Unless the cross-hierarchical nexus is made explicit, one cannot move up the hierarchy, although one can move down by dissecting the connecting nexus even without noticing it as such. A major problem which protobiology sets for itself is how to move up the hierarchy from atoms and molecules to organelles, cells, and organs.

The cross-hierarchical nexus which I shall emphasize is a wide variety of conservation laws. In so far as one understands conservation laws with the reservation that only partially fixed boundary conditions are externally applicable to these laws, the internal manner of fulfilling conservation laws is ubiquitous, irrespective of the hierarchical levels of matter. Conservation laws presume the material process of measuring and upholding these laws. For instance, material flow continuity as an expression of the conservation of matter induces an endogenous process of fulfilling continuity unless the boundary conditions are completely fixed without leaving any room for internal determination. Because of its functional ubiquity, the endogenous process of fulfilling conservation laws can serve as a candidate for the cross-hierarchical nexus connecting the lower level to the upper. Protobiology is an attempt to approach biology via the physics of becoming that postulates an endogenous capacity of fulfilling conservation laws on the part of matter.

Protobiology as a physics of becoming emphasizes the dynamics underlying conservation laws, whereas the physics of being emphasizes the dynamics presupposing conservation laws. Protobiology thus concerns itself with a convoluted problem of how both the law of motion and its boundary conditions develop with time without forgetting that these two are inseparable, in contrast to the physics of being that assumes separability. An empirically infallible frame of reference that we rely upon are the physical conservation laws. I have repeatedly emphasized the law of motion underlying physical conservation laws in the whole book, since the process entailed by and leading to these conservation laws is argued to be fundamental both in protobiology and in biology.

* * *

I have benefited from helpful comments made by Michael Conrad and Giuseppe Sermonti on several points which are discussed in this monograph. Deborah Conrad read each chapter and gave me insightful comments and remarks. Of course, I have full responsibility for any remaining errors. I am indebted to Toshihiro Suzuki for numerical calculations reported in Chapter 1. I should also like to thank Ingrid Betancourt, Ryuhei Hayashi, Eiichi Imai, Etsuji Oku, and Takayuki Tohma for drawings and photos, and Yukiko Matsuno for assuming the lion's share of the typing.

Koichiro Matsuno

THE AUTHOR

Koichiro Matsuno, Ph.D., is currently Professor of Biophysics at the Nagaoka University of Technology, Nagaoka, Japan.

He graduated in 1963 from the University of Tokyo with a B.Sc. in electrical engineering, and obtained the Ph.D. degree in physics in 1971 from the Massachusetts Institute of Technology, Cambridge. From 1982 to 1984 he was a Visiting Research Professor at the Institute for Molecular and Cellular Evolution at the University of Miami, Coral Gables, Florida.

Dr. Matsuno is a member of the Biophysical Society of Japan, the American Physical Society, and the International Society for the Study of the Origins of Life. He serves as a member of the Editorial Board of the journal *BioSystems* and the Board of Reviewers of the *Mathematical Reviews.*

He is the author of more than 70 publications in scientific and technical journals including review articles and several book chapters and an editor of the book *Molecular Evolution and Protobiology* (Plenum Press, New York, 1984). His current major research interests are in the areas of biological information processing, cell motility, origin and evolution of motility, molecular and cellular evolution, protobiology, and especially the theoretical underpinnings of each of these.

TABLE OF CONTENTS

Chapter 1
WHAT IS PROTOBIOLOGY?
I. Introduction ... 1
II. Boundary Conditions ... 1
III. Conservation as a Process ... 6
IV. Mechanics and Biology ... 11
V. Mechanical Motion in Biology ... 15
VI. Protobiology and Its Problems ... 25
VII. Summary ... 27
References ... 28

Chapter 2
INTERNAL MEASUREMENT
I. Introduction ... 31
II. Measurement in Classical Mechanics ... 31
III. Internal Measurement ... 36
IV. Measurement in Quantum Mechanics ... 41
V. Summary ... 53
References ... 54

Chapter 3
EQUILIBRATION
I. Introduction ... 57
II. Force Equilibration ... 57
III. Material Flow Equilibration ... 62
IV. Equilibration and Nonequilibrium Thermodynamics ... 68
V. Boundary Conditions in Nonequilibrium Thermodynamics ... 72
VI. The Origin of Fluctuations ... 73
VII. Summary ... 74
References ... 75

Chapter 4
INTRINSIC IRREVERSIBILITY
I. Introduction ... 77
II. First and Second Laws of Thermodynamics ... 77
III. Entropy ... 79
IV. Time's Arrow ... 85
V. Summary ... 90
References ... 91

Chapter 5
SYMMETRY-BREAKING
I. Introduction ... 93
II. Production Process ... 93
III. Symmetry-Breaking: Protobiological ... 98
IV. Symmetry-Breaking: Biological ... 101
V. Complexity and Complexity Increase ... 105
VI. Summary ... 106
References ... 107

Chapter 6
INFORMATION

I. Introduction ... 109
II. Anthropomorphic Information ... 109
III. Nonanthropomorphic Information ... 112
IV. Transcription ... 122
V. Indefiniteness ... 124
VI. Summary ... 126
References ... 126

Chapter 7
PROTOREPRODUCTION

I. Introduction ... 129
II. Origins of Protoreproduction ... 129
III. Multiplication of Protocells ... 134
IV. Energy Transduction: ATP Process ... 140
V. Molecular Replication ... 144
VI. The Origin of the Genetic Code ... 150
VII. Summary ... 154
References ... 154

Chapter 8
EQUILIBRATION IN DEVELOPMENT

I. Introduction ... 157
II. Internally Regulative Principle ... 157
III. Perpetual Disequilibrium and Equilibration ... 163
IV. Genotype-Phenotype Interdependence ... 167
V. Developmental Symmetry-Breaking ... 168
VI. Cleavage Process ... 169
VII. Growth ... 172
VIII. Synchronic and Diachronic Processes ... 173
IX. Summary ... 173
References ... 174

Chapter 9
EQUILIBRATION IN EVOLUTION

I. Introduction ... 177
II. Reproductive Fidelity ... 177
III. Natural Selection ... 182
IV. Variability ... 187
V. Adaptability ... 194
VI. Summary ... 198
References ... 199

Chapter 10
EQUILIBRATION IN ECOLOGY

I. Introduction ... 201
II. Ecological Niche ... 201
III. Ecological Adaptability ... 204
IV. Ecological Evolution ... 207

V. Summary 208
References..209

Chapter 11
EPILOGUE 211

Index 217

Chapter 1

WHAT IS PROTOBIOLOGY?

I. INTRODUCTION

Protobiology as a physics of becoming that can give birth to biology differs from the physics of being that emphasizes the preservation of invariance and symmetry and rejects emergent properties. The difference is within the manner of how boundary conditions could be materialized in reality and how they could be conceived in theory. Boundary conditions in physics are double-edged. If boundary conditions are fixed so completely and so rigidly that the motion of any degree of freedom turns out to be determinate with the help of a well-defined law of motion, there will be no room for anything unexpected to emerge. Physics of being is guaranteed at the expense of physics of becoming. On the other hand, if boundary conditions are only loosely and partially fixed externally, there remains room for internal determination that is not totally controlled from the outside. Physics of becoming can be saved at the expense of physics of being. Protobiology appreciates partially fixed boundary conditions instead of completely fixed ones. Biological motility provides empirical evidence of the fact that partially fixed boundary conditions are legitimate in the biological realm of the physics of becoming.

II. BOUNDARY CONDITIONS

There are two fundamental ingredients in physics. One is the law of motion, as embodied in Newton's equations of motion in classical mechanics and in Schrödinger's equation of motion in nonrelativistic quantum mechanics. The other is the set of boundary conditions that supplement the law of motion. This dichotomy has been well established.[1] In particular, no argument has existed about the primary importance of the law of motion and its demonstrated validity in classical and in quantum mechanics.[2] However, the boundary condition is not a derivative of the law of motion.[3] It must be reasoned and supported independently of the law of motion.[4]

Newton's equations of motion for celestial mechanics, a prototype of all possible forms of mechanics, take it for granted that every degree of freedom involved, or every variable appearing in the equations of motion, assumes an unambiguous initial value. It is an expression of the boundary conditions set at an arbitrary initial time. This complete fixedness of the boundary conditions is a consequence of employing the equations of motion that do not allow any indefiniteness to intervene. The equations of motion are a scheme of connecting the value of every variable at an earlier time to the later one uniquely and unambiguously. The law of motion embodied in Newton's equations of motion serves as a law of motion of one-to-one mapping type in the sense that the trajectory of every variable is uniquely determined irrespective of whether one proceeds in the forward or in the backward direction in time.

Consequently, although the law of motion and the boundary conditions should be argued independently, as we have already noticed, a law of motion of the one-to-one mapping type imposes a special constraint upon the boundary conditions to which it is subject. The constraint originates in the very nature of one-to-one mapping. Complete fixedness of the boundary conditions is the constraint. The present complete fixedness is a matter of principle and has nothing to do with the technical feasibility of identification in practice so long as one sticks to a law of motion of a one-to-one mapping type. Even if the total number of particles is of the order of Avogadro's number, a law of motion of one-to-one mapping type requires that the boundary conditions for every particle are completely fixable in relation to all of the others without leaving any indefiniteness.

The law of motion of one-to-one mapping type is not restricted to Newton's equations of motion in classical mechanics. Schrödinger's equation of motion in nonrelativistic quantum mechanics also uniquely determines the development of the wave function if its field value is completely known at an intial time along with a complete knowledge about how the wave function is bounded in space. In fact, the later wave function uniquely transforms itself into the earlier one following Schrödinger's equation of motion in the reversed direction as far as the boundary conditions are completely fixed. It is, however, important to notice that what the deterministic one-to-one mapping is applied to is the wave function which is not directly observable, instead of something to be measured directly.

The law of motion of one-to-one mapping type, whether in classical or in quantum mechanics, properly functions only if the boundary conditions for every degree of freedom in motion are completely and unambiguously fixed. If the boundary conditions for some of the variables appearing in the equations of motion are missing for whatever reason, the law of motion does not work even if it remains legitimate in its own right. If it is due simply to a technical difficulty in identification, however, such a hypothetical unavailability of the boundary conditions can easily be overcome by implementing what is missing, at least in theory. As far as the law of motion of one-to-one mapping type is guaranteed, the unique decidability of every degree of freedom involved cannot be affected by a mere technical difficulty in identification in practice.

A theoretician who perceives that the law of motion of one-to-one mapping type is working has every good reason to justify the presence of completely fixed boundary conditions even if one cannot measure it in reality. The law of motion of one-to-one mapping type makes it possible that every degree of freedom involved in motion can be externalized in the form of the law itself and nothing indefinite remains. If something indefinite remains in the motion, one cannot claim a temporal one-to-one mapping anymore because of the lack of unique decidability in the temporal transformation. The boundary conditions have to be unambiguously determinate and completely fixed insofar as the law of motion is of one-to-one mapping type in time.

Nevertheless, the complete fixedness of boundary conditions that any law of motion of one-to-one mapping type intrinsically requires is not as obvious to the experimenter as to a theoretician. If an experimenter who has accepted a law of motion of one-to-one mapping type dares to examine what is obtained in measurements, it would become imperative to literally fully control the experimental boundary condition which one tries to set up. As a rule, the total number of degrees of freedom that one faces is enormous and is far greater than the number one can literally control. A possible best strategy for approaching the complete fixedness of boundary conditions that is compatible with the law of motion of one-to-one mapping type is to design an experimental setup so as to minimize the influences from those degrees of freedom the experimenter cannot control.

One possibility for gaining full control over the experimental boundary conditions is to approach absolute zero in temperature. Here only the zero-point motion predicted by quantum mechanics survives and all of the other motions are frozen out. Another possibility is to abandon full control over every single degree of freedom, and instead to seek only a partial control over those one cannot control in full. What can be controlled in a weaker sense is a statistical variance associated with an ensemble of those individuals, though each member in the ensemble is beyond the experimenter's control.

An experimenter can single out those degrees of freedom one manages to control in practice and can examine whether those variables one chooses on one's own will obey the corresponding law of motion of one-to-one mapping type. This examination is necessarily incomplete in that the boundary conditions are only partially controlled and in that a full consideration has not been given to all of the degrees of freedom in motion.[5]

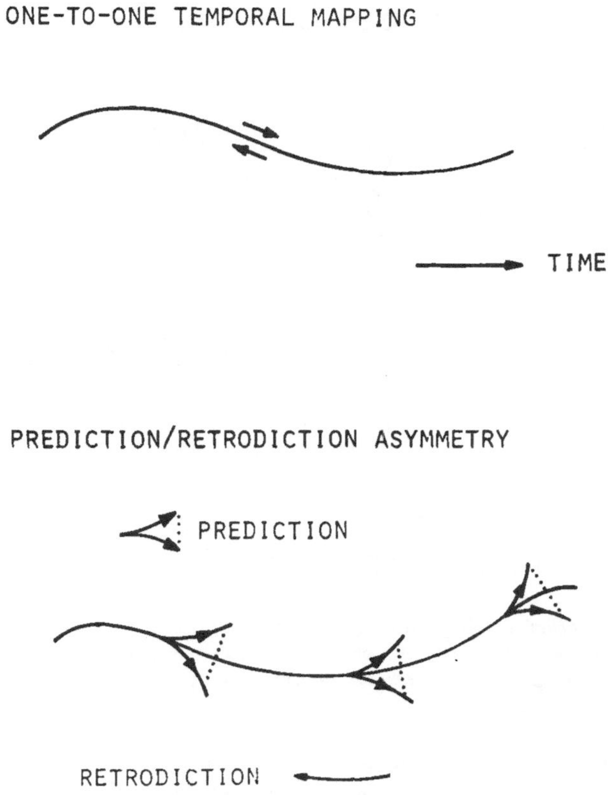

FIGURE 1. A schematic representation of a one-to-one temporal mapping and a prediction/retrodiction asymmetry.

The inevitable partial fixedness of experimental boundary conditions suggests two interpretations. One is that the partial fixedness is due only to the technical difficulty in practical control. If the experimenter shares with a theoretician the view that any degree of freedom in motion obeys a law of motion of one-to-one mapping type, the superficial incompleteness and indefiniteness would simply be a matter of the technical limitation of experiment and would not preclude complete fixedness of boundary conditions in theory. In contrast, another interpretation, although it does not dispense with the first interpretation, is to admit the possibility that partially fixed boundary conditions are not due to technical difficulty. If intrinsic partial fixedness of boundary conditions survives, this would be at variance with the time-honored law of motion of one-to-one mapping type as shown in Figure 1.

Any experimental boundary condition is necessarily only partially controlled and fixed at best. Thus, an examination of whether the partial fixedness of boundary conditions is a matter of principle or simply an artifact will become a serious problem to be investigated in its own right, irrespective of the theoretical conviction the practioner may have.

Intrinsic complete fixedness of boundary conditions and a law of motion of one-to-one mapping type go hand in hand. On the other hand, however, if intrinsic partial fixedness of boundary conditions can gain any reality, the law of motion of one-to-one mapping would have to be given second thought.

One likely symptom which suggests that a law of motion of non-one-to-one mapping type may not be a total absurdity is the asymmetry between prediction and retrodiction[6] of the second half of Figure 1 on the part of an observer who examines the process of, let us say, biological evolution. The observer can clearly and definitely tell in retrospect what has really

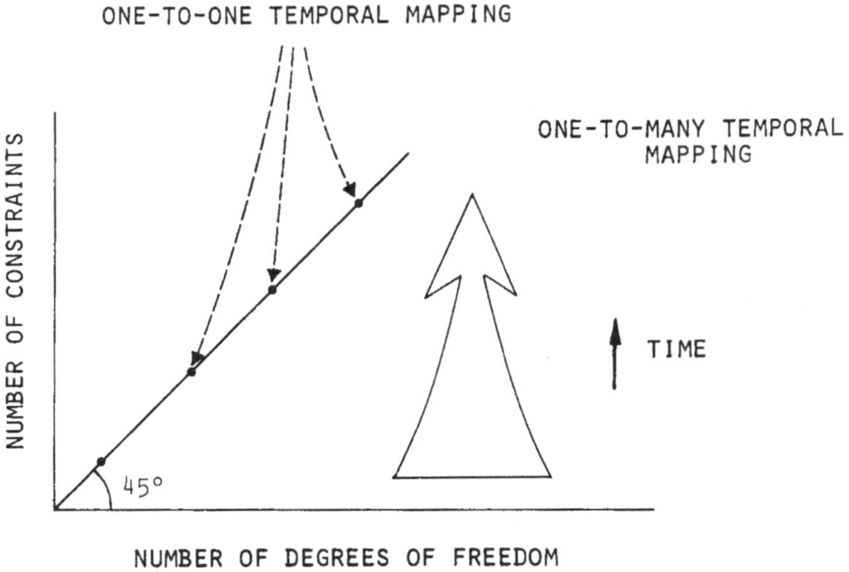

FIGURE 2. Temporal mappings in terms of both the number of degrees of freedom and the number of constraints. One-to-one temporal mapping corresponds to a point on a straight line whose tangential angle is 45°, whereas one-to-many mapping exhibits a movement approaching the line from below.

happened, but such clarity and definiteness are absent when one tries to tell what will happen in the future. The asymmetry between prediction and retrodiction provides an example of the temporal mapping of one-to-many type as exhibited in Figure 2. Many possibilities are anticipated in prediction, but only one alternative has materialized in retrodiction. Of course, we have to accept the charge that a further introduction of an observer as a source of the one-to-many mapping in time may aggravate our physical investigation of the nature of the law of motion, instead of helping us solve the problem. The present one-to-many temporal mapping with regard to an observer by no means logically leads to a law of motion of one-to-many mapping type, because the role of observer must be critically examined in addition to the physical problem under investigation.

However, if we take a view that even an observer exists as a very special mode of matter, the problem of the physical possibility of one-to-many mapping will be recognized as being a more urgent one. We are asked to respond to the question of whether and/or how a law of motion of one-to-many mapping type could be a reality. A rationale lying behind this convoluted reflection is that the very presence of an observer maintaining the prediction/retrodiction asymmetry could not be possible unless a law of motion of one-to-many mapping type can be available in the first place.

If a law of motion of one-to-many mapping type is available, it is irreversible and asymmetric with time.[7] A schematic representation of one-to-many mapping is shown in Figure 3. The temporally forward process constantly necessitates a choice of reality out of many possibilities. In contrast, the backward process requires no choice in tracing in the reversed direction the track that has already been established. The irreversibility or the asymmetry with time refers to the presence or absence of materialistic choice process depending upon the direction of time.[8]

There have already been several theoretical attempts to approach physical irreversibility.[9] The second law of thermodynamics points to an infallible empirical observation on irreversibility, telling us that any system cannot return to its original state without disturbing its surroundings. However, the second law itself stops short of identifying what mechanism is

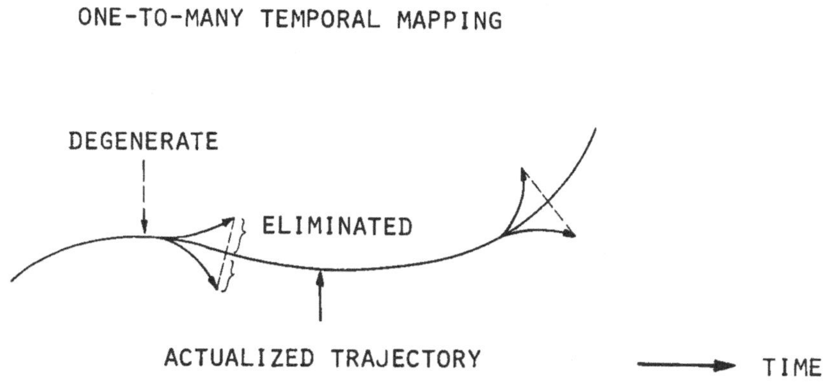

FIGURE 3. A trajectory of one-to-many temporal mapping. The actualized trajectory is degenerate with the possible future trajectories whose degeneracy is constantly dissolved with time.

really involved in securing the claimed irreversibility that is factually irrefutable.[10] The second law does not identify what is microscopically responsible for preventing a thermodynamic system from returning to its orginal state without disturbing its surroundings.

The unique feature of thermodynamics is its silence about the issue of whether the boundary condition for a thermodynamic system is intrinsically completely fixable or partially fixable. It is for this reason that the second law of thermodynamics may be applied either to cases of completely fixed or partially fixed boundary conditions. In fact, thermodynamic quantities such as temperature and thermodynamic entropy do not completely specify how every constituent element of a thermodynamic system behaves. Thermodynamics assumes only a partial fixedness of boundary conditions for the sake of its unique theoretical framework, irrespective of whether the boundary conditions at the most fundamental level of the law of motion are intrinsically completely fixable or only partially fixable in principle.

We must go beyond thermodynamics in order to examine how physical irreversibility can be possible at all. One attempt in this direction is statistical mechanics.[11,12] Statistical mechanics of many particles subject to the law of motion of one-to-one mapping type, whether in classical or in quantum mechanics, sometimes has been thought to give rise to a certain form of irreversibility if full control over the boundary conditions for the microscopic law of motion is abandoned.[13,14] The lack of full control implies that the boundary condition would be only partially fixable. The rationale behind this statistical-mechanical irreversibility, sometimes called Boltzmann's H-theorem, is the presence of hypothetical molecular chaos suggesting that any deviation of the initial conditions even to the slightest degree yields an enormous difference in trajectories later in time.[11,15]

However, the origin of the statistical-mechanical irreversibility is not materialistic, but anthropocentric. The agent who abandons full control of boundary conditions is an external observer, not the particles themselves.[16] The resulting partial fixedness of boundary conditions is an artifact attributed to the observer whose capacity to control boundary conditions is admittedly limited as demonstrated in Figure 4. Statistical-mechanical irreversibility is an outcome of an anthropomorphic intervention, the physical justification of which still remains to be seen. Unless an anthropomorphic artifact intervenes, a many-particle system subject to the law of motion of one-to-one mapping type is reversible in time because of the very nature of the one-to-one temporal mapping. Physical irreversibility, which is factually irrefutable, cannot find its place in the law of motion of one-to-one mapping type. The one-to-one mapping presupposes intrinsic complete fixedness of boundary conditions.

An examination of whether intrinsic partial fixedness of boundary conditions could be

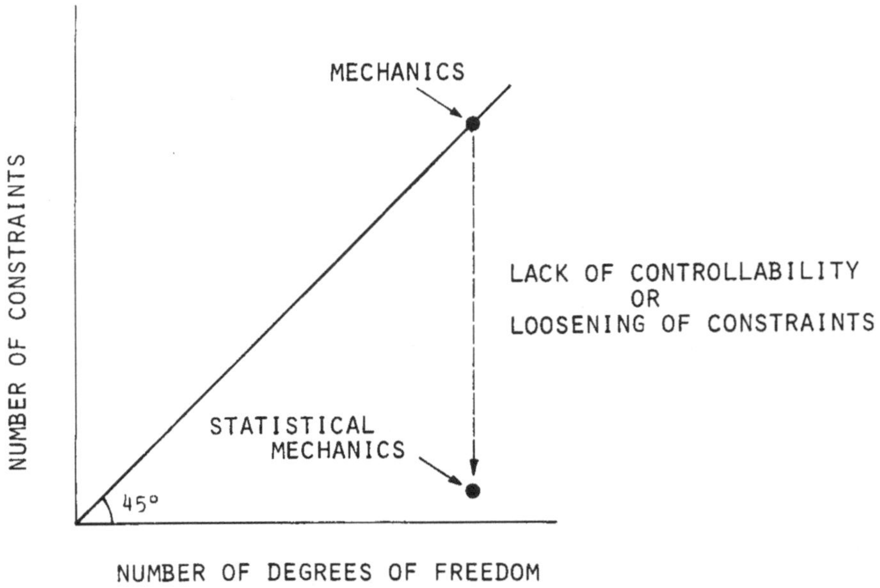

FIGURE 4. A relationship between mechanics and statistical mechanics. Unlimited theoretical controllability of boundary conditions is abandoned in statistical mechanics.

physically possible thus has to be made at the same time as an examination of the physical origin of irreversibility. For this purpose, it is first required to identify the extent to which the law of motion of one-to-one mapping type remains legitimate.

III. CONSERVATION AS A PROCESS

To begin with, one can imagine an isolated many-particle system as following a law of motion of one-to-one mapping type. This is in fact an example of an isolated system which one experiences in classical and in quantum mechanics. An isolated system, though highly theoretical, can serve as a useful, conceptual device to scrutinize the law of motion which makes the presence of the very same system possible.

First of all, isolation imposes on the system the constraint that the total energy is conserved at an invariant level at any moment. However, we have not considered the problem of how the anthropocentric measurement of the total energy would proceed in quantum mechanics. Every degree of freedom involved in the motion is under the constraint of the conservation of the total energy. The law of motion of one-to-one mapping type combined with the associated complete fixedness of boundary conditions renders the value of every degree of freedom in the motion determinate all through the time development in such a manner that the total energy is conserved. There exists a definite one-to-one correspondence between every dynamic variable and its own constraint which specifies uniquely how the value of the variable develops without allowing any indefiniteness. The total number of degrees of freedom in the motion, if countable, is equal to the total number of independent constraints the law of motion incorporates into itself. Even if the number is uncountable, on the other hand, the one-to-one correspondence between every dynamic variable and its unique constraint survives without involving anything indefinite.

The total energy which a theoretician sets for an isolated system is uniquely and unambiguously distributed in the whole system. The way of distribution is such that the value of any dynamic variable is precise and well defined and is, of course, consonant with the conservation of the total energy. In order to see how both the determinate character of each variable involved and its consonance with the conservation of the total energy can be materialized, let us hypothetically imagine such a situation in which the theoretician perturbs an extremely limited local portion of the whole system by supplying a very small amount of energy externally, by whatever hypothetical means, at a given moment. The isolated system that develops by following the law of motion of one-to-one mapping type makes the motion of all its dynamic variables determinate, due to the complete fixedness of boundary conditions, whether we are dealing with classical or quantum mechanics. Insofar as the value of every dynamic variable is determinate (thanks to the complete fixedness), the total decidability of all of the dynamic variables involved in the motion has to be consistent with the conservation of energy at any moment.

This total decidability is such as to make all of the dynamic variables involved vary their values instantaneously in a completely correlated fashion so as to adjust the total energy right at the moment the theoretician introduces perturbation, even if the perturbation is strictly local. Every variable that is uniquely determinate must simultaneously detect all of the others in a manner consistent with the conservation of the total energy as expressed in the law of motion. The present hypothetical situation would imply that the energetic perturbation initiated by the theoretician cannot remain localized in space, no matter how localized the initial perturbation may be. The effect of the localized energy perturbation would have to be propagated in the whole system at an infinite velocity. What makes the propagation of the effect at an infinite velocity superficially possible is a one-to-one law of motion plus completely fixed boundary conditions. Insofar as the consistency between the total decidability of all of the dynamic variables and the conservation of energy is observed, one cannot conceive a case in which some local portion of the isolated system senses a conserved total energy that is different from the total energy that the remaining portions do. Here, simultaneous expression of any interacting physical bodies would necessarily entail similar simultaneous detection of each of them by themselves as they are expressed.

We thus recognize that the law of motion of the one-to-one mapping type forces something to be propagated at an infinite velocity. What is being propagated at an infinite velocity is the "conservedness" of total energy, conservedness having been introduced as a new term in order to distinguish between conservation as a process and as an end result. By conservedness we shall mean the realization of the process entailed by and leading to conservation. Conservation as an end result is a consequence of the process of conservation that operates so as to fulfill conservedness. Unless the process of conservation imputed to conservedness instantaneously prevails in the whole isolated system in a simultaneous manner, the law of motion of the one-to-one mapping type could not properly function right at the moment the theoretician initiates the localized energy perturbation. Nevertheless, no physical process is permitted to propagate at a velocity greater than light velocity. The materialistic processes underlying the actualization of conservedness need to be elucidated.

Conservation laws in physics are unique in having a holistic and synthetic character,[17] rather than being reductionistic and analytic. It is thus legitimate to ask how such a holistic and macroscopic character can be synthesized and materialized on a microscopic basis.[18] There exists a dichotomy between microscopic interaction and macroscopic conservation.[19]

If the principle of conservation were a property to be imposed *a priori*, an intractable problem would arise as to who could impose it and how. On the other hand, if the principle is to be materialized only *a posteriori*, it must be microscopic interactions that are responsible for implementing the macroscopic conservation.[20] In fact, empirical observation attests to no more than an *a posteriori* validity of the conservation principle. It is fundamentally impossible to empirically examine whether the principle of conservation is legitimate *a priori*. Conservation laws are as basic in physics as microscopic interaction is. But, microscopic interaction is more fundamental in its intrinsic capability to implement conservation laws *a posteriori*.

The process of materializing conservedness of energy yields the conservation of energy *a posteriori*, in which the process refers to that entailed by and leading to conservation of energy. In this regard, as far as the law of motion of one-to-one mapping type is concerned, the complete fixedness of its initial condition as a particular form of boundary condition has already satisfied in itself the conservedness of the total energy at the very moment the initial condition is specified. The law of motion preserves the conservedness of the total energy during development. The difference between preserving and materializing the conservedness of the total energy vanishes for the theoretician who sets the level of the total energy to be conserved and sees a complete fixedness of boundary conditions. Such a theoretical imposition letting the law of motion be of one-to-one temporal mapping makes the internal communication of conservedness proceed at an infinite velocity, though only in theory, and not in reality.

The internal communication of conservedness is a particular mode of materialistic interaction implementing the principle of conservation *a posteriori*. A theoretician who would impose (in the form of completely fixed boundary conditions) how the conservation of energy is implemented, for instance, is to take over the role of the materialistic activity that the communication process of conservedness intrinsically assumes.

Materialistic interactions involved in the conservation of energy assume two roles. One is as an energy carrier and the other is as a signaling agent to communicate how the carried energy is implemented so as to satisfy conservedness. A convoluted character latent in these two roles is that the communication process of conservedness also carries a certain amount of energy along with it. Intrinsic partial fixedness of boundary conditions can accommodate the internal and endogenous capacity of determining how materialistic interactions are actualized through both the processes of implementing and communicating the conservedness of energy. On the other hand, completely fixed boundary conditions which a theoretician perceives dispense with the materialistic process of communicating conservedness, leaving only the process of preserving conservedness. The communication of conservedness under completely fixed boundary conditions is a theoretical artifact at its best and not materialistic. The conservedness is simply an imposition.

The process materializing conservedness assumes the propagation of changes in interaction in the medium because the energy carrier of the isolated system is the whole body that embodies interaction. The real physical adjustment of interaction does not let the materialistic communication of conservedness proceed at a velocity exceeding light velocity. Thus, the theoretical imposition letting the communication velocity of conservedness diverge can survive in reality at most approximately and only when all of the other velocities of interest are far less than the real communication velocity of conservedness which cannot exceed light velocity. The nonrelativistic regime of classical and quantum mechanics can make the law of motion of one-to-one mapping type practically feasible as far as the real communication velocity of conservedness remains at or near light velocity. Yet, there is no guarantee whatsoever that the real communication velocity of conservedness always remains around its possible maximum value. We need to scrutinize what would happen if we really wish to appreciate the finiteness of the communication velocity of conservedness.

One well known example that indicates the communication of conservedness proceeding at a finite velocity is hydrodynamics.[21] Sound waves demonstrate in fact a process embodying how the conservedness of momentum is propagated in the medium.[22,23] But, hydrodynamics is not well suited to our purpose of examining the extent to which the law of motion of one-to-one mapping type remains legitimate. The boundary condition for hydrodynamic quantities such as the velocity field of fluids is only partially fixable because of the macroscopic dimensions. The boundary condition for every microscopic element constituting a hydrodynamic quantity is only partially fixable. An issue at this point is whether the partial fixedness of boundary conditions in hydrodynamics is an artifact or an intrinsic reality.

Hydrodynamics by itself does not answer this question. We are thus led to inquire more microscopically about how the real communication of conservedness proceeds, especially in relation to whether or not intrinsic partial fixedness of boundary conditions is the case. The present inquiry is motivated by the observation that the law of motion of one-to-one mapping type accompanied by completely fixed boundary conditions causes the communication velocity of conservedness to diverge, since the communication of conservedness to be accomplished right at the moment the initial condition is identified.

Intrinsic partial fixedness of boundary conditions is the only alternative to account for the finiteness of the communication velocity of conservedness. If a theoretician still claims that every degree of freedom in motion has to be unambiguously identified and externalized in the form of the law of motion in theory, the resulting complete fixedness makes the finiteness of the communication velocity of conservedness untenable at least in principle. The intrinsic partial fixedness requires that no matter how much effort may be paid in theory, let alone in experiment, there always remain some indefinite degrees of freedom. As a consequence, no theoretician can insist on a complete specification of their boundary conditions unambiguously in an externalized and expressible form. The impossibility of complete specification of boundary conditions is not a matter of technical difficulty, but a matter of principle.

A theoretician can think of an agent with an infinite capability for identifying the boundary condition of every degree of freedom in motion, such as a Maxwell's demon,[24] that can distinguish every molecule from all of the others even though their total number is of the order of Avogadro's. In contrast, intrinsic partial fixedness of boundary conditions, if tenable, remains as such even if the intervention of such an agent assumed to have an infinite capability in identifying boundary conditions is conceived theoretically. The intrinsic partial fixedness of boundary conditions is the necessary price we have to pay in order to save the communication of conservedness proceeding at a finite, not at an infinite, velocity.

One might think that an appreciation of something indefinite, or more precisely, indefinite boundary conditions, in theory would pave the way for a kind of theoretical defeatism on the grounds that any theory that can be verifiable or falsifiable must be definite. Strictly speaking, any theoretical attempt that includes something indefinite is hard to verify in an externalized and expressible form as an invariant theoretical artifact. It is not possible to identify and to examine what is left over as indefinite in such theorizing.

Intrinsic partial fixedness of boundary conditions is incompatible with the established rule of physical theorizing that requires a practicing theoretician to choose only definitive quantities as being relevant to the theory to be constructed. The theoretician should be advised not to choose any quantity which cannot be controlled definitively, in spite of the fact that the presence of indefiniteness is not totally denied in reality. But, once such definitive theorizing is attempted, anything indefinite which is rejected from participating in theory is supposed to be irrelevant, according to the rule. An objective of the verification or falsification of the attempted theory is to examine how and to what extent those variables and parameters rejected from the theory as being indefinite remain factually irrelevant in reality.

If an attempted theory turns out to be unsatisfactory in some respect, the next step to follow is to try another theory by rescuing something definite that was previously discarded as being embedded in what has been thought indefinite and by embedding something that has been appreciated as being definite in what remains indefinite. The presence of indefiniteness in reality is appreciated only through an indefinite sequence of theorizing of definitive character, its falsification and renewed theorizing of the same defintive character.

An advantage of definitive theorizing is that the theoretician controls (in theory) all of the degrees of freedom involved in the motion and can claim complete fixedness of boundary conditions. Its inevitable disadvantage is to make the communication of conservedness of whatever quantity to proceed at an infinite velocity (in theory). The intrinsic partial fixedness of boundary conditions that we are trying, on the other hand, is an attempt to save the finite communication velocity of conservedness at the expense of the theoretical clearness of definitive character.

An appraisal of the intrinsic partial fixedness of boundary conditions is in the observation that it recognizes the difference between conservation as an end result and the process entailed by and leading to conservation. Conservation in the mode of having fulfilled and preserving conservedness is a consequence or a product of the process materializing conservedness that is an actualization of interactions yielding conservedness *a posteriori*. The recorded conservation of energy as an *a posteriori* consequence is irrevocable and remains as such, since what has already been recorded remains frozen and intact, not in motion. All of the dynamic variables that can be conceived have fixed their values in the record. As far as the record is concerned, the boundary conditions for every dynamic variable have been unambiguously fixed. Complete fixedness of boundary conditions survives at least in the record. Nevertheless, it is important to notice that the complete fixedness of boundary conditions in the past record does not necessarily guarantee the same complete fixedness in the future. If one claims that the same complete fixedness of boundary conditions applies to the future as well as to the past, the law of motion that connects the future to the past has to be of one-to-one temporal mapping, otherwise the complete fixedness would fail. The outcome is again a denial of the communication of conservedness proceeding at a finite velocity. We thus need to recognize that the boundary conditions imposed on the process yet to come are not necessarily of the same quality as those measured in the record, especially with regard to the extent of being fixed and identifiable.

Intrinsic partial fixedness of boundary conditions does not apply to the recorded past, but can apply to the process yet to come. To clarify the matter, let us imagine a situation in which the process materializing conservedness can potentially allow a diversity of trajectories yet to come. In the past record, however, only one trajectory has been actualized. What is responsible for the trajectory reduction is the self-constraining actualizaton process which eliminates those trajectories that can be connected to the one already actualized, but that do not comply with the *a posteriori* principle of conservation during the very process of materializing conservedness. The actualized trajectory is always degenerate or condensed with those trajectories to be eliminated later. We can thus recognize the intrinsic partial fixedness of boundary conditions in the degeneracy of trajectories to be constantly dissolved because the extent of and the bounded nature of the degeneracy cannot be completely specified in advance.

The intrinsic partial fixedness of boundary conditions implies that boundary conditions are intrinsically indefinite and possess a certain extent of degeneracy to be dissolved. Consequently, the one-to-many temporal mapping, by definition, lets the preceding one be degenerate with the subsequent many, and the associated actualization process assumes successive dissolution of the degeneracy. In contrast, the one-to-one temporal mapping lets the entire subsequent trajectory be condensed into the preceding one, and the associated actualization process of trajectory is a unique extension of the one already actualized.

The law of motion of one-to-one mapping type maintains every degree of freedom in motion constrained so that the trajectory may be uniquely definable. The total number of degrees of freedom in motion is equal to the total number of independent constraints involved in the law of motion. On the other hand, the law of motion of one-to-many mapping type always keeps generating constraints so as to dissolve the degeneracy, but maintains the total number of degrees of freedom in motion above the number of those accumulated independent constraints that have been actualized to dissolve the degeneracy. The law of motion of one-to-many mapping type survives until the number of accumulated independent constraints approaches the total number of degrees of freedom in motion. A physical condition that makes a law of motion of one-to-many mapping type possible is the intrinsic partial fixedness of boundary conditions.

The relationship between the intrinsic partial fixedness combined with the self-constraining actualization process and the law of motion of one-to-many mapping type is circular. One of the two always serves as a raison d'être for the other. This circularity, however, is by no means of a vicious type because the emphasis is limited strictly to their mutual consistency.

Finiteness of the communication velocity of conservedness in the medium is within the process that keeps yielding one trajectory out of an indefinite number of possible alternatives. In fact, the self-constraining process of trajectory is unique to the law of motion of one-to-many mapping type and has nothing to do with the law of one-to-one mapping type that forces the communication of conservedness to proceed at an infinite velocity. The intrinsic partial fixedness of boundary conditions assumes that boundary conditions are always degenerate with those in the future. The extent of degeneracy is dissolved with time. And those boundary conditions already dissolved in the record are fixed in the sense that they are irrevocable.

We have seen that there exist at least two modes of coping with boundary conditions in physics. One is to recognize the presence of completely fixed boundary conditions in theory, combined with a law of motion of one-to-one mapping type. The symmetry of time survives in this case because of the one-to-one temporal mapping.[25,26] This mode of thinking has prevailed in mechanistic models of motion such as Newton's equations of motion in classical mechanics and Schrödinger's equation of motion in quantum mechanics. The other way of coping is to emphasize intrinsic partial fixedness of boundary conditions along with a law of motion of one-to-many mapping type. Time's arrow is saved because of the asymmetry between the future that is degenerate within the present when viewed from the present and the past that has already been dissolved of its degeneracy.[27] The present asymmetry clearly differs from the anthropocentric asymmetry between prediction and retrodiction on the part of an observer in spite of its superficial similarity. The materialistic origin of the asymmetry in the one-to-many mapping is in the intrinsic partial fixedness of boundary conditions on the part of matter.

A candidate for using physics in biology, especially in an attempt to give birth to biology, seems to be one that assumes the intrinsic partial fixedness of boundary conditions, at least because of its capacity for physical irreversibility and asymmetry with time. Further investigation is in order on the nature of intrinsic partial fixedness of boundary conditions and the law of motion of one-to-many mapping.

IV. MECHANICS AND BIOLOGY

Intrinsic partial fixedness of boundary conditions and the communication of conservedness proceeding at a finite velocity are compatible with each other. Likewise, intrinsic complete fixedness of boundary conditions and the communication of conservedness proceeding at an infinite velocity are also compatible with each other.

A clue for discovering whether boundary conditions could be completely fixable or only partially fixable is in the quantity which is to be conserved. A most representative quantity to be conserved is mechanical force defined in mechanics. The third law of mechanics claims no net contribution from those forces of action and reaction. The first law defining inertia as a continuation of invariant state in the absence of a force to act upon and the second law defining force as an impetus to induce changes in the state defined in the first law, when combined together, yield the third law of mechanics stating the null-conservation of forces of action and reaction.

The null-conservation of forces means simply that there is no net contribution from those forces actualized inside an isolated system, otherwise the isolation would be violated. In fact, insofar as one makes a system large enough to include in it every pair of elements to act upon and to be acted upon in an exhaustive manner, the null-conservation of forces holds irrespective of whether the system is nonbiological or biological. The null-conservation of forces as an expression of the third law of mechanics holds whatever mechanisms are involved in deriving and actualizing forces. We therfore notice that the null-conservation of forces is more fundamental than, for instance, what is implied by a conservative force defined as a derivative from another quantity to be conserved, such as the resultant of kinetic and potential energies.

The null-conservation of forces is a consequence of the process materializing null-conservedness of forces. Accordingly, it is required to examine how the process of materializing null-conservedness proceeds in reality. One hypothetical possibility is that the null-conservedness of forces would be communicated in the medium at an infinite velocity. Such a hypothetical communication would necessarily assume the propagation of changes in force proceeding at an infinite velocity because it involves internal adjustment of forces themselves.

The propagation of changes in force at an infinite velocity could materialize in an action-at-a-distance, such as the classical Coulomb force. Forces actualized in action-at-a-distance are completely in equilibrium at any moment in the sense that the null conservation of forces is fulfilled instantaneously and simultaneously in the whole medium. Those forces, once actualized, have not admitted in themselves any room for internal determination, allowing for further endogenous movement aiming at the null-conservedness of forces. Its preservation is guaranteed at any moment. Henceforth, the difference between preserving and materializing null-conservedness of forces disappears. Force equilibrium as an expression of the null-conservedness of forces is always there.

The communication of null-conservedness of forces proceeding at an infinite velocity that is conceivable in the case of action-at-a-distance is unfailingly compatible with intrinsic complete fixedness of boundary conditions. For an instantaneous establishment of force equilibrium makes those forces involved in the motion unambiguously determinate at any moment without leaving anything indefinite. Those determinate and definite forces now drive the corresponding mechanical motion in a unique manner according to the second law of mechanics. Such unambiguously determinate forces are, in fact, possible only when boundary conditions are completely fixable, at least in theory, since forces are undoubtedly an attribute of boundary conditions. If initial conditions are only partially fixable in principle and admit an inclusion of something indefinite, one cannot claim in theory a complete specification of forces that drive the later mechanical motion, for the effect of initial indefiniteness is left unspecified. The present indefiniteness contradicts the presence of force equilibrium that allows no ambiguity in identifying those forces to act upon and to be acted upon.

Newton's equations of motion in classical mechanics, or more precisely, in celestial mechanics, assume action-at-a-distance. Accordingly, they cause the communication velocity of null-conservedness of forces to diverge. The boundary conditions for Newtonian mechanics have to be completely fixable as a matter of principle. The resulting trajectory is unambiguously determinate. Similarly, Schrödinger's equation of motion in nonrelativistic quantum mechanics uniquely determines the trajectory of its wave function, though not how to measure it. The determinate character of a quantum-mechanical wave function perforce requires complete fixedness of its boundary condition. Schrödinger's equation of motion requires those forces acting upon the wave function to be unambiguously determinate at any moment like Newton's equations of motion. As far as force equilibrium as a manifestation of the null-conservedness of forces being guaranteed at any moment is concerned, the completely determinate character of both the trajectory and its boundary conditions applies to Schrödinger's as well as to Newton's equations of motion. The difference between classical and quantum mechanics, however, will arise if one examines how the process of measurement intervene in these two theories, though which we have not yet elaborated.

Action-at-a-distance has been identified as yielding the process preserving null-conservedness of forces in which the very communication of null-conservedness is to be propagated at an infinite velocity, at least in theory. An alternative to action-at-a-distance is the one in which the communication of null-conservedness of forces is propagated in the medium at a finite velocity not exceeding light velocity. The present second alternative is more physical than the first in appreciating that no action can be propagated beyond light velocity. The communication of null-conservedness of forces at a finite velocity materializes in an action-through-medium that allows only those changes in force that are propagated at a finite velocity.

Action-through-medium cannot yield an instantaneous fulfillment of force equilibrium in the whole medium in a simultaneous manner, contrary to the case of action-at-a-distance. What is being actualized instead is the process of force coming to equilibrium, namely, the process of force equilibration. Force equilibration appreciates the physical fact that the process of force coming to equilibrium initiated in a local region cannot instantaneously reach every local part of the whole medium, whereas force equilibrium in action-at-a-distance assumes that any process for the null-conservedness of forces initiated in an arbitrary local region simultaneously reaches every other local part of the whole medium. Force equilibration thus makes it a rule, not an exception, to allow its indefinite repercussions in the whole medium because the preceding force equilibration at one place never fails to spill over and to induce the subsequent equilibrations in its neighborhood.

Force equilibration being propagated at a finite velocity cannot specify its own boundary condition in full even in theory. If the initial conditions for one and the same process of force equilibration were completely fixed and identified without leaving any indefiniteness at two separate time points, these two conditions would have to be uniquely concatenated in time, otherwise the complete fixedness would fail. Moreover, if the complete fixedness of initial conditions survived, the underlying law of motion would have to be of one-to-one temporal mapping. It cannot be of one-to-many mapping. The resulting law of motion of one-to-one mapping type accompanied by the associated complete fixedness of initial conditions would render any macroscopic quantity to be conserved explicitly expressible at any moment. This would mean the communication of conservedness proceeds at an infinite velocity. We thus recognize a contradiction that force equilibration as a process for null-conservedness of forces being propagated at a finite velocity would reduce to the process being propagated at an infinite velocity. The contradiction originates in the false assumption that the boundary conditions, or more specifically, the initial conditions, be completely fixable.

Boundary conditions have to be only partially fixable even in principle in order to save force equilibration that proceeds at a finite velocity. One examination to check whether boundary conditions are completely fixable or only partially fixable is to measure the propagation velocity of force equilibration. If it turns out that the propagation velocity of force equilibration remains finite, the imposition of completely fixed boundary conditions would be surely incorrect even in theory, let alone in experiment.

We already know in continuum mechanics that the process of conserving momentum is propagated as are sound waves in the medium at a finite velocity. However, the present process of conserving momentum being propagated at a finite velocity does not allow us to reach such a hasty conclusion that intrinsic partial fixedness of boundary conditions and, accordingly, their indefiniteness could survive in continuum mechanics. We must first pay attention to the fact that the process of conserving momentum assumes an infinitely large, homogeneous space because the conservation of momentum is equivalent to the invariance of motion against its spatial translation. An infinitely large space conceived as a theoretical device, by definition, is an imprecise description of reality and remains indefinite about how the space is bounded. A mere theoretical imposition of indefiniteness does not work as a materialistic basis of intrinsic indefiniteness yielding boundary conditions fixable only partially. In continuum mechanics, however, such a theoretically imposed indefiniteness admittedly underlies a justification of the process of conserving momentum being propagated at a finite velocity.

A fragility of the above logic will become obvious if one examines how the conservation of momentum could materialize in a bounded space whose boundary condition is completely specified. The bounded space contradicts the presence of spatial translational invariance as the very basis of the conservation of momentum. If one claims a complete fixedness of boundary conditions, there is no room left for the process of conserving momentum in continuum mechanics. Conversely, although the lack of complete fixedness of boundary conditions can make feasible the process of conserving momentum being propagated at a finite velocity (at least in theory), such theorizing does not tell us anything specific about whether intrinsic partial fixedness of boundary conditions could be possible materialistically.

One more likely source of partially fixed boundary conditions are fluctuations that one experiences in equilibrium and nonequilibrium thermodynamics.[28-31] It is of course legitimate to say that fluctuations are only imprecisely definable in anticipation, otherwise fluctuations would cease to be fluctuations. But, the origin of such an indefiniteness differs depending upon the viewpoint one takes.

An experimenter can associate with fluctuations those over which one cannot claim a full control. This confession of the lack of full control capability suggests two possibilities. One is that what were measured as fluctuations have been identified as such because of the technical difficulty in identification on the part of the experimenter, while fluctuations are completely identifiable of themselves in principle. The other is that fluctuations represent a character of indefiniteness intrinsic to matter. The present dichotomy, however, would not bother the experimenter very much if one maintains that the lack of full control in experiment renders what looks like fluctuations to be as such. This reasoning is admittedly modest insofar as it addresses inevitable limitation of any experiment.

A theoretician, on the other hand, takes a rather different view. If the law of motion of one-to-one mapping type is chosen as a most fundamental frame of reference, such as Newton's equations of motion on the classical trajectory or Schrödinger's equation of motion on the quantum-mechanical wave function, the inevitable complete fixedness of boundary conditions enables the theoretician to control every single degree of freedom in motion, at least in theory, no matter how large their number may be. Nothing uncontrollable is left behind when viewed from the eyes of the theoretician. Consequently, it turns out to be a self-contradiction that the theoretician who bases one's theory upon the law of motion of one-to-one mapping type in one occasion observes the working of fluctuations as an indication of the lack of full controllability in another.

The lack of completely fixed boundary conditions observed in statistical mechanics does not support the presence of intrinsic partial fixedness in a many-particle system. The lack of complete fixedness which the theoretician sees in statistical-mechanical theorizing is again anthropomorphic in its origin in spite of the underlying law of motion to the contrary. Theoretically, we have only two choices, either to fully rely upon the law of motion of one-to-one mapping type supplemented by completely fixed boundary conditions without allowing anything like fluctuations, or to appreciate the law of motion of one-to-many mapping type supplemented by intrinsically partially fixed boundary conditions. In the latter, fluctuations are viewed as something pointing to the intrinsic partial fixedness.

The partial fixedness of boundary conditions conceived in continuum mechanics is a theoretical artifact that, in reality, cannot be reduced to intrinsic partial fixedness. Similarly, the lack of completely fixed boundary conditions in statistical mechanics is another theoretical artifact, being contrary to the presence of completely fixed boundary conditions at the level of mechanics underlying statistical mechanics. In order to examine whether intrinsic partial fixedness of boundary conditions could have any reality, continuum mechanics and statistical mechanics are not the right subject matter. This is because of their theoretical double standards that are intended to save partially fixed boundary conditions in spite of the evidence to the contrary on the more microscopic level, though this negative assessment by no means depreciates the proved capability of these mechanics in their own right.

One theoretical discipline that appreciates the communication of conservedness proceeding at a finite velocity is relativistic mechanics, either classical or quantum. Any type of motion proceeds at almost, but not exceeding, light velocity. The communication of conservedness is no exception. However, biology and the physics which gives birth to a biology function in the nonrelativistic regime, though it is not denied that those building blocks of biology such as atoms are admittedly the evolutionary products of the relativistic Big-Bang cosmology.[32] One objective is to examine whether and how the intrinsic partial fixedness of boundary conditions, or equivalently the communication of conservedness proceeding at a finite velocity, could persist in the physics of becoming that can lead to biology.

We note that mechanics per se can serve as a theoretical basis of physics, giving birth to biology. Of course, if mechanics were associated with the law of motion of a one-to-one mapping type, such as Newton's equations of motion in classical mechanics, one would be allowed to have only a limited interpretation of the third law of mechanics. The process of force equilibration underlying the null-conservedness of forces would be propagated at an infinite velocity and that force equilibrium would aways be actualized as being supplemented by completely fixed boundary conditions. Both an instantaneous communication of null-conservedness of forces and completely fixed boundary conditions would have been a matter of theory. In fact, possible laws of motion that mechanics would admit are not strictly restricted to the type of one-to-one temporal mapping. Even if boundary conditions are only partially fixable as a matter of principle, the third law of mechanics survives and can yield the process of force equilibration proceeding at a finite velocity. We shall examine a possibility for intrinsic partial fixedness of boundary conditions within the framework of mechanics.

V. MECHANICAL MOTION IN BIOLOGY

The third law of mechanics states no net contribution from forces of action and reaction. This no-net contribution admits two possibilities, either force equilibrium completed or force equilibration in the making. The third law has almost, but not quite, been equated with force equilibrium as a holistic fulfillment of the null conservation of forces in the medium in a simultaneous manner. In particular, what is implied by the third law of mechanics is more than force equilibrium. It also allows the process of force equilibration to be propagated at a finite velocity. Force equilibration can thus extend the capability of mechanics per se.

As a matter fact, biological motility which attests to the presence of mechanical motion in biology can serve as a test to examine how extended mechanics as a candidate of the physics leading to biology could prove itself.[33,34] A reason behind this assessment is that biological processes observed in evolution and development are factually irreversible and asymmetric with time, and that the reversible law of motion of one-to-one mapping type plus the necessarily completely fixed boundary conditions would meet with difficulty in dealing with such factual irreversibility. Instead, a law of motion of one-to-many mapping type supplemented by intrinsically partially fixed boundary conditions could emerge as an alternative, with the inevitable consequence of force equilibration being propagated at a finite velocity. Complete fixedness of boundary conditions and force equilibration proceeding at a finite velocity are incompatible, as already noted. For complete fixedness unfailingly lets the underlying law of motion be of one-to-one temporal mapping that makes force equilibrium, instead of force equilibration, inevitable.

A particular example of mechanical motion in biology is seen in the flagellar movement of eukaryote.[35] Hiramoto and Baba have recorded the flagellar movement of spermatozoa of starfish *Asterias amurensis.*[36] Their result is reproduced in Plate 1a (Plate 1 follows page 22) with the framing rate of 450 per second, in which the spermatozoon was identified to be swimming in a trough far below, about 300 μm below, the coverslip in the normal seawater at temperature of 22°C. The observed flagellar movement was found to be planar as demonstrated. For comparison, we also demonstrate the forced bending motion of a flexible but inelastic metallic chain in Plate 1b with the framing rate of 33 per second.

Our objective is to measure the propagation velocity of force equilibration, both for cases of biological and nonbiological motion. If the propagation velocity is measured as remaining finite for the flagellar movement of starfish spermatozoa and as diverging for the forced bending motion of a metallic chain, we can hope to proceed further along the line we have already set. Otherwise, we would have to stop here.

It has already been established that the bending motion of the flagellum originates in the sliding force of the axoneme along which the active site of generating the force is traveling. The sliding force is identified as an internal tensile force generated along the axoneme. Consequently, the motive force for the flagellar movement understandably comes from the two sources. One is the internal tensile force generated both biologically and nonbiologically. The other is the resistive force due nonbiologically to the fluid medium, seawater in the present case, external to the flagellum.

The third law of mechanics states that the resistive force acting upon the flagellum is counterbalanced by the force of reaction from the flagellum. Similarly, the third law implies that the internal tensile force to pull is counterbalanced by the similar force to pull back. The internal communication of null-conservedness of internal tensile forces to pull and to be pulled is propagated along the flagellar axoneme. Mechanical displacements due to the second law of mechanics are also propagated along the axoneme.

One thus recognizes that force equilibration communicating the null-conservedness of forces of the third law is immediately followed by mechanical displacement because of the inevitable involvement of the second law. At the same time, there also exist those mechanical displacements that are propagated more slowly than the communication of null-conservedness. The slower process is immediately followed by force equilibrium of the third law or instantaneous communication of the null-conservedness of forces. The problem which now comes up is: suppose we have measured the bending displacement of a flagellar axoneme. Which is the source of the observed displacement, the communication process of null-conservedness of internal tensile forces in force equilibration or the propagation of mechanical displacement in force equilibrium?

Contribution from those internal tensile forces generated along the axoneme is conceptually schematized in Figure 5, in which the entire flagellum is divided into many small segments of equal length. Each segment is pulled by two adjacent segments. If the strengths of internal tensile force at both ends of each segment are known, one can specify the contribution of internal force to the bending moment of each local segment around the one end of it that is moving.

There also exists a contribution to the bending moment from the resistive force imputed to the surrounding medium. The third law of mechanics asserts that the resistive force to act upon is counterbalanced by the force to be acted upon from the flagellar axoneme. The communication of null-conservedness of forces is to take place on and across the contact surface between the axoneme and the surrounding fluid. Consequently, the actualization of the third law with the resistive force will be instantaneous even if the communication proceeds at a finite velocity. As far as the resistive force is concerned, the third law reduces it to a form of force equilibrium. On the other hand, the third law along the flagellar axoneme would yield force equilibration, instead of force equilibrium, if the communication of null-conservedness turns out to proceed along it at a finite velocity.

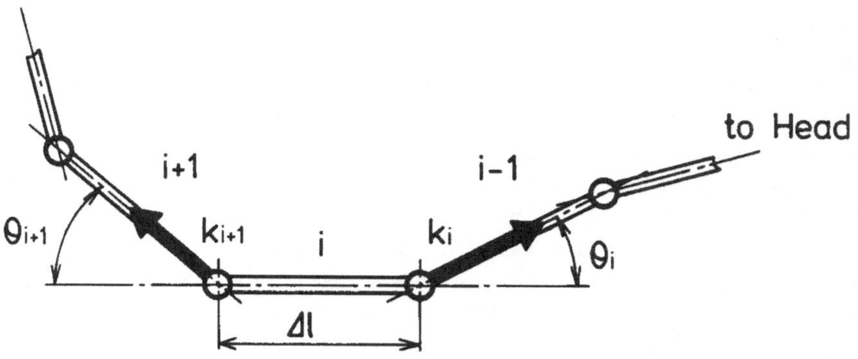

FIGURE 5. A schematic representation of internal tensile force K_i and bending angle θ_i of local segment i of length $\Delta\ell$ along a filament.

We can think of two possibilities for how the bending moment around the moving end determines its own value and how the third law of mechanics is implemented for those internal tensile forces. One possibility is that if internal tensile forces fulfill the third law of mechanics at every moment along the entire axoneme without assuming any underlying process to actualize it, the bending moment of each flagellar segment around its moving end has to vanish. All of the available forces in force equilibrium that the third law uniquely determines are involved at the same time in the acceleration motion, according to the second law of mechanics. If the bending moment around a certain moving end is observed not to vanish by an observer sitting on the moving end, it would give rise to the contradiction that the realized bending motion that has been viewed as a consequence of acceleration, due to all of the forces available on the one hand in fact lacks some of them on the other. In fact, as far as forces are definable uniquely, the second law of mechanics specifies how these forces are involved in the acceleration of motion without leaving any indefiniteness.

One more possibility is that if internal tensile forces fulfill the third law of mechanics through the underlying communication of null-conservedness of forces, the bending moment of a flagellar segment around its moving end also has to vanish but, this time, through the underlying force equilibration. The internal motion that allows the bending moment to vanish is not the type of motion attributed to the second law of mechanics subject to the third law in the form of force equilibrium, but the one attributed to the third law in the form of force equilibration subject to the second law. Internal tensile forces are made to vary in a propagative manner with the progression of the communication of null-conservedness of forces along the flagellar axoneme. A corollary of the present internal motion is that the internal motion imputed to cause the vanishing of the bending moment of one local segment around its moving end successively serves as an impetus to induce the similar motion in the neighborhood segments.

More specifically, internal tensile force at one segment can function as an impetus to induce changes in internal tensile forces in the neighborhood segments in a temporally propagative manner. This sequential propagation of changes in internal tensile force is a form of what we have called the process of force equilibration. Force equilibration takes it for granted that force functions as an impetus to induce changes in the force in a propagative manner in space without invoking the second law of mechanics as an intermediary, while the second law varies forces only through changes in mechanical displacement. The second law would become the sole intermediary for concatenating temporal changes in force only at the limit that the communication velocity of null-conservedness of forces would diverge. The third law at that limit would be deprived of the internal dynamics of its own.

The bending movement of flagellar axoneme indicates a form of mechanical displacement.

The dynamics is either in the communication process of null-conservedness of forces followed by the associated mechanical displacement or in the propagation of mechanical displacement followed by instantaneous communication of the null-conservedness. We are thus required to measure the propagation velocity of force equilibration, especially with regard to internal tensile force associated with the recorded flagellar bending movement.

Let us suppose that we divide the entire flagellar axoneme into n segments of equal length $\Delta\ell$, and that the effect of internal motion causing the bending moment of an arbitrary segment around its moving end to vanish reaches its neighboring adjacent segments after time interval of Δt. The internal motion assumes changes in the strength of internal tensile force at both ends of each segment under the constraint that the bending angles of each segment follow those angles recorded on the real flagellar movement. The internal motion of adjusting the bending moment of each segment subsequently spills over into its adjacent segments at every time interval of Δt. This sequence of spillover, or the process of force equilibration, continues to survive indefinitely. The method of identifying the real propagation velocity of force equilibration is to search for the optimum value of $\Delta\ell/\Delta t$ that can most faithfully reproduce both internal tensile forces and the observed bending angles of the flagellum, the values of the latter of which are read from the record presented in Plate 1a.

In particular, the bending moment due to the internal tensile forces K_i and K_{i+1} of segment i (cf. Figure 5) around its moving right-most end gives

$$\Delta M_i = K_i\Delta\ell\theta_i - K_{i+1}\Delta\ell\theta_{i+1}$$

Internal dynamics of the third law imputed to the communication of null-conservedness of forces functions so as to cause ΔM_i to vanish by endogenously adjusting forces K_i and K_{i+1}, whereas the bending angles θ_i and θ_{i+1} follow those recorded. The only additional assumption in obtaining this expression of the bending moment is that the mass distribution over each segment is longitudinally uniform at least locally. The internal adjustment of internal tensile forces K_i and K_{i+1} admits some arbitrariness in fulfilling the null condition $\Delta M_i = 0$, because the third law by itself does not specify how the communication of null-conservedness of forces proceeds internally. We determine the values of K_i and K_{i+1} by a random sampling at every time interval of Δt.

It is thus noticed that if the adjustable parameter $\Delta\ell/\Delta t$ is set equal to the real propagation velocity of force equilibration, the most faithful retracements will be obtained among those traced through random samplings. As the parameter $\Delta\ell/\Delta t$ deviates from the real propagation velocity of force equilibration, on the other hand, the retracings will become less faithful because of the nature of the search based upon random samplings.

Furthermore, if the propagation velocity of force equilibration along the flagellar axoneme remains finite, the underlying law of motion will have to be of one-to-many temporal mapping supplemented by intrinsically, partially fixed boundary conditions as we have already noticed. For if there exists a unique one-to-one correspondence between mechanical displacement and force to induce it, the third law of mechanics would have to be observed at the same time, thus dispensing with the communication of null-conservedness of forces proceeding at a finite velocity. Force assumes two roles. One is to communicate the null-conservedness, and the other is to induce mechanical displacement. Intrinsic indefiniteness about the communication process is implicated in the lack of externalized explicitness about how force participates in these two roles at the same time.

One characteristic which is unique to the law of motion of the one-to-many mapping type is its self-constraining capability in that only those internal forces that fulfill the null conservation of forces *a posteriori* are able to be realized and others are eliminated. If the real propagation velocity of force equilibration is fixed, those internal forces that would be

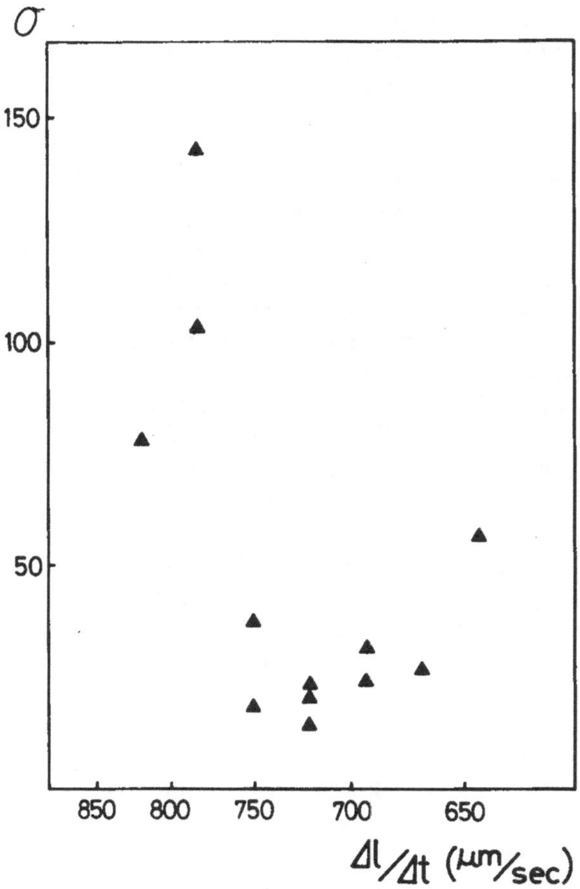

FIGURE 6. Standard deviation σ of internal tensile forces among those retracings run independently against the adjustable propagation velocity $\Delta\ell/\Delta t$ of force equilibration, in which the entire flagellar axoneme is divided into 25 segments of equal length. Tensile forces internal to the flagellum were measured in arbitrary units. Measurement of standard deviation σ was made at 24.2 ms after the top frame in Plate 1a. Indentification of the real propagation velocity of force equilibration remained almost indifferent to the choice of the adjustable parameter of length $\Delta\ell$.

accompanied by the hypothetical force equilibration proceeding at velocities either greater or smaller than the real one are necessarily eliminated from those realized forces. Accordingly, when the adjustable parameter $\Delta\ell/\Delta t$ is set equal to the real propagation velocity of force equilibration, the variance of internal tensile forces that are able to be realized would become a minimum there at least locally.

Identification of the propagation velocity of force equilibration was made by referring to the standard deviation of internal tensile forces over the similar retracements run independently through random sampling. Denoting by σ_i the standard deviation of internal tensile force K_i over retracements run independently, we demonstrate in Figure 6 the resultant standard deviation $(\sigma = \sigma_1^2 + \sigma_2^2 + \ldots + \sigma_n^2)^{1/2}$ against the adjustable velocity parameter $\Delta\ell/\Delta t$. Here, we have divided the entire flagellar axoneme into 25 segments of equal length. Internal tensile forces have been identified at 24.2 ms after the top frame in Plate 1a. The propagation velocity of force equilibration for the internal tensile forces generated along the flagellar axoneme of starfish spermatozoa has been found to be 720 ± 30 μm/s.

The resultant standard deviation σ does not vanish at its minimum. This indicates that tensile forces internal to the bending flagellum are not uniquely definable against a given mechanical displacement, being consistent with the law of motion of one-to-many mapping type supplemented by intrinsically partially fixed boundary conditions. When the communication velocity of null-conservedness of forces remains finite, those forces involved in communicating the null-conservedness but not in inducing mechanical displacements attest to the absence of a unique one-to-one correspondence between mechanical displacement and force.

Of course, many intricate molecular mechanisms are involved in the observed bending motion. We have not explicitly taken into account the role of these microscopic mechanisms in reaching a finite propagation velocity of force equilibration. Nevertheless, the present result suggests that the law of motion at a microscopic level would also be of one-to-many mapping type. If the microscopic law of motion were of a one-to-one mapping type, the resulting macroscopic law as a summation of those microscopic motions would also have to be of a one-to-one type. Conversely, the one-to-many mapping at a macroscopic level cannot be reduced to the microscopic one-to-one mapping. We have examined the rather abstract and theoretical problem of estimating the dynamic pattern of internal tensile forces needed to retrace the observed planar bending of a string. Our conclusion has been that the flagellar bending displacement and internal tensile force acting there maintains a one-to-many correspondence.

Following a similar procedure, as with the case of flagellar movement, we can also measure the propagation velocity of force equilibration associated with the forced bending motion of an inelastically flexible metallic chain recorded in Plate 1b. It is, however, evident even before the actual measurement that a form of classical equation of motion that is admittedly of one-to-one temporal mapping governs the bending motion of the nonbiological string. The law of motion of one-to-one mapping type supplemented by completely fixed boundary conditions uniquely determine both displacement of the bending chain and the internal tensile force action along it, letting the propagation velocity of force equilibration diverge in effect. The uniqueness of force lets the variance of internal tensile forces that can reproduce the same bending motion actually vanish. We thus recognize that the propagation of force equilibration proceeding at an infinite velocity is consistent with the vanishing variance of forces that can reproduce the same movement.

We demonstrate in Figure 7 a relationship between the resultant standard deviation $\sigma = (\sigma_1^2 + \sigma_2^2 + \ldots + \sigma_n^2)^{1/2}$ of internal tensile forces acting in the forced bending chain recorded in Plate 1b and the adjustable velocity parameter $\Delta\ell/\Delta t$, in which the measurement procedure remains the same as the one we have employed for the flagellar bending motion. The adjustable velocity parameter $\Delta\ell/\Delta t$ that minimized the resultant standard deviation of internal tensile forces is to give the real propagation velocity of force equilibration. In fact, the present result suggests that the resultant standard deviation decreases its value toward zero as the adjustable velocity parameter $\Delta\ell/\Delta t$ diverges. An experimental consequence derivable from this observation is that the propagation velocity of force equilibration in the bending metallic chain would diverge, implying the presence of a unique one-to-one correspondence between mechanical displacement and force. One thus notes that the law of motion of one-to-one mapping type supplemented by completely fixed boundary conditions does unfailingly function in the nonbiological motion of metallic string.

Flagellar movement is one piece of evidence suggesting that mechanics can be extended so as to include the law of motion of one-to-many mapping type supplemented by intrinsically partially fixed boundary conditions. In fact, this is a case of biological motility, but the flagellar movement is not the sole evidence exhibiting biological motility.

When two different proteins, actin and myosin, are complexed into a sarcomeric structure in the presence of ATP(adenosine triphosphate) as an energy currency in biology, they serve as an agent for generating contractile force as we observe in muscle. Yanagida et al. have recorded the movement of actin filaments both in their rigor and in the ATP-activated state in the presence of mysion as an ATPase,[37] as demonstrated in Figure 8.

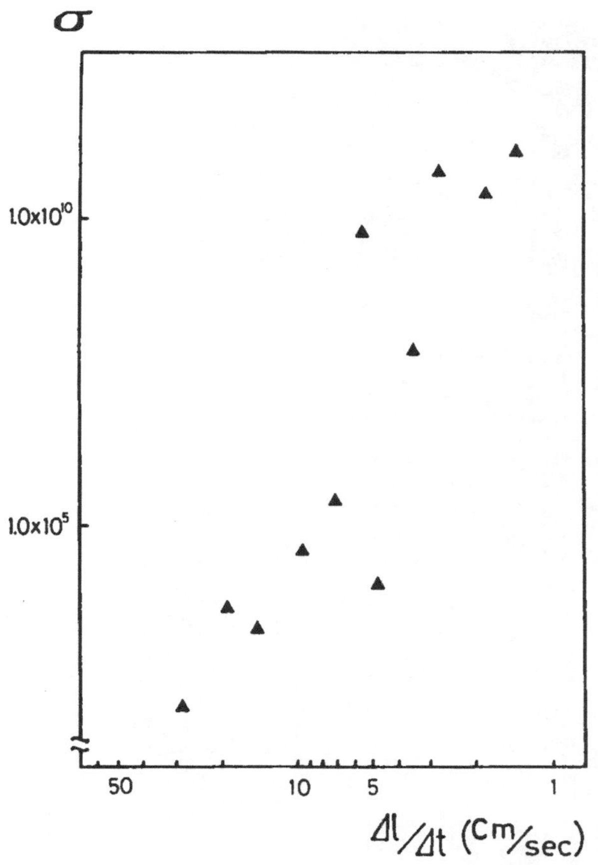

FIGURE 7. Standard deviation σ of internal tensile forces among those retracings run independently against the adjustable propagation velocity $\Delta\ell/\Delta t$ of force equilibration, in which the entire metallic string of Plate 1b is divided into 25 segments of equal length. Tensile forces internal to the metallic string were measured in arbitrary units. Measurement of standard deviation σ was made at the 25th time step in units of Δt from the initial time frame of Plate 1b. The monotonous decrease of σ with the increase of $\Delta\ell/\Delta t$ remained invariant irrespective of the choices of $\Delta t, \Delta\ell$, and the time point at which σ was measured. Both extrapolation and interpolation of the observed data were employed for the estimation.

Insofar as it is in the rigor state, the actin filament moves only nonbiologically. The actin filament in its rigor state fluctuates in the medium due to thermal motion of fluid particles. Since the force acting upon the actin filament in the rigor state originates in the surrounding particles in thermal motion, the resulting Brownian motion understandably follows the well-established law of motion of the one-to-one mapping type in the nonbiological realm of physics. On the other hand, when the actin filament is ATP activated in the presence of myosin that can transform the chemical binding energy stored in ATP into the kinetic energy of mechanical motion through hydrolysis, the process involved resembles that of muscle contraction as a representative example exhibiting biological motility. This observation turns out to be obvious if one compares the two different types of actin movement. The ATP-activated actin filament moves more flexibly and more rapidly compared with the one in the rigor state, contrary to the commonly held view on Brownian motion of polymers stating that the more flexible they become, the less rapidly they move.

A more quantitative examination of the nature of the bending motion in both the rigor and the ATP-activated states can be accomplished by measuring the propagation velocity of force equilibration with respect to internal tensile force generated along its longitudinal axis.

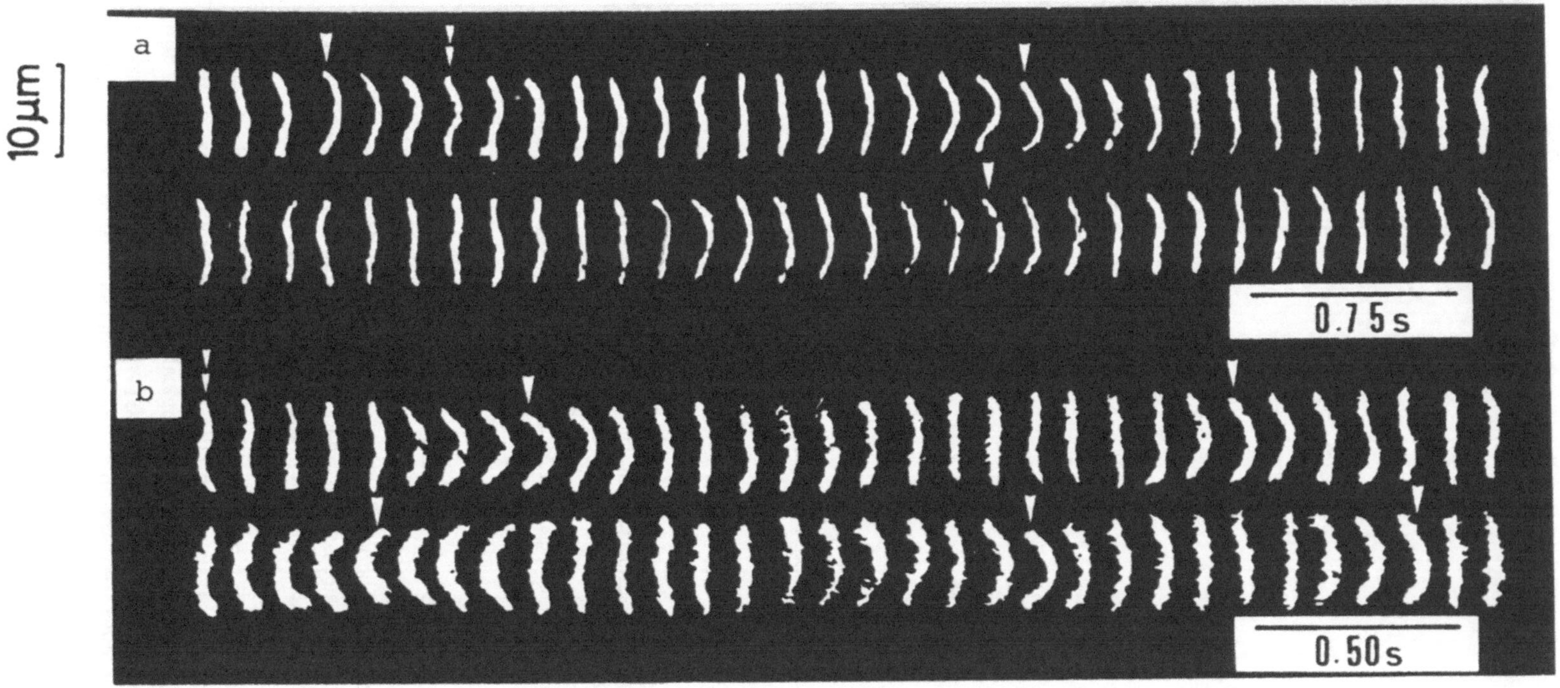

FIGURE 8. Sequence of bending actin filaments in (a) the rigor state and in (b) the ATP-activated state in the presence of myosin. The contour length of filaments was about 10 μm. The framing rates of successive pictures were 0.15 s for (a) and 0.1 s for (b). (Modified from Yanagida, T., Nakase, M., Nishiyama, K., and Oosawa, F., *Nature (London)*, 307, 58, 1984. With permission.)

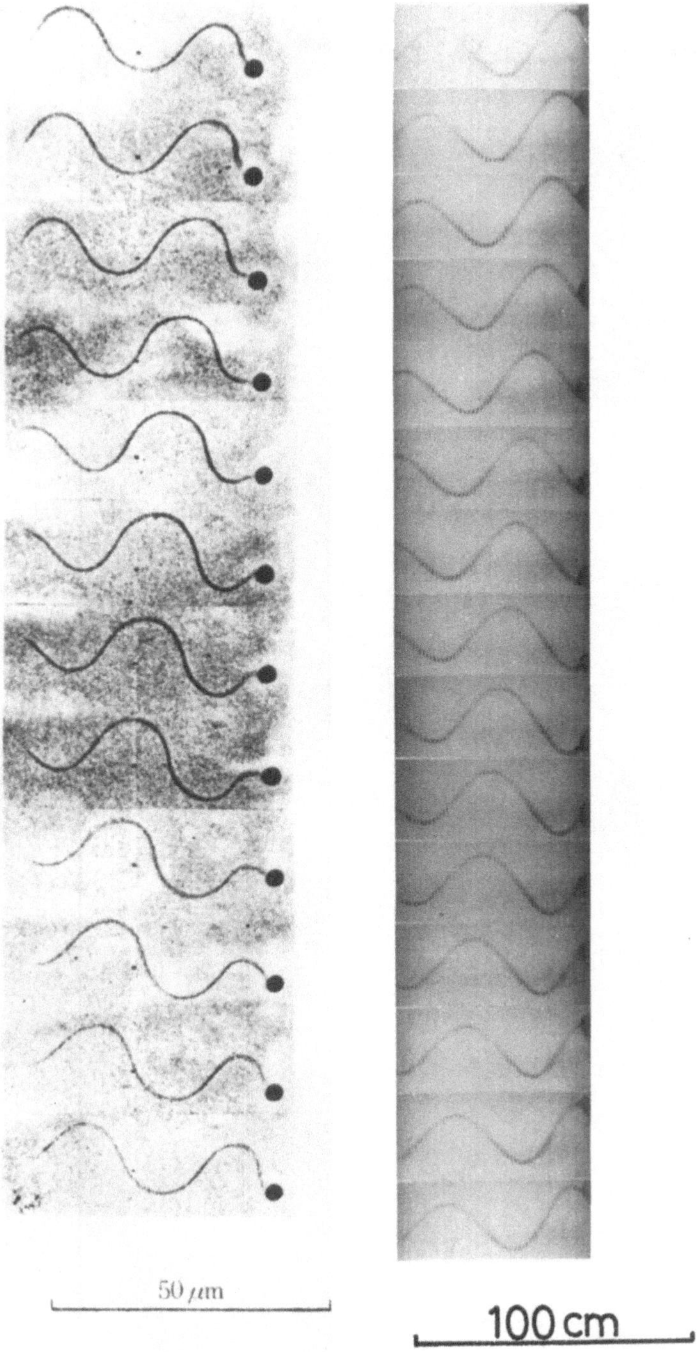

a b

PLATE 1. (a) Flagellar movement of a starfish spermatozoon swimming about 300 μm away from coverslip in a trough 1 mm in depth of normal sea water at 22°C. The time interval between successive picture from top to bottom is 2.2 ms. (From Hiramoto, Y. and Baba, S. A., *J. Exp. Biol.*, 76, 85, 1978. With permission.) (b) Forced bending motion of a flexible but inelastic metallic chain with the framing rate of 33 per second.

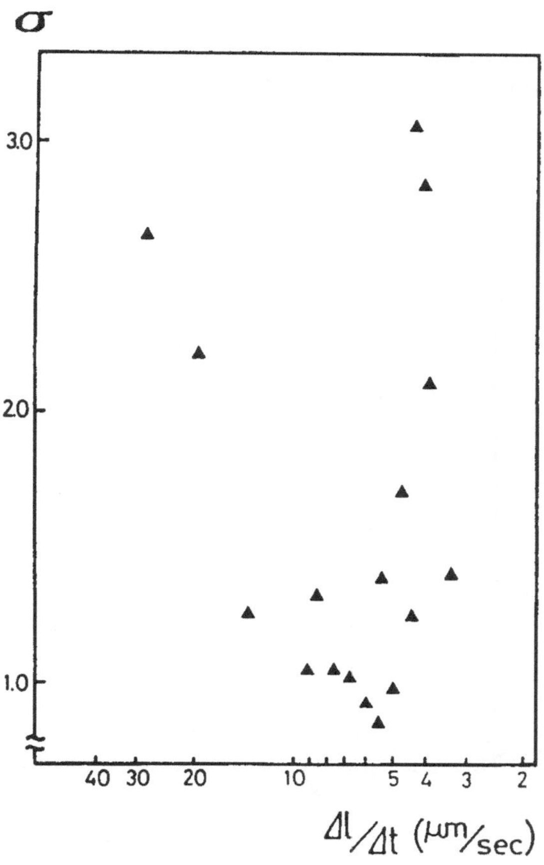

FIGURE 9. Standard deviation σ of internal tensile forces generated along the actin filament in the ATP-activated state in the presence of myosin of Figure 8b against the adjustable velocity parameter $\Delta\ell/\Delta t$ of force equilibration, in which the entire filament was divided into 15 segments of equal length and σ was measured among those retracings run independently. Tensile forces internal to the actin filament were measured in arbitrary units. Measurement of standard deviation σ was made at about 2.7 s after the initial frame of Figure 8b. Similar measurements were repeated for the filament divided into 13 (and 17) segments of equal length at about 2.7 s after the initial frame and for the same filament divided into 15 segments of equal length at about 3.2 s (and about 3.7 s) after the initial frame. The concave character of standard deviation σ remained invariant irrespective of the choices of Δt, $\Delta\ell$, and the time point at which σ was measured.

The filament movement has been found to be planar to the extent demonstrated in the record in Figure 8. The measurement method remains the same as with the case of the flagellar movement. The resultant standard deviation σ of internal tensile forces generated along the molecular longitudinal axis against the adjustable velocity parameter $\Delta\ell/\Delta t$ is plotted in Figure 9 for the actin filament ATP activated in the presence of myosin, and in Figure 10 for the filament in its rigor state.

The ATP-activated actin filament clearly exhibits a character unique to biological motility, as demonstrating the propagation of force equilibration proceeding at velocity 9 ± 4 μm/s. In contrast, the actin filament in its rigor state unmistakably exhibits a character intrinsic to a nonbiological mechanical process that would let the propagation velocity of force equilibration diverge. It preserves a unique one-to-one correspondence between mechanical displacement and force as being subject to completely fixed boundary conditions. The ATP-activated movement of actin filament joins the flagellar movement of starfish spermatozoa in experimentally confirming the law of motion of one-to-many mapping type supplemented

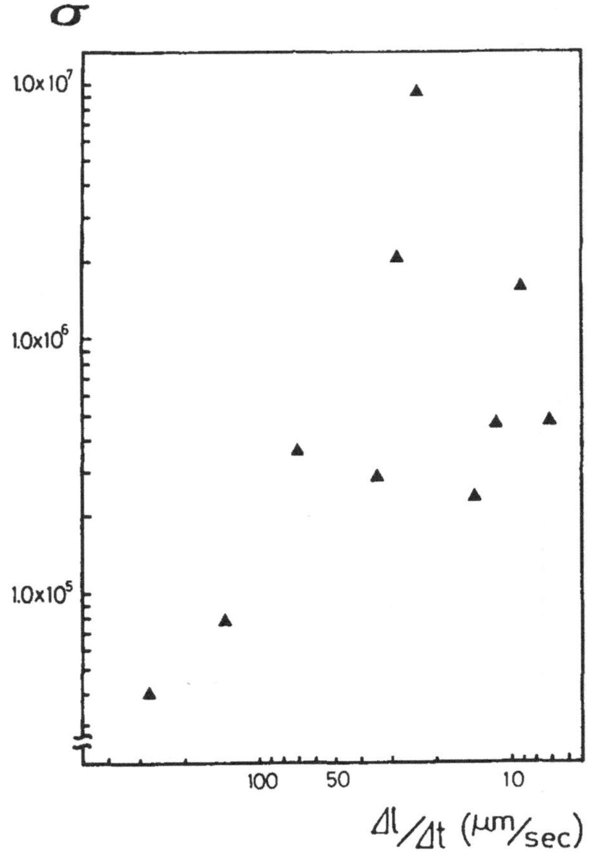

FIGURE 10. Standard deviation σ of internal tensile forces generated along the actin filament in the rigor state of Figure 8a against the adjustable velocity parameter $\Delta\ell/\Delta t$ of force equilibration, in which the entire filament was divided into 15 segments of equal length and σ was measured among those retracings run independently. Tensile forces internal to the actin filament were measured in arbitrary units. Measurement of standard deviation σ was made at the 25th time step in units of Δt from the initial frame of Figure 8a. Both extrapolation and interpolation of the observed data were employed for the estimation. The monotonous decrease of σ with the increase of $\Delta\ell/\Delta t$ remained invariant irrespective of the choices of Δt, $\Delta\ell$, and the time point at which σ was measured.

by intrinsically partially fixed boundary conditions. Experimental examination of biological motility is evidence for the law of motion of the one-to-many mapping type, that is to say, the communication of null-conservedness of forces proceeding at a finite velocity.

Our identification of the propagation velocity of force equilibration for the flagellar movement of starfish spermatozoa and for the bending motion of ATP-activated actin filaments has been factual, but indirect in that what has been directly measured is the angular displacement instead of the internal tensile force itself. It is therefore desirable to examine the process of force equilibration being propagated at a finite velocity by measuring internal force directly.

The propagation of force equilibration identified with the flagellar movement of starfish spermatozoa and with the bending motion of ATP-activated actin filaments assumes mechanochemical energy transformation. Force equilibration carries with itself a certain amount of energy. If force with its intensity f is carried by the process of force equilibration proceeding in the medium at velocity v, the work fv done by the mediate process of force comes from the chemical binding energy stored in energy-rich chemical complexes such as ATP. One thus observes an identity

$$fv \equiv -\frac{\partial E}{\partial t}$$

in which the right hand side expresses the amount of chemical binding energy released per unit time. This identity holds irrespective of the mechanism involved in the energy release from energy-rich chemical complexes.

In fact, a few parameters appearing in the physical identity have already been available to the flagellar movement of sea-urchin spermatozoa. Kamimura and Takahashi have measured internal tensile (sliding) force generated along the flagellar axoneme, with $f \sim 10^{-7}$ dyne per dynein arm.[38] Dyneins are positioned every 24 nm apart on the filament of another protein named tubulin along the axoneme. The propagation velocity of force equilibration along the sea-urchin flagellum is identified indirectly by measuring the propagation velocity of the bending wave. Ishijima and Hiramoto have reported the propagation velocity $v \sim 10^{-3}$ cm/s.[39] On the other hand, the number of ATP molecules hydrolyzed within one beat cycle of the flagellum is about two per dynein arm and the beat cycle is 10^2/s.[35,36] Since the energy released during hydrolysis of ATP to ADP (adenosine diphosphate) is 7.3 kcal/mol,[40] the chemical binding energy released in the ATP-activated flagellar movement is $-(\partial E/\partial t) \sim 10^{-10}$ erg/s per dynein arm. One thus recognizes an empirical confirmation of the equality $fv \sim -(\partial E/\partial t) \sim 10^{-10}$ erg/s per dynein arm at least as far as the order of magnitude is concerned.

Consequently, force equilibration is found to serve as a means of transforming the chemical binding energy directly to the kinetic energy of mechanical motion although each transformation requires its own specific energy releaser such as a specific ATPase like dynein or myosin. Mechanochemical couplings in biological motility are unique in that there exists a process of mechanical force being propagated with the energy-releasing chemical reactions.[41] The spatially distributed chemical reactions generating mechanical force affect themselves at least in a manner consistent with the third law of mechanics stating the null-conservation of forces of action and reaction. A mere simultaneous expression of these distributed chemical reactions alone does not lead to the similar simultaneous detection of each of them by themselves as they are expressed. Each chemical reaction takes time in detecting others, and the endogenous detection proceeds in a manner such that the infallible third law may be observed *a posteriori*. The process effecting the third law, or force equilibration, is to proceed with the propagation of the energy-releasing chemical reactions.

Biological motility exhibits a type of mechanical process that would allow force equilibration to be propagated at a finite velocity. In fact, biological motility shows that physics holds within itself room for the law of motion one-to-many mapping type supplemented by intrinsically partially fixed boundary conditions.

VI. PROTOBIOLOGY AND ITS PROBLEMS

The time-honored mechanistic model based on the law of motion of one-to-one mapping type supplemented by completely fixed boundary conditions is competent in coping with the physics of being. For one-to-one temporal mapping uniquely determines the future development of a trajectory insofar as the initial conditions are unambiguously specified. However, intrinsic partial fixedness of boundary conditions making the law of motion of one-to-many mapping type inevitable is also admissible in physics as demonstrated in biological motility. The law of motion of one-to-many mapping type renders what is to be transformed less degenerate with the elapse of time.

Constant dissolution of the previous degeneracy with time itself constitutes a generative source of changing and becoming that can serve as a basis of the physics of becoming. In fact, the propagation of force equilibration proceeding at a finite velocity provides factual evidence for protobiology as a physics of becoming that can yield and give birth to biology. Insofar as one understands that the process basic to both contemporary biology and its emergence persistently survives in the physics of becoming, force equilibration can serve as a process fundamental to protobiology.

What makes protobiology unique compared with other branches of physics is its appreciation of the endogenous and final capability of aiming at infallible conservation laws on the part of matter.[42] If the law of motion were of one-to-one temporal mapping, its completely specified initial condition will exogenously guarantee an instantaneous embodiment of those conservation laws right at the moment the condition is set. On the other hand, the materialistic endogenous capability for conservation laws that is unique to the law of motion of one-to-many mapping type necessitates the internal process of both identifying what has been accomplished and realizing what will be in accord with these laws. Internal identification and realization are in fact a principal attribute of the process named measurement.

Of course, measurement is a notorious problem in physics and it is not our intention to exhaustively clarify what measurement is all about. We only wish to emphasize a common denominator of measurements, namely, measurement as identification and realization of a particular pattern of interaction between an object to be measured and a means of measurement. Although it is customary to provide a means of measurement externally from an anthropocentric perspective, measurement understood as identification and realization of a particular pattern of interaction is ubiquitous as far as any pair of interacting bodies are available irrespective of their origin. Either component of the interaction bodies can serve as a means of measuring the other.

Protobiology can be seen as an instance of internal measurement that accompanies the law of motion of one-to-many mapping type. The situation is, however, quite different from the law of motion of one-to-one mapping type. In the latter, the completely specified initial condition unambiguously determines *a priori* how both internal identification and realization of a particular pattern of interaction would proceed. The later development following the one-to-one temporal mapping would simply be a deterministic extension of what has been identified in an exhaustive manner. The law of motion of one-to-many mapping type, on the other hand, does not specify in advance what will be identified and what will be realized. Internal measurement is responsible for what has been identified and realized.

Force equilibration has been argued to be a process unique to protobiology. It does demonstrate a functioning of the law of motion of one-to-many mapping type. But, it does not specify how internal measurement proceeds. Internal measurement is a problem in protobiology that has to be further elaborated. The problem of measurement in quantum mechanics, which we have not yet elaborated on, will become unavoidable in clarifying internal measurement in protobiology.

Internal measurement in protobiology is not restricted to force equilibration as an agent materializing the null-conservedness of forces which the third law of mechanics implies. Internal measurement as the process materializing conservedness also applies to the conservation of matter, or more specifically to material flow continuity, and to the conservation of energy. In view of the fact that biological and protobiological systems are open to material flow during their evolution and development, the endogenous and materialistic capability for actualizing material flow continuity plays a major role in protobiology. Emergence and evolution of cellular material aggregates that are open to material flow constitute a problem that protobiology sets for itself.

A microscopically materialistic capability for materializing both the third law of mechanics and material flow continuity would not need to be called for if the law of motion is of one-to-one mapping type as with Newton's equations of motion in classical mechanics. However, if the law of motion of one-to-many mapping type deserves its due attention as demonstrated in force equilibration underlying biological motility, its function turns out to be self-constraining in that only those motions that fulfill conservation laws *a posteriori* are realizable.

The law of motion of one-to-many mapping type, when combined with the associated internal measurement, becomes informational, because one attribute of information is in the process of constraining the extent of possibilities.[43,44]

An intriguing aspect of the origin of protobiological and biological information is in the process that makes what would have looked less constrained become more constrained with the elapse of time.[45-48] If one wishes to model the informational process by a law of motion of a one-to-one mapping type supplemented by completely fixed boundary conditions, the origin of information has to be sought in an external agent that exogenously controls boundary conditions.[49] Of course, those exogenous origins of biological information cannot be denied purely on a logical basis. However, the factual evidence of force equilibration in the physics of becoming implies that the law of motion of one-to-many mapping type is unfailingly functioning in reality and is endogenous in its capacity of constraining.

The origin of protobiological and biological information can be endogenous.[50,51] Conversely, the endogenous origin of protobiological and biological information necessitates the law of motion of one-to-many mapping type. The associated intrinsic partial fixedness of boundary conditions assures that the materialistic endogenous constraint would by no means destroy the very partial fixedness no matter how much it may be constrained. The partial fixedness is intrinsic in the sense that it cannot be identified externally in an exhaustive manner.

Information in evolution refers to both the evolutionary process and its products. Protobiology is an attempt to cope with the evolutionary problem of processes and products from the perspective of the law of motion of one-to-many mapping type supplemented by intrinsically partially fixed boundary conditions. Interplay between internal measurement and the process of equilibration thus becomes inevitable.

Information does evidently contrast with what is real with what could be possible. Moreover, it is a common practice in physics to associate what is possible with entropy. A problem unique to entropy is how to define it. If entropy could not be defined, its link with information would be lost. Entropy is in fact definable insofar as a probability distribution of events is available. Statistical-mechanical entropy, however, does not fulfill the purpose in protobiology, for the law of motion of one-to-one mapping type underlies statistical mechanics, whether in classical or in quantum mechanics. A major problem for protobiology is to examine whether entropy is intrinsic to the law of motion of one-to-one mapping type supplemented by completely fixed boundary conditions. If the basis upon which entropy could be definable is not evident, protobiology would have to dispense with entropy, no matter how useful it may be for other purposes. The definitional problem of entropy must be settled before we enter details on entropy in protobiology and in biology.

Our protobiology is thus a commitment to scrutinize measurement or internal measurement, equilibration, information, and entropy on the basis of the fact that the law of motion of one-to-many mapping type supplemented by the intrinsic partial fixedness of boundary conditions paves the way to the physics of becoming that can yield and give birth to biology. Protobiology then examines the functional origin of full-blown biological processes such as natural selection, mutation, and adaptation on the physical basis of internal measurement, equilibration, information, and entropy.

VII. SUMMARY

Protobiology is an attempt to identify a physics of becoming that can yield and give birth to biology. Mechanistic models built on the law of motion of a one-to-one mapping type supplemented by completely fixed boundary conditions do not fulfill this purpose. Being contrary to the one-to-one temporal mapping in Newton's equations of motion of the classical trajectory and in Schrödinger's equation of motion of the quantum-mechanical wave function, the law of motion in protobiology has to be of one-to-many temporal mapping supplemented by intrinsically partially fixed boundary conditions. Biological motility observed in contemporary biology exhibits the law of motion of one-to-many mapping type in operation, demonstrating that the process materializing the third law of mechanics is propagated in the medium at a finite, not infinite, velocity. Intrinsic, partial fixedness of boundary conditions that has nothing to do with the technical limitation on the part of an observer is identified in mechanical motion in biology. The partial fixedness is factually observed in the experimental evidence that force is not uniquely definable even if mechanical displacement in motion is identified. The third law of mechanics yielded by and leading to the process of becoming, counterbalanced between acting and reacting forces, serves as a regulative principle in protobiology and biology as maintaining itself to be a form of temporal transformation of a one-to-many mapping character, in contrast to the third law in the context of asking the state of being, counterbalanced between these acting and reacting forces at any moment.

The law of motion of one-to-many mapping type, supplemented by intrinsically partially fixed boundary conditions that we see in protobiology, motivates us to take a new look at measurement or internal measurement, equilibration, information, and entropy as a functional basis of the physics of becoming that can give birth to biology.

REFERENCES

1. **Wigner, E.,** Events, law of nature, and invariance principles, *Science,* 145, 995, 1964.
2. **Mehlberg, H.,** *Time, Causality, and The Quantum Theory,* Vol. 1 and 2, (Parts 2 and 3), Reidel Dordrecht, Boston, 1980, chap. 9 and 4.
3. **Polanyi, M.,** Life's irreducible structure, *Science,* 160, 1308, 1968.
4. **Mercer, E. H.,** *The Foundation of Biological Theory,* John Wiley & Sons, New York, 1981.
5. **Leggett, A. J.,** The superposition principle in macroscopic systems, in *Proc. Int. Symp. on Foundations of Quantum Mechanics,* Kamefuchi, S., Ezawa, H., Murayama, Y., Namiki, M., Nomura, S., Ohnuki, Y., and Yajima, T., Eds., Physical Society of Japan, Tokyo, 1984, 74.
6. **Watanabe, S.,** Time and the probabilistic view of the world, in *The Voices of Time,* Fraser, J. T., Ed., Braziller, New York, 1966, 527.
7. **Denbigh, K. G.,** *Three Concepts of Time,* Springer-Verlag, Berlin, 1981, chap. 7 and 8.
8. **Prigogine, I.,** *From Being to Becoming: Time and Complexity in the Physical Sciences,* Freeman, San Francisco, 1980.
9. **Prigogine, I. and Stengers, I.,** *Order out of Chaos,* Bantam, New York, 1984.
10. **Peacocke, A.,** Thermodynamics of life, *Zygon,* 19, 395, 1984.
11. **Boltzmann, L.,** *Vorlesungen über Gastheorie,* Barth, Leipzig, 1896-98; *Lectures on Gas Theory* (transl.), Brusch, S. G., University of California Press, Berkeley, 1964.
12. **Gibbs, J. W.,** *Elementary Principles in Statistical Mechanics,* Scribners, New York, 1902.
13. **Penrose, O.,** *Foundations of Statistical Mechanics,* Pergamon Press, London, 1970.
14. **Brusch, S. G.,** *Statistical Physics and the Atomic Theory of Matter, from Boyle and Newton to Landau and Onsager,* Princeton University Press, Princeton, New Jersey, 1983.
15. **Ehrenfest, P. and Ehrenfest, T.,** *The Conceptual Foundations of the Statistical Approach in Mechanics* (transl.), Moravcsik, M. J., Cornell University Press, Ithaca, N.Y., 1959.
16. **Martin-Löf, A.,** *Statistical Mechanics and the Foundations of Thermodynamics,* Springer-Verlag, Berlin, 1979.
17. **Rogers, B.,** Probabilistic causality, explanation, and detection, *Synthesis,* 48, 201, 1981.
18. **Slobodkin, L. B.,** Listening to a symposium—a summary and prospectus, in *Biotic Crises in Ecological and Evolutionary Time,* Nitecki, M. H., Ed., Academic Press, New York, 1981, 269.
19. **Matsuno, K.,** Material self-assembly as a physicochemical process, *BioSystems,* 13, 237, 1981.
20. **Matsuno, K.,** A theoretical construction of protobiological synthesis: from amino acids to functional protocells, *Int. J. Quantum Chem.,* QBS9, 181, 1982.
21. **Landau, L. D. and Lifshitz, E. M.,** *Fluid Mechanics,* Addison-Wesley, Reading, Mass., 1959.
22. **Kadanoff, L. P. and Martin, P. C.,** Hydrodynamic equations and correlation functions, *Ann. Phys.,* 24, 419, 1963.

23. **Martin, P. C., Parodi, P., and Pershan, P. S.,** Unified hydrodynamic theory for crystals, liquid crystals, and normal fluids, *Phys. Rev.*, A6, 2401, 1968.
24. **Porod, W., Grondin, R. O., Ferry, D. K., and Porod, D.,** Dissipation in computation, *Phys. Rev. Lett.*, 52, 232, 1984.
25. **Whitrow, G. J.,** *The Natural Philosophy of Time*, 2nd ed., Clarendon Press, Oxford, 1980.
26. **Davies, P. C. W.,** *The Physics of Time Asymmetry*, Surrey Univeristy Press, Guildford, 1974.
27. **Prigogine, I.,** The rediscovery of time, *Zygon*, 19, 433, 1984.
28. **Glansdorff, P. and Prigogine, I.,** *Thermodynamic Theory of Structure, Stability and Fluctuations*, John Wiley & Sons, New York, 1971.
29. **Nicolis, G. and Prigogine, I.,** *Self-Organization in Nonequilibrium Systems: from Dissipative Structures to Order Fluctuations*, John Wiley & Sons, New York, 1977.
30. **Haken, H.,** Cooperative phenomena in systems far from thermal equilibrium and in non-physical systems, *Rev. Mod. Phys.*, 47, 67, 1975.
31. **Haken, H.,** *Synergetics*, Springer-Verlag, New York, 1977.
32. **Papagiannis, M. D.,** Life-related aspects of stellar evolution, *Origins Life*, 14, 43, 1984.
33. **Purcell, E. M.,** Life at low Reynolds number, *Am. J. Phys.*, 45, 3, 1977.
34. **Berg, H. C. and Turner, L.,** Movement of microorganisms in viscous environments, *Nature (London)*, 278, 349, 1979.
35. **Brokaw, C. J.,** Flagellar movement: a sliding filament model, *Science*, 178, 455, 1972.
36. **Hiramoto, Y. and Baba, S. A.,** A quantitative analysis of flagellar movement in echinoderm spermatozoa, *J. Exp. Biol.*, 76, 85, 1978.
37. **Yanagida, T., Nakase, M., Nishiyama, K., and Oosawa, F.,** Direct observation of motion of single F-actin filaments in the presence of myosin, *Nature (London)*, 307, 58, 1984.
38. **Kamimura, S. and Takahashi, K.,** Measurement of sliding forces along the flagellar microtubule under the condition of constant length, *Biophysics (Japan)*, 22, S56, 1982.
39. **Ishijima, S. and Hiramoto, Y.,** Direct measurement of bending forces of spermatozoa, *Biophysics (Japan)*, 23, S57, 1983.
40. **Lehninger, A. L.,** *Principles of Biochemistry*, Worth, New York, 1982, chap. 17.
41. **Yanagida, T., Arata, T., and Oosawa, F.,** Sliding distance of actin filament induced by a myosin crossbridge during ATP hydrolysis cycle, *Nature (London)*, 316, 336, 1985.
42. **Matsuno, K.,** Protobiology: a theoretical synthesis, in *Molecular Evolution and Protobiology*, Matsuno, K., Dose, K., Harada, K., and Rohlfing, D. L., Eds., Plenum Press, New York, 1984, 433.
43. **Jaynes, E. T.,** Information theory and statistical mechanics, *Phys. Rev.*, 106, 620, 1957.
44. **Brillouin, L.,** *Science and Information Theory*, Academic Press, New York, 1962.
45. **Wicken, J. S.,** A thermodynamic theory of evolution, *J. Theor. Biol.*, 87, 9, 1980.
46. **Wiley, E. O. and Brooks, D. R.,** Victims of history—a nonequilibrium approach to evolution, *Syst. Zool.*, 31, 1, 1982.
47. **Brooks, D. R. and Wiley, E. O.,** Evolution as an entropic phenomenon, in *Evolutionary Theory: Paths into the Future*, Pollard, J. W., Ed., John Wiley & Sons, New York, 1984, chap. 6.
48. **Brooks, D. R.,** Information and entropy in a simple evolution model, *J. Theor. Biol.*, 109, 77, 1984.
49. **Frautschi, S.,** Entropy in an expanding universe, *Science*, 217, 593, 1982.
50. **Layzer, D.,** The arrow of time, *Sci. Am.*, 233, 56, 1975.
51. **Layzer, D.,** Information in cosmology, physics and biology, *Int. J. Quantum Chem.*, 12(Suppl. 1), 185, 1977.

Chapter 2

INTERNAL MEASUREMENT

I. INTRODUCTION

A process underlying the physics of becoming that is fundamental to protobiology is the one that actualizes the conservedness as evidenced in the equilibration of forces between two types of force, action and reaction. Inherent to the actualization of conservedness is the process of measurement that is to identify and then to implement what is required for fulfilling the principle of conservation *a posteriori*. Measurement is the realization of a particular pattern of interaction between an object to be measured and a means of measurement. Measurement is ubiquitous in any pair of interacting bodies in the sense that either one of the two serves as a means of measuring the other, though it is rather common from an anthropocentric perspective to provide a means of measurement externally.

The law of motion of the one-to-one mapping type supplemented by completely fixed boundary conditions makes the process of measurement superfluous, since what will be expected from measurement is already latent unambiguously in the law of motion and the associated boundary conditions. On the other hand, if the law of motion is of the one-to-many mapping type supplemented by intrinsically partially fixed boundary conditions, the process of measurement that proceeds internally is responsible for identifying which one out of many possible alternatives is actually chosen in time. The law of motion of one-to-many mapping type can indicate by itself what alternatives could be possible. But, the law of motion cannot identify what will be actualized in reality unless it is supplemented by internal measurement that is intrinsically selective in its operation. Quantum mechanics naturally allows internal measurement, though the quantum-mechanical development of wave function following Schrödinger's equation of motion of the one-to-one mapping type dispenses with such an internal measurement. The matter of internal measurement in quantum mechanics becomes extremely important when we examine how the boundary condition imposed on the quantum-mechanical wavefunction is provided and how it could be measured. Internal measurement in quantum mechanics prepares the ground for protobiology in such a manner that the law of motion is of a one-to-many mapping type necessarily accompanied by an endogenous selection process acting upon those boundary conditions that are intrinsically partially fixable.

II. MEASUREMENT IN CLASSICAL MECHANICS

We have already observed that biological motility presupposes the law of motion of the one-to-many mapping type supplemented by intrinsically partially fixed boundary conditions as evidenced in the flagellar movement of starfish spermatozoa and in the bending motion of ATP-activated actin filaments in the presence of myosin. However, the one-to-many temporal mapping is not specific enough by itself to identify how the trajectory develops with time. The trajectory of motion that has been unique in the record in retrospect always serves as a source of possible trajectories yet to come. But the one-to-many mapping by itself does not tell us in anticipation which trajectory will really be selected. Something that concatenates a wide variety of anticipated possibilities with the recorded uniqueness in retrospect is missing in the one-to-many mapping. What is missing is an agent to constrain the extent of possible trajectories when the real trajectory comes into existence.

It is of course difficult to exhaustively list all of the characteristics of such a constraining agent. Nevertheless, one can identify one characteristic common to any of those agents.

This is the process materializing conservedness underlying infallible conservation laws. The actualization of conservedness assumes an elimination of what would violate the principle of conservation when recorded. Such an elimination comes to enhance the extent of constraint on what carries the very process of materializing conservedness. We thus recognize that measurement is involved in the actualization of conservedness, to say the least. For the materialistic communication of conservedness assumes a process of measurement taking place internally that identifies what is implied by the conservedness.

In fact, measurement is an agent for realizing and indicating a particular pattern of interaction between an object to be measured and a means of measurement.[1] Measurement is ubiquitous in any pair of interacting bodies in spite of its heavily anthropocentric connotation. Either one of the two bodies can serve as a means of measuring the other. Above all, measurement exhibiting a particular pattern of interaction between arbitrary interacting bodies underlies the process materializing conservedness, since the latter is actualized only through materialistic interactions. Further elucidation of measurement per se will be in order for the purpose of specifying the process materializing conservedness.

Measurement is classified into at least two subgroups, internal and external. Internal measurement is inherent in realization of a particular pattern of interaction between an arbitrary pair of interacting bodies in the absence of an external observer. Such an internal measurement is ubiquitous in any material system.

In contrast, external measurement that assumes an intervention of an external observer is very common in physical experiments in spite of its very exceptional role as embodied in the frequently quoted dictum saying that measurement always perturbs what is to be measured. Physical experiment usually presumes that the effect of the measurement apparatus on what is to be measured could be kept nil or could be maintained on the lowest possible level at its best. It is only in this sense of nonvanishing effects from the measurement apparatus that the very process of measurement intrinsically disturbs measurement itself. It is one thing to call the effect of measurement apparatus an inevitable disturbance on measurement from an anthropocentric perspective, but quite another to recognize the presence of mutual interference between what is doing the measuring and what is being measured. External measurement is an anthropocentric artifact of regarding the effect of the measurement apparatus as irrelevant or at most not substantial by referring to it simply as disturbance, however inevitable the presence is. Internal measurement, on the other hand, is inherent in the process of fully mutual interference between what is doing the measuring and what is being measured.

Classical mechanics incorporates both internal and external measurements in a unique manner. One-to-one temporal mapping in classical mechanics, whose representative example is Newton's equations of motion of the classical trajectory, deprives internal measurement of the capacity of selecting a particular trajectory out of many possible alternatives. The one-to-one temporal mapping that is consistent with the presence of completely fixed boundary conditions lets the past trajectory unambiguously determine the future trajectory and vice versa. Internal measurement thus turns out to be superfluous in classical mechanics. As far as the law of motion of the one-to-one mapping type supplemented by completely fixed boundary conditions is guaranteed, internal measurement that is endogenous and self-constraining need not be invoked anymore. Internal measurement underlying the actualization and communication of conservedness is to proceed at an infinite velocity in classical mechanics. For the effect of internal measurement observing the principle of conservation *a posteriori* is already latent in the boundary conditions once they are fixed completely. Hypothetical internal measurement proceeding at an infinite velocity does not occupy the status of measurement in practice. Instead, external measurement claims its reality in classical mechanics.

Objectives of an external observer are multifaceted when the observer measures how the trajectory develops with time as following the law of motion of one-to-one mapping type, such as Newton's equations of motion. One objective is to explicitly determine the law of motion underlying the measured trajectory. One more objective is to identify the boundary conditions to which the measured trajectory is subjected.

Empirical or experimental identification of the law of motion assumes measurement of the trajectory at at least two different time points. Both the coordinate of the trajectory at a given moment and its temporal displacement are prerequisites for the identification of the underlying law of motion. However, one can recognize a queer disparity between an empirical confirmation of and a theoretical affirmation of the law of motion that is supposed to be of the one-to-one mapping type. As far as it is proposed as a theoretical artifact, the law of motion of the one-to-one mapping type is a simultaneous expression of both the coordinate of the trajectory and its temporal displacement. Both are conceived only theoretically and no measurements are involved.

In contrast, as far as their external measurements are concerned, both the coordinate of the trajectory at a given moment and its temporal displacement cannot be identified simultaneously. Both are measurable only sequentially, one by one. When the coordinate of the trajectory is measured, its displacement could not be allowed to intervene. Likewise, when the displacement along the trajectory is measured, its coordinate could not be allowed to be free from its displacement. External measurement alone cannot tell us whether or not the law of motion could be of the one-to-one mapping type. Unless it is supplemented by an independent theoretical premise stating that the law of motion has to be of the one-to-one mapping type, mere external measurements of both the coordinate of the trajectory and its temporal displacement cannot be good enough to confirm the presence or the absence of such a one-to-one temporal mapping.

Nevertheless, it may be possible to examine an experimental validity of the theoretically proposed law of motion of the one-to-one mapping type through external measurement if measurement refers only to those quantities measurable at each moment. Initial conditions as a particular form of boundary conditions are the ones that are measurable by definition at a given initial moment. The proposed one-to-one temporal mapping claims that once an initial coordinate of the trajectory is determined by whatever means, its later development will simply be a unique temporal extension of what has been specified at the initial time point. Such a unique temporal extension excludes those initial conditions that could remain ambiguous in their later development.

External measurement can determine whether an experimentally imposed and identified initial condition could guarantee a unique time development of the trajectory if its later coordinate is externally measurable independently of the manner of its identification. However, one must pay attention to the fact that there inevitably exists an experimental limitation upon the resolution power for any type of external measurement. Experimental controllability does not properly function below the resolution level that any measurement apparatus inevitably succumbs to in its own way. If the two sets of initial conditions that could not be identified as being separate due to the inevitable limitation of experimental controllability yield those two trajectories whose separation could become measurable later, this would manifest a lack of the capacity for preparing initial condition in a precise manner. If the resolution of measurement is not fine enough, an experimenter can design a measurement apparatus with finer resolution. Still, this sequence of achieving the finer resolution cannot proceed without limit. No matter how sophisticated the apparatus may be, there does exist the resolution limit due to the inevitable limitation of experimental controllability. The problem will become more serious if one is asked to examine on the basis of experiment whether the initial condition could guarantee a unique later development of the trajectory.

If the inevitable experimental limit of resolution masks the difference between two sets of initial conditions which could later yield two different trajectories being measurable

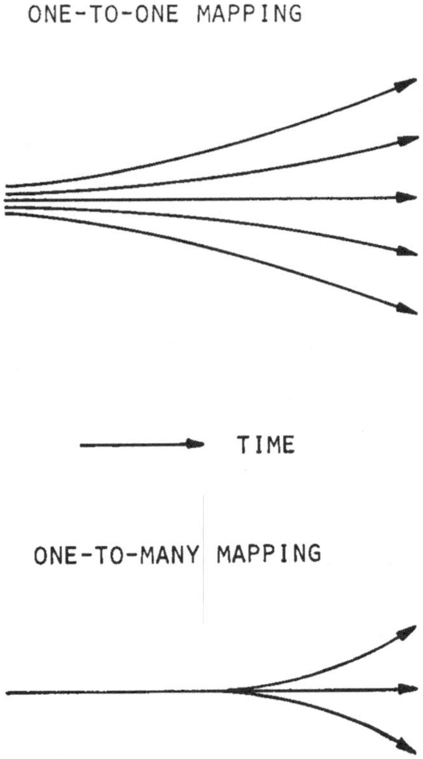

FIGURE 1. A qualitative difference between one-to-one and one-to-many temporal mappings. One-to-many mapping assumes an intrinsic degeneracy of trajectories, whereas one-to-one mapping does not allow such a degeneracy under the scheme of unlimited theoretical controllability of boundary conditions even though the trajectory may seem chaotic.

separately, one cannot answer on the basis of experiment the question of whether the underlying law of motion be uniquely of the one-to-one mapping type. In fact, there is a class of one-to-one temporal mapping such that the slightest amount of separation in the initial conditions can later develop an apparently appreciable difference.[2,3] A unique retracing of such a chaotic trajectory is beyond limited experimental controllability. However, insofar as one claims unlimited theoretical controllability of initial conditions, any type of one-to-one temporal mapping can be saved as such, at least theoretically.

It is possible to theoretically imagine two initial conditions that are separated only infinitesimally and to deduce some theoretical consequence on the later difference between the two trajectories from such a premise. But, this line of argument does not work in experiment. External measurement based upon limited experimental controllability cannot serve as a means to empirically confirm the one-to-one temporal mapping. It cannot cope with unlimited theoretical controllability of boundary conditions in the case of chaotic trajectories. Furthermore, if a one-to-many temporal mapping is the case, even theoretical controllability of boundary conditions would remain intrinsically limited in the sense that it is not possible to identify in an exhaustive manner how the future developments are degenerated in the present. Experimental controllability alone cannot distinguish whether the observed trajectory could develop as following one-to-one temporal mapping supplemented by completely fixed boundary conditions or if it could develop as following one-to-many temporal mapping,[4] as depicted in Figure 1.

The law of motion of the one-to-one mapping type claims in its own right unlimited theoretical controllability of boundary conditions even if the temporal mapping happens to be chaotic in the sense that the slightest amount of deviation in the initial conditions yields an appreciable difference later. Otherwise, the uniqueness of trajectory would be destroyed even in theory.

However, there has been some attempt to lessen the extent of controllability even on theoretical grounds because any theory must be checked against its corresponding experiment and because experimental controllability of boundary conditions is inevitably limited. Limited controllability of initial conditions can transform the one-to-one temporal mapping that allows chaotic trajectories into a form of one-to-many temporal mapping that necessarily lacks the uniqueness of its development.

Statistical mechanics is just such a theoretical attempt to abandon unlimited controllability of boundary conditions, being contrary to the fact that theoretical controllability does remain unlimited as far as the underlying law of motion is of the one-to-one mapping type. Whether such an artificial lessening of controllability could survive in theory is totally dependent upon how well or poorly the corresponding experiment can reproduce what the theory predicts. Only when the artificial lessening of theoretical controllability coincides with the limitation of its experimental counterpart can statistical mechanics also survive in experiment.

One-to-one temporal mapping in classical mechanics is very theoretical because it assumes unlimited theoretical controllability of boundary conditions in its own right. Although statistical mechanics is a theoretical compromise to appreciate the practical impossibility in controlling boundary conditions in an unlimited manner, the one-to-one mapping, by definition, lets boundary conditions be completely fixable as a matter of principle. One-to-one temporal mapping supplemented by completely fixed boundary conditions is thus consistent with unlimited theoretical controllability of boundary conditions.

In theory, any one-to-one temporal mapping guarantees a unique and deterministic development of its trajectory no matter how many degrees of freedom may be involved. Even if external measurement is not capable of tracing the unique development of a trajectory due to its limited controllability of boundary conditions, this incapability would neither negate nor justify the unique development of trajectory. Unless an artificial lessening of theoretical controllability is applied to the one-to-one temporal mapping as experienced in statistical mechanics, the theoretical uniqueness of a trajectory remains invulnerable to any external measurement.

One can observe in statistical mechanics a curious coexistence of both an artificial lessening of theoretical controllability over boundary conditions and its inevitable limitation in experiment. This coexistence is understandable as far as the underlying law of motion remains of the one-to-one mapping type. Of course, if it is not necessary to invoke an artificial lessening of theoretical controllability, the unique and deterministic development of a trajectory survives at least in theory. However, the situation will turn out to be completely different if one-to-one temporal mapping is not evident even on theoretical grounds.

If the law of motion is of a one-to-many mapping type, theoretical controllability of boundary conditions will be intrinsically limited. Complete identification of the past trajectory does not lead to a unique specification of the future trajectory. One-to-many temporal mapping always renders the past trajectory degenerate with the future ones. But a new problem would arise if one wishes to examine on the basis of experiment whether the law of motion could be of the one-to-many mapping type. As far as external measurement is concerned, experiments alone cannot distinguish between an artificial lessening of unlimitedness of theoretical controllability pertinent to one-to-one temporal mapping and its intrinsic limitedness unique to one-to-many mapping. Experimental controllability of boundary conditions is intrinsically limited irrespective of whether the underlying law of motion is of the one-to-one or of the one-to-many mapping type. As a result, limited experimental controllability could be compatible with both artificial lessening and intrinsic limitedness of theoretical controllability over boundary conditions.

Further clarification of one-to-many temporal mapping will require due examination of the role of internal measurement that is involved in the process of transforming anticipated possible trajectories into the recorded single one. Classical mechanics in the form of the law of motion of the one-to-one mapping type dispenses with the involvement of such an internal measurement. Measurement in classical mechanics is restricted exclusively to external measurement. The one-to-one temporal mapping is thus conceivable in theory even without invoking measurement, either external or internal. Accordingly, classical mechanics is destined to boast unlimited theoretical controllability over boundary conditions beyond the bound that any experiment sets for itself.

One-to-one temporal mapping, complete fixedness of boundary conditions, and unlimited theoretical controllability over those conditions constitute a consistent tripartite circle. However, once one pays equal attention to the inevitable limitedness of experimental controllability over boundary conditions, it is by no means evident whether the consistent tripartite circle could remain intact as before. Classical mechanics is very exceptional in maintaining the consistent tripartite circle purely on theoretical grounds. The only footnote it would admit is that limited experimental controllability over boundary conditions could be equated with an artificial lessening of theoretical controllability whose capacity remains unlimited as a matter of principle. Intrinsic limitedness of theoretical controllability is excluded from classical mechanics. The problem of measurement at issue is whether limited experimental controllability over boundary conditions could reduce to an artificial lessening of theoretical controllability or to its intrinsic limitedness. We must go beyond classical mechanics in order to answer this question.

III. INTERNAL MEASUREMENT

The consistent tripartite circle among a one-to-one law of motion, completely fixed boundary conditions, and unlimited theoretical controllability over them would become vulnerable altogether if any one of the three is attacked. Completely fixed boundary conditions require that the initial condition applied to the law of motion of the one-to-one mapping type is fully specified and guarantees the unique later development of trajectory. Any quantity that is conserved during development, such as energy and mass, remains invariant from the moment the initial condition is specified. This implies that complete specification of the initial condition and the principle of conservation are compatible. The law of motion of one-to-one mapping type witnesses that the conserved quantity develops with time as maintaining its invariant level. In spite of this, the quantity that is conserved in reality assumes the communication process of conservedness. Interactions are propagated at less than light velocity inside the carrier of what is conserved. This, however, is self-contradictory.

Take, for instance, the principle of the conservation of energy. Insofar as changes in interaction are propagated in the medium at a finite velocity, the interaction energy varies its magnitude with the propagation of interaction changes. The communication process of conservedness of energy is also accompanied by its own energy, though its magnitude is left unspecified. If one tries to save both the law of motion of the one-to-one mapping type and completely fixed boundary conditions, the propagation velocity of interaction changes would have to be infinite. For the conservation of energy to be inherent to the law of motion would require an instantaneous adjustment of interaction in the whole medium in order to maintain its invariant level. Nevertheless, no physical processes are allowed to be propagated beyond light velocity. The law of motion of the one-to-one mapping type supplemented by completely fixed boundary conditions is thus incompatible with the finiteness of the propagation velocity of interaction changes.

As stated previously, the finiteness of the propagation velocity of interaction changes necessitates the process communicating conservedness of energy. Such a process is not found in the law of motion of the one-to-one mapping type, in the latter of which the surviving process is the one transforming what has been conserved. The process materializing conservedness does not find its place in the law of motion of the one-to-one mapping type supplemented by completely fixed boundary conditions.

In addition, unlimited theoretical controllability over boundary conditions takes it for granted that one can specify, at least in theory, the initial condition upon the process preserving conservedness of energy at a predetermined level. Unlimited theoretical controllability can maintain its own claim as far as boundary conditions are completely fixable in principle. Such an unlimited controllability is inherent to the process preserving conservedness that transforms in time what has been conserved. The process preserving conservedness is reducible to a one-to-one temporal mapping supplemented by completely fixed boundary conditions. In contrast, the process materializing conservedness that fundamentally differs from the process preserving conservedness cannot claim its unlimited theoretical controllability over boundary conditions. For the law of motion underlying the process materializing conservedness and its boundary condition are not separable. In the following, we shall rephrase the inseparability in a slightly different form.

Insofar as the propagation velocity of interaction changes is understood to remain finite, the conservation of energy viewed as a particular attribute of boundary conditions already includes in itself a law of motion that yields the process materializing conservedness. The initial condition admitting the conservation of energy is inseparable from the law of motion. Unlimited theoretical controllability of the initial condition is thus not feasible, otherwise it would yield the unacceptable consequence that one could theoretically control the law of motion.

The law of motion that remains incapable of detaching itself from its initial condition cannot be of the one-to-one mapping type, since in the latter the initial condition is definable in its own right independently of the law of motion. The initial condition for the law of motion underlying the process materializing conservedness can be only partially fixable in principle. The corresponding law of motion thus turns out to be intrinsically partially fixable. When one classifies laws of motion into three distinctive groups of one-to-one, one-to-many, and many-to-one[5-7] temporal mappings, the law of motion that is only partially fixable belongs to the class of one-to-many mapping. Many-to-one temporal mapping is precluded because it can suggest a unique goal even if the initial condition remains indefinite to a certain extent.

One-to-many temporal mapping supplemented by intrinsically partially fixed boundary conditions thus underlies the process materializing conservedness, in which the finiteness of the propagation velocity of interaction changes is rightly observed. The tripartite circle among the law of motion of one-to-one mapping type, complete fixedness of boundary conditions, and unlimited theoretical controllability could remain consistent only if the approximation that allows the propagation velocity of interaction change to diverge persists for some reason. In contrast, once it is admitted that the propagation velocity does remain finite, the tripartite circle ceases to be consistent. An alternative to the previous tripartite circle is a new one among the law of motion of one-to-many mapping type, intrinsic partial fixedness of boundary conditions and limited theoretical controllability.

The law of motion of one-to-many mapping type supplemented by intrinsically, partially fixed boundary conditions is derived from the process materializing conservedness. Both the law of motion and the associated boundary conditions are indefinite in their implication because of their intrinsic partial fixedness. Nevertheless, these two are constrained by the *a posteriori* principle of conservation. The constraining agent is measurement that proceeds internally. Internal measurement serving as an internal watchdog is necessary for observing and communicating the conservedness when both the law of motion and its boundary conditions are only partially identifiable.

Internal measurement underlying the process materializing conservedness of energy is

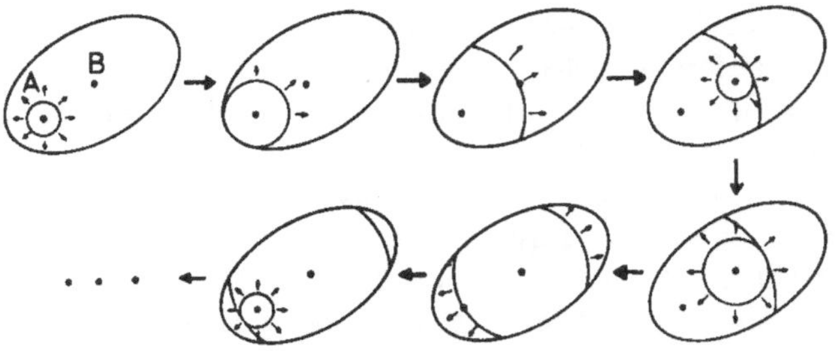

FIGURE 2. A schematic representation of indefinite repercussions of interaction changes and their propagation underlying the process materializing conservedness. When an interaction change originating in arbitrary point A reaches another arbitrary point B, a successive interaction change will be induced in the latter so as to fulfill the principle of conservation *a posteriori*. And again, when the interaction change imputed to point B reaches point A, further interaction change will follow in the latter. This sequence continues to survive indefinitely.

ubiquitous in any material system.[8] Moreover, differing from external measurement, internal measurement is an instance of exhibiting a mutual constraint between an arbitrary pair of interacting bodies. Internal measurement is thus capable of not only recognizing but also reacting upon others, while external measurement assumes only the capacity of recognition. The mutual constraining inherent to internal measurement allows either one of the interacting bodies to influence the other. Then, what has been influenced comes to react upon what has influenced while both are involved in the process materializing conservedness altogether. This form of mutual influence survives indefinitely. Both internal recognition and reaction are associated with internal measurement.

In fact, any material aggregate has the capacity of recognizing its exterior through material interaction. To say that matter can recognize its exterior implies simply that it has a capacity for materialistic interaction with the exterior. A consequence of recognizing the exterior serves as a sequential cause for the process of how the interaction with the exterior could be altered. The pattern of interaction between the material aggregate and its exterior is actualized and maintained through materialistic mutual recognition, followed then by successive renewal of the pattern as demonstrated in Figure 2. The renewed interaction configuration may remain the same as previously or not. What is required in this renewal process of interaction configuration is that the communication of conservedness of energy must be observed internally. Thus the process materializing conservedness of energy internally controls the pattern of materialistic interaction.

As the interaction configuration develops with time, the materialistic constraint which originated in the process materializing conservedness accumulates in a temporally multiplicative manner. This implies that the pattern of materialistic interaction observed in the material aggregate cannot develop arbitrarily. It is endogenously regulated due to the process materializing conservedness acting internally. If one calls the internal materialistic process of an arbitrary material aggregate recognizing others in its neighborhood "internal structural identification" and the following process of renewing the interaction configuration "internal structural realization", the sequence of structural identification and realization must alternate in a successive manner.

An impetus for successive alternations between structural identification and realization is the process materializing conservedness. The internal communication of conservedness assumes both the capacities of identifying and implementing what is implied by the conservedness on materialistic grounds. In view of the fact that interaction changes are propagated in the medium at a finite velocity, the effect of interactive change for fulfilling the conservation of energy at one place subsequently reaches other places, inducing successive changes for fulfilling the conservation of energy there. Internal structural identification is the process recognizing the need for a change in the interaction configuration so as to maintain the conservation of energy, and internal structural realization is the process implementing the recognized change in the interaction configuration. Internal structural identification and realization, however, do not converge with each other simultaneously, because identification of what has been implemented proceeds in a sequential manner, instead of in a simultaneous manner.

If internal structural identification and realization converge with each other, one can define the structure of interaction unambiguously. If every atom and molecule involved in material aggregates maintains an invariant pattern of materialistic recognition of others, the structure of interaction configuration among themselves would also remain unaltered. Once an unaltered structure of interaction is established, it would not be necessary to consider how every interacting element recognizes others and how it adjusts the way of interaction with others accordingly. However, this does not mean that interacting elements are deprived of their capacity of recognizing and reacting upon others. Unaltered interaction configuration is simply a matter of theoretical artifact that allows the elimination of internal active agents from interacting material aggregates. The problem which remains to be settled is whether such an elimination of active agents is a good approximation. The price of eliminating active agents is a denial of evolutionary processes as endogenous structural transformations because the structure of interaction configuration would remain unaltered unless external agents intervene.

Every interacting atom and molecule is an agent for recognizing and reacting upon others. The presence of such internal agents implies no more than that interaction changes are propagated in the medium at a finite velocity. Recognition on the part of an arbitrary interacting element is the process of sensing the manner of interaction with others, and the reaction is the process of readjusting the manner of interaction with others. Internal structural identification and realization are intrinsic to every interacting element. If we call the difference between internal structural identification and realization a disequilibrium, unaltered structure of interaction configuration will be said to be in equilibrium in the sense of vanishing disequilibrium. An impetus that causes changes in an equilibrium structure has to be sought in external agents.

The problem of whether a system of interacting elements can establish its equilibrium configuration, however, must be examined for its own sake. One possible solution is to search for relevant empirical facts. If an unaltered structure of interaction configuration is empirically confirmed, one could accept the theoretical assumption that interacting elements in the system may finally reach the state where the difference between structural identification and realization vanishes. The process materializing conservedness would reduce to the process preserving conservedness. Internal measurement that distinguishes between materializing and preserving conservedness would become superfluous.

If the structure of interaction configuration changes endogenously, on the other hand, one cannot define the structure except in conjunction with the mechanism of inducing those internal changes. The structure in equilibrium lacks this internal mechanism of changing itself. We thus arrive at an irony. If an external observer claims that the presence of structure can be identified *a posteriori,* this statement would automatically imply that the structure must be in equilibrium. For the external identification of structure presumes the convergence between internal structural identification and realization on the part of interacting material elements, otherwise an externally observable structure could not be definable. Still, if the observer maintains that the identified structure changes endogenously with time, it would lead to a self-contradiction.

Complete separation between the law of motion and its boundary condition is consistent with the assertion that structure is identifiable externally, where structure points to both the law of motion and the associated boundary condition. It is this complete separation that renders both the law of motion and the boundary condition unambiguously definable independent of each other. There is no need for invoking the internal measurement characteristic to the law of motion that is inseparable from its boundary conditions. External identification is concerned exclusively with external measurement. Insofar as this sort of separation is maintained, an impetus to change the equilibrium structure has to be external to the structure. However, such a complete separation is exceptional and possible only when the disequilibrium between internal identification and realization vanishes. Disequilibrium between internal identification and realization cannot vanish unless the propagation velocity of interaction change is forcibly set to diverge.

In the presence of disequilibrium between internal structural identification and realization, external measurement fails to completely and unambiguously identify the structure of interaction configuration because of the interference from internal measurement. Internal measurement underlying the process materializing conservedness continues to survive indefinitely. Accordingly, it leaves both the law of motion and the associated boundary condition inseparably interwoven. One is not allowed to theoretically control the structure of interaction configuration in the presence of disequilibrium because of the inability to measure the structure externally.

Internal structural identification and realization may sound anthropomorphic. However, their origin is undoubtedly materialistic, and it is found in the difference between before and after the process materializing conservedness. Internal identification and realization are no more than a particular attribute of materialistic interaction. Internal measurement is thus imputed to the nonanthropomorphic disequilibrium between internal identification and realization.

It should also be noted that internal measurement is finalistic in the sense that the process materializing conservedness is internally aiming at the *a posteriori* principle of conservation. Nevertheless, indefinite repercussions of the process are inevitable because of the lack of an agent to coordinate the whole interaction configuration in a simultaneous manner. The goal orientedness of the process materializing conservedness is in letting disequilibrium between internal structural identification and realization vanish *a posteriori* internally, though the process leaves the spillover of disequilibrium inevitable. Imparted to internal measurement is a limited active role of raising what has been identified as need for fulfilling the principle of conservation *a posteriori*. Disequilibrium between internal identification and realization intrinsically implies continual equilibration between the two in such a manner that the internal dynamics of eliminating disequilibrium successively leaves further disequilibria to be eliminated.

Disequilibrium and equilibration are compatible. Disequilibrium refers to the difference between internal structural identification and realization, though not measurable as such externally, whereas equilibration refers to the process eliminating such a difference. Unless internal identification and realization proceed simultaneously, disequilibrium will persist and necessitate the process of equilibration. The process of equilibration is thus associated with internal measurement.

The process of equilibration is self-constraining in its operation because internal structural realization is limited only to those interaction configurations that can meet the internal requirement for fulfilling the principle of conservation *a posteriori*. Other hypothetical configurations are eliminated from the process of equilibration. In particular, the impossibility of complete separation between the law of motion and its boundary condition renders the material carrier of internal measurement to be the source of inseparability. Internal identification of both the law of motion and its boundary condition is not however comprehensible to an external observer, while the inseparability is variable as being subject to the process materializing conservedness. What makes the variability possible is the presence of inevitable indefiniteness about the separation between the law of motion and its boundary condition.

The extent of indefiniteness is constrained as being subject to the process of equilibration. Development of internal measurement lessens the extent to which the separation between the law of motion and its boundary condition would remain indefinite.

Internal measurement makes its material carrier indefinite about its own specification, whether it may be subatomic, atomic, molecular, or supramolecular. Every atom and molecule remains indefinite with regard to its own internal measurement.[9,10] We are thus motivated to experimentally examine how legitimate the theoretical consequence of the indefiniteness associated with internal measurement could be. Experimental examination however assumes external measurement.[11] Accordingly, the question arises about how the process of internal measurement could be measured externally, especially from an anthropomorphic perspective.

Interference between internal and external measurements becomes a serious matter in quantum mechanics, because the law of motion in quantum mechanics in the form of Schrödinger's equation of motion refers to the wave function that is not directly measurable.[12] Classical mechanics as represented by Newton's equations of motion makes internal measurement superfluous, but on the other hand, completely distinguishes between the law of motion of the one-to-one mapping type and its boundary conditions. External measurement can identify the law of motion based upon unlimited theoretical controllability of its boundary condition at the limit that experimental controllability would approach its theoretical counterpart. In contrast, quantum mechanics first has to address itself to how unlimited theoretical controllability of boundary conditions could survive in internal measurement and then to how internal measurement would interfere with experimental controllability of boundary conditions.

IV. MEASUREMENT IN QUANTUM MECHANICS

Classical mechanics has a long tradition of maintaining both unlimited theoretical controllability of boundary conditions and external measurement that could disturb an object to be measured only infinitesimally, if at all. One thus naturally comes to ask how both unlimited theoretical controllability of boundary conditions and external measurement could apply to quantum mechanics,[13] since quantum mechanics undoubtedly serves as a foundation of the material sciences including protobiology and biology. Quantum mechanics is divided into two classes, nonrelativistic and relativistic. We shall limit ourselves only to nonrelativistic quantum mechanics, since our principal objective is protobiology and biology in the nonrelativistic regime.

The law of motion in the nonrelativistic regime of quantum mechanics is one of the one-to-one mapping type if Schrödinger's equation of motion and its wave function are the case. Unlimited theoretical controllability of boundary conditions lets Schrödinger's equation of motion unambiguously determine how the wave function develops with time. This is possible because Schrödinger's equation of motion and its boundary condition are separable in principle. Complete fixedness of boundary condition is tenable at least in theory.[14] Such a wave function and its derivatives become objects to be measured externally in quantum mechanics.

External measurement in quantum mechanics however differs from its counterpart in classical mechanics. The wave function that can be transformed through the one-to-one temporal mapping is not directly measurable. The quantum-mechanical wave function assuming complex-number values is measurable only indirectly through identification of its absolute value, namely, its probability amplitude. Once the boundary conditions in general, and the initial conditions in particular, of the wave function are completely identified, the underlying Schrödinger's equation of motion unambiguously determines the probability amplitude that could be measured later externally.

External measurement of the quantum-mechanical wave function is its projection onto a space that is more contracted as compared with the space in which it originally belongs. Being a projection, external measurement is intrinsically a many-to-one mapping from the wave function to the corresponding probability amplitude. The nature of such a many-to-one mapping associated with external measurement in quantum mechanics has raised a controversy, especially with regard to the appropriateness of the Copenhagen interpretation of the quantum-mechanical wave function viewing only the probability amplitude as being a reality.[15] Nevertheless, external measurement in quantum mechanics has established the view that if measurements are repeatable under the same invariant boundary condition, the relative frequency of events can be equated with the probability density that the corresponding wave function would imply. The correspondence between the relative frequency of events and the probability amplitude has been confirmed irrespective of whatever interpretation may be taken about the nature of many-to-one mapping from the wave function to the probability amplitude.

As far as the quantum-mechanical law of motion and its boundary condition are separable (as demonstrated in Schrödinger's equation of motion), what external measurement can identify is the boundary condition. Schrödinger's equation is a means of transforming the boundary condition through its own one-to-one temporal mapping. External measurement of the quantum-mechanical wave function at a later time is entirely dependent upon its initial condition set at an earlier time.

One characteristic aspect of the boundary condition upon the quantum-mechanical wave function is that if initial conditions are additive, the corresponding wave functions are additive during their quantum mechanical development. This is what one calls linear superposition of the quantum-mechanical wave functions.

Consider, for instance, the case that boundary condition (A or B) is set at time t initially and remains invariant from then on. And further suppose that the quantum-mechanical wave function that develops under boundary condition (A or B) is $\psi_A(x,t)$ [$\psi_B(x,t)$] in which x represents an aggregate of coordinates and t refers to time. Initial condition $\psi_A(x,t_{initial})$ [$\psi_B(x,t_{initial})$] is externally controlled. Then, if both boundary conditions A and B are additive, that is to say, if the wave funciton $\psi_{A+B}(x,t)$ subjected to

$$\psi_{A+B}(x,t_{initial}) = \psi_A(x,t_{initial}) + \psi_B(x,t_{initial})$$

is externally definable, the underlying Schrödinger's equation of motion yields

$$\psi_{A+B}(x,t) = \psi_A(x,t) + \psi_B(x,t) \qquad \text{for } t > t_{initial}$$

External measurement of the superimposed wave function $\psi_{A+B}(x,t)$ refers to the probability amplitude $|\psi_{A+B}(x,t)|$ $(= |\psi_A(x,t) + \psi_B(x,t)|)$. Identification of the quantum-mechanical coherence pointing to the difference between the probability amplitude of linearly superimposed wavefunctions $|\psi_A(x,t) + \psi_B(x,t)|$ and the resultant of each probability amplitude $(|\psi_A(x,t)|^2 + |\psi_B(x,t)|^2)^{1/2}$ is unique to external measurement in quantum mechanics. However, such an external measurement that demonstrates characteristics pertinent to quantum mechanics is entirely up to the boundary condition one can prepare and controll externally. Additiveness of boundary conditions underlies linear superposition of the quantum-mechanical wave functions.

One can further recognize that even external measurement constitutes an attribute of the boundary condition controlled externally.[16,17] We shall see a reason lying behind this below.

Let us imagine a controlled experiment on measuring three spin-coordinates, S_x, S_y, and S_z of a polarized particle, in which the eigen-wave function of upward oriented spin along the z-direction is ψ_z^+ satisfying

$$S_z\psi_z^+ = \frac{\hbar}{2}\psi_z^+ \qquad \text{and similarly} \qquad S_z\psi_z^- = -\frac{\hbar}{2}\psi_z^-$$

for downward oriented spin, where $\hbar$ is Planck's constant divided by 2π. Accordingly, linear superposition of the wave functions ψ_z^+ and ψ_z^- will give the eigen-wave functions of spin polarized along the x and y directions such as

$$S_x(\psi_z^+ + \psi_z^-) = \frac{\hbar}{2}(\psi_z^+ + \psi_z^-)$$

$$S_x(\psi_z^+ - \psi_z^-) = -\frac{\hbar}{2}(\psi_z^+ - \psi_z^-)$$

and

$$S_y(\psi_z^+ + i\psi_z^-) = \frac{\hbar}{2}(\psi_z^+ + i\psi_z^-)$$

$$S_y(i\psi_z^+ + \psi_z^-) = -\frac{\hbar}{2}(i\psi_z^+ + \psi_z^-)$$

If a beam of polarized particles is externally measured with regard to the z-component spin and if the meter reading gives a result of being polarized 100% upward, the spin-wave function that has been identified will consist only of the eigen-wave function ψ_z^+, and none of ψ_z^-. Moreover, if we imagine a situation such that external measurement of the z-component spin is replaced by that of the x-component spin while maintaining all of the other conditions to be the same as previously, the present interchange of measurement apparatus will affect the controllability of experiment only minimally, if at all. This interchange of measurement apparatus will let the eigen-wave functions $(\psi_z^+ + \psi_z^-)$ and $(\psi_z^+ - \psi_z^-)$ for the x-component spin to measurable externally, instead of ψ_z^+ and ψ_z^- for S_z. Accordingly, identity

$$\psi_z^+ \equiv \frac{1}{2}(\psi_z^+ + \psi_z^-) + \frac{1}{2}(\psi_z^+ - \psi_z^-)$$

tells us that external measurement of the x-component spin will give 50% of being polarized upward and 50% downward along the x direction.

Let us further suppose that the x-component measurement is followed by the z-component measurement. The spin-wave function to be fed into the apparatus measuring the z component is either in the form of $(\psi_z^+ + \psi_z^-)$ or $(\psi_z^+ - \psi_z^-)$ because of the preceding measurement of the x component. The subsequent measurement of the z component will thus give the meter reading of 50% of being polarized upward and 50% downward along the z direction.

We thus come to observe a superficial paradox or discrepancy such that a beam of polarized particles that has been measured to be polarized 100% upward along the z direction can be found polarized only 50% upward and 50% downward along the same z direction if external measurement of the x-component spin intervenes and precedes, as pictorially demonstrated in Figure 3. External measurement of the z-component spin will give different results depending upon whether or not another external measurement of the x-component spin interferes. But, the difference is more than superficial. External measurement of the z-component spin implicitly sets such a boundary condition that the spin-wave function is a linear superposition of those that are eigen-wave functions of the spin operator S_z, instead of those that are eigen-wave functions of S_x or S_y. Likewise, external measurement of the

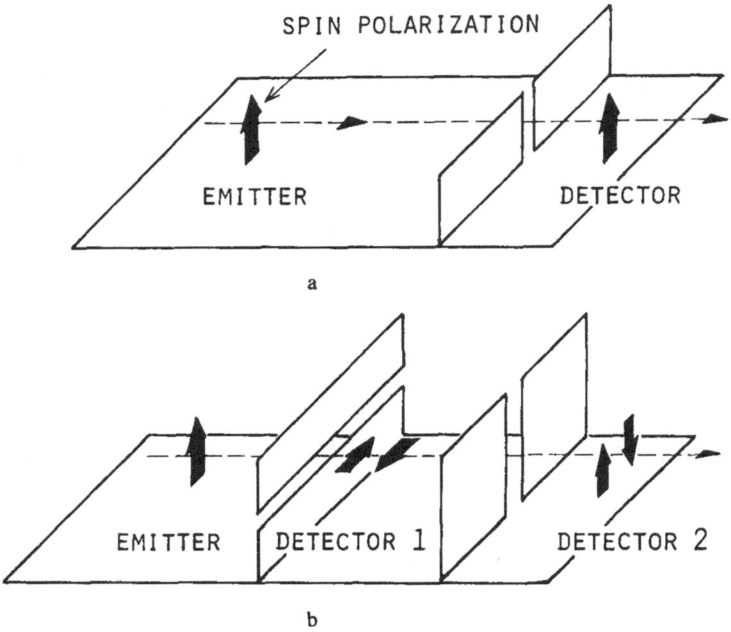

FIGURE 3. A conceptual scheme of measuring the spin-polarization vector of those particles which are prepared in a manner of being 100% polarized upward along the z axis initially. (a) When the measurement is so contrived as to detect the z component, the detector reading will give 100% polarization upward along the z direction. (b) On the other hand, if the detector measuring the x-component polarization is set in between, the final z-component reading will give 50% upward and 50% downward polarizations.

x-component spin of the same polarized beam lets the eigen-wave function of the z-component spin operator S_z to be a linear superposition of those eigen-wave function of the x-component spin operator S_x. This illustrates that the presence of successive external measurements affects what each external measurement identifies. External measurements determine a part of the boundary condition under which the quantum-mechanical process to be measured proceeds, especially with regard to the nature of the wave functions that are linearly superimposable.

Interference among external measurements becomes evident even if these measurements proceed simultaneously in different spatial locations. If two external measurements on the z-component and on the x-component spins of the same polarized beam proceed simultaneously in different locations, it would necessarily occur that a portion of the wave function that has been subject to either one of the two measurements is fed into the remaining other. This is due simply to the presence of the wave function being reflected at the measurement apparatus. The z-component measurement lets a part of the incoming wave function be reflected and lets the remaining other part decompose itself into a linear superposition of those eigen-wave functions of the spin operator.

The incoming wave function fed into the apparatus measuring the x-component spin now consists of three different parts, the one that has not yet been either subjected to nor reflected at any measurement apparatus, and the two others that have been either subjected to or reflected at the z-component measurement. The wave function contributed from the one that was subject to the z-component measurement has already been linearly decomposed into the eigen-wave functions of the spin operator S_z, the latter of which again have to be decomposed into the eigen-wave functions of the spin operator S_x in the process of measuring the x-component spin. The z-component measurement has provided such a boundary condition that the wave function is decomposed into a linear superposition of the eigen-wave functions of S_z. The linear decomposability of the remaining two parts of the wave function also depends upon the character of the boundary condition set for the measurements altogether.

The factor that determines and selects a particular type of linear decomposability of the wave function out of many alternatives is controlled entirely externally. It is the characteristic of the boundary conditions that specifies the nature of the wave function, especially with regard to its linear decomposability. Addition or deletion of external measurements could disturb what is to be measured only infinitesimally or minimally, but can appreciably alter how the wave function is linearly decomposed. The remark on the interference among simultaneous external measurements, that has been due originally to Einstein-Podolsky-Rosen, does not raise a real paradox, but points to a fundamental characteristic unique to quantum mechanics, especially about its boundary condition.[18,19]

An overly simplified, exaggerated, but still pedagogical example indicating an intriguing interference between the boundary condition upon a quantum-mechanical system and its external measurement is that of Schrödinger's cat.[15] Suppose such a quantum-mechanical system containing a cat, a canned poisonous material, and a radiative atom, in which the whole system is designed by an experimenter so that an emitted particle from the radiative atom will let the can open and let the cat die. An external observer can identify whether the cat is alive or dead by inspecting the system externally. The quantum-mechanical wave function of the cat, ψ_{cat}, could be identified as either in the alive state ψ_{alive} or in the dead state ψ_{dead}.

The quantum-mechanical wave function of the cat could be conceivable theoretically even if external measurement intervenes. In particular, if the external observer has not yet finished measurement, the wave function of the cat might be described as a linear superposition of both the alive and the dead states, $\psi_{cat} = \psi_{alive} + \psi_{dead}$ because of the linearity of the underlying Schrödinger's equation of motion with regard to its wave function. One thus recognizes that external measurement would make the wave function contracted either to ψ_{alive} or to ψ_{dead}. It is, however, conceptually pretty difficult at first sight to imagine such a sudden and abrupt contraction of the quantum-mechanical wave function due to a mere external inspection, because external measurement is so designed as to disturb an object to be measured only minimally.

In fact, the origin of such an artificial contraction of the wavefunction can be traced back to the assertion for the presence of linearly superimposed wavefunction $\psi_{cat} = \psi_{alive} + \psi_{dead}$ in the absence of external measurement. The boundary condition for the wavefunction ψ_{alive} is that the radiative atom has not yet decayed, and similarly the boundary condition for ψ_{dead} is that the radiative atom has already decayed. We have already noted that linear superposition of the quantum-mechanical wave functions necessitates the similar linear superposition of the corresponding boundary conditions. Consequently, an intriguing characteristic of Schrödinger's cat will become clearer if one examines whether the boundary condition letting the linear superposition $\psi_{cat} = \psi_{alive} + \psi_{dead}$ be possible could be consistent with the boundary condition the experimenter has set. At this point, it is important to notice that the boundary condition the experimenter has set is by no means definite about when the radiative atom decays. In contrast, the boundary condition for the wave function $\psi_{cat} = \psi_{alive} + \psi_{dead}$ is a linear superposition of two definite conditions. One is that the radiative atom has not yet decayed, and the other is that the atom has already decayed.

Crucial to our present examination is the recognition that the boundary condition that is indefinite about when the radiative atom decays cannot be substituted by a linear superposition of the two definite boundary conditions accompanied by the not-yet decayed and the already decayed radiative atom. Even if a physicist requires a superposition of the two definite boundary conditions, a biologist would dismiss the case as charging it simply as a matter of artifact that has nothing to do with the biological reality.

The original boundary condition the experimenter has set does not allow such a linear decomposition of the wave function as $\psi_{cat} = \psi_{alive} + \psi_{dead}$ because of the impossibility of decomposing the indefinite boundary condition into the two definite ones without leaving any residue. External measurement can tell us that the wave function is decomposable as $\psi_{cat} = \psi_{alive} + \psi_{dead}$. However, one cannot say for sure that such a decomposition is acceptable even before actual measurement. What is responsible for determining how the wave function could be decomposed is its boundary condition. A sudden change in the boundary condition such as an interference from external measurement can induce the corresponding abrupt change in the linear decomposability of wave function, instead of its hypothetical contraction. A serious lesson we can learn from an intriguing character of Schrödinger's cat is that external measurement does induce a change in the boundary condition on how the wave function develops as following Schrödinger's equation of motion.

External measurement plays a decisive role in determining how the wave function is linearly decomposed. But, an exhaustive clarification of the character of external measurement could be difficult to attain because of its inevitable anthropomorphic connotation. What we can expect to do instead is to examine a common denominator of the process called measurement in general and external measurement in particular.

External measurement as a specific instance of measurement in general is the realization of a particular pattern of interaction between an object to be measured and a means of measurements. Quantum-mechanically, external measurement decomposes the wave function to be measured in a manner unique to itself. This implies at the least that given an arbitrary pair of interacting bodies, either one of the two serves as a boundary condition upon the other, especially with regard to how the wave function associated with the interacting pair is decomposed. Realization of a particular pattern of interaction that is ubiquitous between arbitrary pairs of interacting bodies inevitably provides internal boundary conditions upon the wave function with each other.

We thus come across a difficulty again in distinguishing between the law of motion identified as Schrödinger's equation of motion and its boundary condition. One way out of this difficulty, though overly theoretical, is to dismiss internal boundary conditions altogether. If one can claim unlimited theoretical controllability of boundary conditions, what looks like an internal boundary condition upon a local interacting body could be theoretically derived as a specific attribute of the corresponding total wave function and there would be no need for invoking the separability of internal boundary conditions and internally localized wave functions. Unlimited theoretical controllability of boundary conditions makes those boundary conditions completely determinate exogenously and leaves out the endogenous capacity of their internal determination.

Such an unlimited theoretical controllability assumes that the initial condition of the total wave function is completely identifiable at a given initial moment along with the remaining other part of boundary conditions. Schrödinger's equation of motion is further supposed to be fully responsible for the transformation of the given initial condition with time. However, the claim for unlimited theoretical controllability of boundary conditions will raise another formidable difficulty when it is asked how the present claim could be justified in practice especially with regard to its realization, let alone its identification. Intrinsically limited, experimental controllability of boundary conditions could not approach its theoretical counterpart of unlimited controllability except for the cases such as low temperature limit and complete alignment of polarized particles. One has to admit that unlimited theoretical controllability of boundary conditions cannot apply to interacting atoms and molecules that constitute protobiology and biology, in spite of the fact that Schrödinger's equation of motion keeps requiring completely fixed boundary conditions in order to save its wave function.

An interacting many-molecule system makes every molecule capable of serving internally as an endogenous apparatus measuring other molecules interacting with it. In essence, there is no fundamental difference between external and internal measurement in exhibiting a particular pattern of interaction between interacting bodies. Internal measurement intrinsic to interacting molecules influences the linear decomposability of the wave function to be measured internally as much as external measurement does. However, internal measurement is neither clear nor definite about the separation between the boundary condition and the bounded quantum-mechanical motion, being different from the case of external measurement in which the demarcation is forcedly made clear. Internal measurement maintaining the process materializing conservedness as a common denominator, as we have already noticed, renders both the law of motion and its boundary condition inseparable. Endogenous and internal separation between the bounded quantum-mechanical motions and their boundary conditions now emerges as a major problem with internal measurement.

Consider, for instance, three molecules interacting with each other. If the boundary condition upon the three molecules is completely identified without leaving any indefiniteness, the corresponding Schrödinger's equation of motion will unambiguously determine how the wave function of the three molecules develops with time. External measurement, though requiring a unique linear decomposition of the wave function, can tell us how the wave function develops in reality. The boundary condition is thus completely controllable and externalized. In contrast, if the boundary condition on the three molecules remains indefinite because of intervening internal measurement that necessarily makes both the law of motion and its boundary condition inseparable, one could conceive the theoretical alternative that one molecule moves under the internal boundary condition provided by the other two molecules. The present theoretical alternative is undoubtedly incomplete in reflecting the impossibility of thinking of completely fixed boundary conditions. Of course, one cannot deny that there has been a long and strong tradition in physics demanding the separability between the law of motion and its boundary conditions. Internal measurement that is physically unavoidable now urges us to examine how both the boundary condition and the bounded motion could be separable in a practical sense in spite of its intrinsic and undeniable inseparability.

Endogenous separation between the boundary conditions and the bounded motions, if possible in some practical sense, can be either stationary or variable with time. If the separation remains stationary on average, one can hope to cope with this particular class of internal measurement within the framework of quantum statistical mechanics. Internal measurement lets every molecule join in constituting the boundary condition upon other molecules interacting with it and at the same time lets it move under the boundary condition constructed by those molecules interacting with it. Each local boundary condition upon the locally bounded quantum-mechanical motion determines how the associated wave function is linearly decomposed. In other words, the endogenous local boundary condition carries with itself a coherent linear superposition of local quantum-mechanical wave functions but does not guarantee such a quantum coherence beyond its own boundary, both in space and in time. This recognition makes it evident that the endogenous separation between the boundary condition and the bounded motion is internally accomplished to the extent the local quantum coherence is maintained.

The endogenous quantum coherence, though inevitably localized, is a consequence of the endogenous separation between the boundary condition and the bounded motion, not vice versa. However, once the extent of local quantum coherence is identified, the underlying many-molecule system can be viewed as a collection of or an ensemble of the quantum-mechanical wave functions, in which the emergence of quantum coherence beyond its local limit is prohibited due to its premise. The situation is completely different from an interacting many-molecule system under unlimited theoretical controllability of boundary conditions. In the latter, the quantum coherence that guarantees a linear superposition of wave functions can be extended to the entire system because its boundary conditions are completely fixed.

Both limited and unlimited quantum coherences are theoretical in their origins. Their difference, on the other hand, is in whether theoretical controllability of boundary conditions is limited or unlimited. Theory alone cannot answer this question. An appropriate answer could come only from a closer examination of what looks like boundary conditions in the real world in general, or in physical experimentation in particular.

Quantum statistical mechanics is one theoretical attempt that admits the impossibility of claiming unlimited theoretical controllability of boundary conditions. Pure quantum states that would presuppose that the quantum coherence extends over all of the interacting particles are abandoned.[20,21] Instead, mixed quantum states that could maintain their quantum coherence only locally both in space and in time are saved in quantum statistical mechanics. It is internal measurement that is responsible for giving birth to mixed quantum states. Although both pure and mixed quantum states can be conceived theoretically, the distinction between the two different kinds of quantum state does require its raison d'être. In fact, internal measurement serves this purpose.

Internal measurement is a double-edged sword in quantum statistical mechanics. Both local boundary conditions and bounded quantum-mechanical motions owe their genesis to internal measurement, whereas internal measurement intrinsically prevents the law of motion and its boundary conditions from being separable. Mixed quantum states in quantum statistical mechanics are in fact derived from an overemphasis of one side of the same reality named internal measurement. Accordingly, as far as endogenous separation between local boundary conditions and bounded motions remains stationary for some reason, quantum statistical mechanics that apparently differs from quantum mechanics could survive as a theoretical tool to cope with an interacting, many-particle system. A principal way that quantum statistical mechanics differs from quantum mechanics is in its denial of unlimited theoretical controllability over boundary conditions. Abandonment of unlimited theoretical controllability is admittedly more than just theoretical. It makes it possible to properly appreciate the functioning of internal measurement. In addition, internal measurement incorporates in itself more than what quantum statistical mechanics can tell. One further possibility of internal measurement is to vary, especially to enhance the extent of local quantum coherence beyond the limit intrinsic to quantum statistical mechanics.

An examination of how internal measurement can vary the extent of local quantum coherence requires a more detailed study of the endogenous separation between boundary conditions and bounded motions. We have already recognized that the process materializing conservedness underlies internal measurement serving as an internal agent to realize a particular pattern of interaction between interacting bodies. The endogenous separation is always under the influence of the process materializing conservedness. The process makes it inevitable that variations will be successfully generated through the internal repercussion of equilibration to eliminate disequilibrium in order to satisfy the principle of conservation *a posteriori* internally. Locally bounded quantum-mechanical motions generate local boundary conditions upon the subsequent bounded motions in a successive manner.

Successive genesis of local boundary conditions is, however, not arbitrary, but constrained internally so as to eliminate those that would violate the principle of conservation *a posteriori*. This amounts to saying that the intrinsic indefiniteness about the separation between the law of motion and its boundary conditions is internally constrained with time. The process materializing conservedness eliminates those hypothetical separations that have been degenerated with the preceding ones but that would violate the principle of conservation *a posteriori*. Quantum statistical mechanics in fact corresponds to the case in which the indefiniteness associated with the separation between local boundary conditions and bounded motions could remain stationary. The stationary indefiniteness can thus be equated with those variances within a stationary statistical ensemble.

In contrast, internal measurement per se decreases the extent of indefiniteness about the separation between the law of motion and its boundary condition in a self-constraining

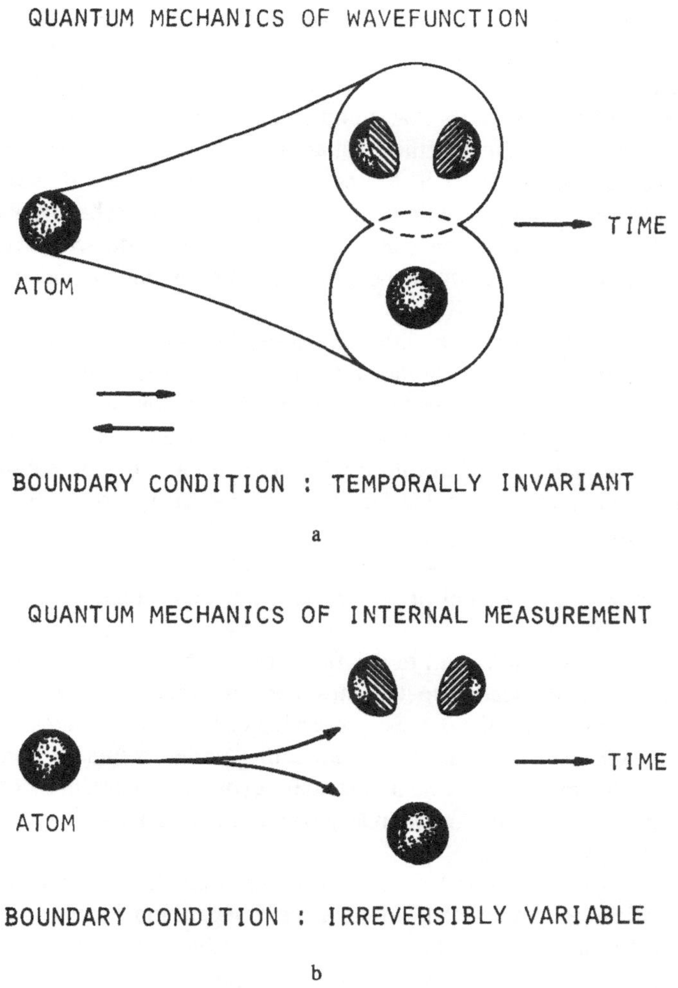

FIGURE 4. A qualitative difference between (a) quantum mechanics of wave function and (b) quantum mechanics of internal measurement. The quantum mechanical equation of motion of its wave function preserves its boundary condition during development even if a radiative atom is the case. However, measurement constantly reshapes boundary conditions because of its characteristic of constraining. The process of internal measurement is an agent of varying boundary conditions in quantum mechanics internally. One way to approach a superficial paradox of Schrödinger's cat is to examine which could be more appropriate, whether quantum mechanics of wave function under invariant boundary conditions or quantum mechanics of internal measurement.

manner. Enhancement of internal constraint endogenously generates further *de novo* boundary conditions upon those that have already been acting and accordingly, lessens the extent of independence among the preceding local boundary conditions. Each local boundary condition on each bounded quantum-mechanical motion can enhance its extension both in space and in time with the increment of endogenous internal constraint. Thus, the enhancement of quantum coherence naturally follows because the boundary condition upon the bounded quantum-mechanical motion determines the extent of quantum coherence. Figure 4 represents a characteristic difference between quantum mechanics of wave function and its internal measurement.

Internal measurement is an agent enhancing the extent of quantum coherence, whereas quantum statistical mechanics admits a consequence of internal measurement only to a limited extent. It is also contrasted with quantum mechanics under unlimited theoretical controllability of boundary conditions that admits the quantum coherence extending over the whole system, in spite of its practical difficulty in materializing such a controllability. As a matter of fact, if the quantum coherence in which we are interested remains stationary, one can expect that either quantum mechanics that admits pure states subject to complete quantum coherence or quantum statistical mechanics that admits mixed states subject only to limited quantum coherence can be appliable in a practical sense. However, if variable quantum coherence is the case, one can no longer rely upon either quantum mechanics under unlimited theoretical controllability of boundary conditions or quantum statistical mechanics based upon a fixed separability between local boundary conditions and bounded quantum-mechanical motions.

Internal measurement that can cope with variable quantum coherence basically refers to the dynamics of variable boundary conditions of endogenous origin, in contrast to quantum mechanics that refers to the dynamics under given boundary conditions of exogenous origin. Quantum mechanics is valid,[22] but not complete as the fundamental basis of materialistic processes.[23] The incompleteness of quantum mechanics rests upon its inability to specify the boundary conditions under which the quantum-mechanical law of motion identified as Schrödinger's equation of motion operates. Insofar as it is erroneously taken for granted that the law of motion is completely separable from its boundary conditions, the origin of the boundary conditions could be out of our reach and would remain intact only by being imposed.

However, once due attention is paid to the finiteness of the propagation velocity of any interaction change or, equivalently, to the difference between the process materializing conservedness in the making and the process preserving conservedness in the record, internal measurement is identified as a form of dynamics that varies boundary conditions in an endogenous manner. Internal measurement thus goes beyond the quantum mechanics of one-to-one temporal mapping in letting its boundary conditions be endogenous, though accompanied by inevitable indefiniteness.

Quantum mechanics under unlimited theoretical controllability of boundary conditions and quantum statistical mechanics that admit a fixed separation between local boundary conditions and locally bounded quantum-mechanical motions, are two opposite extremes. In between are allowed protobiology and biology that can enhance the extent of quantum coherence with time, since biological material aggregates and their evolutionary precursors exhibit those molecules that are quantum-mechanically coherent among themselves.[24] The presence of quantum coherence in the biological realm has been satisfactorily evidenced by those biological molecules such as DNA, whose quantum-mechanical underpinning comes to light when a specific boundary condition is provided.[25] What is more, the emergence of such specific boundary conditions on protobiological and biological molecules is a consequence of the process varying boundary conditions endogenously, that is to say, internal measurement.

Internal measurement that can yield protobiology and biology lets the law of motion of every interacting molecule inseparable from its boundary condition, and thus has the law of motion vary the boundary condition on the subsequent law of motion. Every molecule as a carrier of internal measurement keeps constraining the boundary condition under which it moves as interacting with other molecules. What internal measurement has identified is the boundary condition on its subsequent development, whereas what internal measurement is doing is to keep constraining its own indefiniteness. This character is markedly different from external measurement where the boundary condition, once identified, remains invariant if it is free of any interference from internal measurement.

External measurement in reality, however, is a form of internal measurement that could vary the boundary condition only to a possible minimum extent, because the measurement apparatus is no more than a material unit interacting with what is to be measured. If one considers a hypothetical external measurement of the wave function that develops as following Schrödinger's equation of motion while maintaining completely fixed boundary conditions unchanged, the notorious problem of the contraction of the wave function would arise; namely how the measurement can identify for sure what the quantum-mechanical law of motion could predict only probabilistically. One would come to face the problem of what is responsible for raising the contraction of wave function, when measured. The origin of the contraction cannot be found within Schrödinger's equation. At the same time, it is also maintained that unless boundary conditions could be varied so as to appreciate the intrusion of a measurement apparatus, there would be no agent that could contract the wave function.

In fact, external measurement is an indication of how the accompanying measurement apparatus changes the whole boundary condition compared with the one in effect in its absence. External measurement is irreversible only in the sense that the intrusion of a measurement apparatus induces an irreversible change in the boundary condition. However, the irony of external measurement is that one cannot identify the difference of the boundary condition before and after the intrusion of a measurement apparatus, because one cannot measure externally what is there without using a measurement apparatus. In contrast, internal measurement has to cope with this type of irony rather seriously and traces how the preceding internal measurement influences the subsequent ones.

Measurement in quantum mechanics in its strict sense refers to identification of the boundary conditions under which the quantum-mechanical motion proceeds. For the underlying Schrödinger's equation of motion that has been confirmed preserves its boundary condition with the elapse of time because of its one-to-one temporal mapping character. Hypothetical external measurement could admittedly point to the possibility of externally controllable boundary conditions if unlimited theoretical controllability of boundary conditions is invoked so as to eradicate a likely interference from internal measurement. On the other hand, if only limited controllability is available both in theory and in experiment, as it should be, internal measurement identifies and then varies boundary conditions in an endogenous manner. We thus come to recognize that internal measurement shares something in common with a decision-making process under imperfect knowledge, though the latter is a bit tinted with an anthropocentric connotation.[26]

Decision-making under imperfect knowledge assumes that what the decision-maker has done constitutes the boundary condition upon the subsequent decisions to be made and that the boundary condition always remains indefinite about its future development, as being constrained with time. Henceforth, every interacting molecule can be equated with a decision-making agent under imperfect knowledge only to the extent that internal measurement constitutes the boundary condition upon its subsequent development, while its extent of indefiniteness is internally constrained with time.

Crucial to the superficial similarity between internal measurement and decision-making is the observation that molecules lessen their future indefiniteness with internal measurement. The source of indefiniteness is in the inseparability between the law of motion and its boundary conditions. By indefiniteness we shall mean this inseparability on material grounds.

Endogenous constraint upon the indefiniteness goes with the process materializing conservedness proceeding at a finite velocity. Although quantum mechanics under unlimited theoretical controllability of boundary conditions leaves the probabilistic interpretation of wave function tenable, the wave function does not serve as the source of indefiniteness to be constrained endogenously. For the boundary condition to which the wave function refers is controlled totally exogenously.

Interacting atoms and molecules as building blocks of protobiology and biology incorporate in themselves the capacity of recognizing and then reacting upon other atoms and molecules purely on materialistic grounds. Moreover, during successive internal identification and realization, these atoms and molecules eliminate through their interaction those interaction configurations that would violate the principle of conservation *a posteriori*. Such an elimination is associated with the transition from the possible to the actual,[27] the latter of which again serves as the possible for the subsequent transition of a similar nature. One notes that the process of elimination inherent in the transition from the possible to the actual is ubiquitous in biology, especially in evolution and in morphogenesis.[28] What we have recognized is that even atoms and molecules assume the self-constraining capacity of transforming the possible into the actual by eliminating the impossible. As far as the self-constraining capability of matter is concerned, there is no discontinuity between physics and biology. The extent of self-constraining varies with time only in degree because what has been constrained serves as an agent of constraining in a successive manner.

Physics in general and quantum mechanics in particular have been blamed for not being able to construct biology, in spite of the fact that atoms and molecules as material units of biology are undoubtedly things physics can cope with. This negative charge against physics, however, can be circumvented if one notices the difference between physics per se and the way physics has been practiced. There has been a long tradition that the quantum-mechanical law of motion as the most fundamental basis of materialistic processes is separable from its boundary conditions. This separation has been a theoretical imposition, that still remains to be examined in its own right. In fact, infallible conservation laws that might be conceived as an attribute of boundary conditions necessitates the process materializing conservedness. Physics of course admits that the process materializing conservedness has to be distinguished from the process preserving conservedness. Thus, the problem is reduced to how one can implement the process materializing conservedness in physics. One way to solve it is to deny unlimited theoretical controllability over boundary conditions.

It is one thing to think of unlimited controllability, but quite another to justify it in reality. Internal measurement fills in the gap caused by the absence of unlimited controllability. Quantum mechanics of internal measurement is one way of practicing physics that can save biology by disproving the charge that quantum mechanics is incompetent to construct biology.

Quantum mechanics of internal measurement will make its role more evident if one considers, for instance, the primordial soup of reacting chemicals as an evolutionary matrix, either natural or simulated. Quantum statistical mechanics maintains a stationary separation between local boundary conditions and locally bounded quantum-mechanical motions of molecules, thus precluding the possibility of letting the preceding local boundary conditions be more interdependent with time. Protobiological molecular evolution can be traced within the framework of quantum statistical mechanics only to the extent that imposed local boundary conditions that remain stationary for some reason are realistic. The quantum-mechanical problem with the evolutionary origin of protobiology cells in the primordial sense, for instance is reduced to how the local boundary conditions upon locally bounded quantum-mechanical motions can be formed and transformed internally.

Locally bounded quantum-mechanical motion obeys Schrödinger's equation of motion under the corresponding local boundary condition. Moreover, the local boundary condition varies with the process materializing conservedness, since the communication of conservedness that yields the principle of conservation *a posteriori* is a common denominator of variable boundary conditions. One thus notices that boundary conditions in quantum mechanics are a materialistic expression of the way the principle of conservation has been implemented. If it is claimed that the boundary condition upon the whole primordial soup could be conceived in its every detail in advance, this theoretical claim would lead to a tentative conclusion suggesting that the way every interacting molecule participates in the process materializing conservedness be unambiguously externalized in the same boundary condition. Then, a formidable difficulty would arise as to how one could expect an extremely unlikely coincidence between the theoretically imposed boundary condition and the one generated endogenously. One rational way out of the present impasse is to restrain oneself from asking for unlimited theoretical controllability of boundary conditions on the primordial soup, and to properly appreciate the endogenous mechanism of generating boundary conditions through the process materializing conservedness.

Measurement in quantum mechanics is dynamic in changing the boundary condition, but the origin of such a dynamic change has to be sought outside of Schrödinger's equation of motion. Anthropocentric intervention originating in an external measurement apparatus of course induces a change in the boundary condition. Nevertheless, external measurement is not the only cause for the change in the boundary condition. The process materializing conservedness that is undoubtedly nonanthropocentric, can also induce dynamic changes in the boundary condition, though the process is not imputed to Schrödinger's equation of motion. This mutual interference gives rise to the quantum mechanics of internal measurement, that is to say, an interplay between the quantum-mechanical motion subject to Schrödinger's equation of motion and the process materializing conservedness. The interference can be understood as internal measurement inducing changes in the boundary condition on Schrödinger's equation of motion if one takes the view that the quantum-mechanical motion is influenced by the process materializing conservedness. In contrast, if the process materializing conservedness is viewed as being influenced by the quantum-mechanical motion, the interference can be understood as a materialization of quantum coherence.

The protobiology that we are developing is an attempt at a quantum mechanics of internal measurement.

V. SUMMARY

Although it is difficult to identify what measurement is all about because of its inevitable anthropocentric connotation, a common denominator is that measurement is realization of a particular pattern of interaction between an arbitrary pair of interacting bodies. Measurement is thus ubiquitous in any system of interacting bodies. What is more, every interacting atom and molecule can serve as an internal agent of measuring others interacting with it. Any cell in the body serves as an apparatus measuring what is all about the whole body, as much as the whole body is altogether measuring how its every constituent cell behaves.

Interaction between an object to be measured and a means of measurement is an instance of applying a particular type of boundary condition of each other, since either one of the two influences the other. Unless the combined boundary condition is completely identifiable, the complex interplay between measuring and being measured could not be shown to follow the quantum-mechanical law of motion identified as Schrödinger's equation of motion. If a measurement apparatus is provided externally and if its influence over an object to be measured is maintained infinitesimal, ideal measurement would proceed. Classical mechanics just happens to be this case. The classical law of motion of one-to-one mapping type supplemented by completely fixed boundary conditions allows external measurement that can identify its boundary conditions completely and unambiguously. Classical mechanics under unlimited theoretical controllability of boundary conditions will be externally measurable as such if the influence from external measurement is kept infinitesimal and if intrinsically limited experimental controllability of boundary condition approaches its theoretical counterpart of unlimited control.

However, the law of motion of the one-to-one mapping type supplemented by unlimited theoretical controllability of boundary conditions breaks down in reality at two points. One is that the process materializing conservedness proceeding at a finite velocity makes complete separation between the law of motion and its boundary conditions untenable because the principle of conservation as a form of boundary condition includes in itself the dynamic process of materializing conservedness. The boundary condition to be measured internally remains indefinite. The other is that the introduction of a measurement apparatus irrevocably induces changes in the boundary condition upon an object to be measured. This abrupt and irrevocable change in the boundary condition is evident in the change in the linear decomposability of the wave function in quantum mechanics in response to the introduction of the measurement apparatus, in spite of the fact that the quantum-mechanical law of motion is of the one-to-one mapping type with regard to the wave function as supplemented by completely fixed boundary conditions, at least in theory.

By means of external measurement, one cannot approach the law of motion of the one-to-one mapping type supplemented by unlimited theoretical controllability of boundary conditions. An alternative is the acceptance of internal measurement inherent to any interacting atoms and molecules. Internal measurement is an interplay between the quantum-mechanical motion subject to Schrödinger's equation of motion and the process materializing conservedness that keeps varying the boundary condition upon the former. A consequence of internal measurement is an enhancement of quantum coherence among interacting atoms and molecules. This enhancement of quantum coherence is significant from the viewpoint of protobiology and biology, since both evolution and morphogenesis in the latter are a specific instance of exhibiting particular molecular coordinations in which every molecular element unfailingly follows quantum mechanics.

Further discussion of how internal measurement is physically implemented will follow.

REFERENCES

1. **Wigner, E. P.,** The problem of measurement, *Am. J. Phys.* 31, 6, 1963.
2. **Misra, B. and Prigogine, I.,** Time, probability and dynamics, in *Long-Time Prediction in Dynamics,* Horton, C. W., Reichl, L. E., and Szebehely, A. G., Eds., John Wiley & Sons, New York, 1982, 21.
3. **Courbage, M. and Prigogine, I.,** Intrinsic randomness and intrinsic irreversibility in classical dynamical systems, *Proc. Natl. Acad. Sci. U.S.A.,* 80, 2412, 1983.
4. **Prigogine, I.,** *From Being to Becoming: Time and Complexity in the Physical Sciences,* Freeman, San Francisco, 1980.
5. **van Hove, L.,** Quantum mechanical perturbations giving rise to a statistical transport equation, *Physica,* 21, 369, 1955.
6. **Nakashima, S.,** On quantum theory of transport phenomena, *Progr. Theor. Phys.,* 20, 948, 1958.
7. **Zwanzig, R. J.,** Ensemble methods in the theory of irreversibility, *J. Chem. Phys.,* 33, 1338, 1960.
8. **Matsuno, K.,** How can quantum mechanics of material evolution be possible?: symmetry and symmetry-breaking in protobiological evolution, *BioSystems,* 17, 179, 1985.
9. **Bohm, D. and Hiley, B. J.,** Measurement understood through the quantum potential approach, *Found. Phys.,* 14, 255, 1984.
10. **Bohm, D. J., Dewdney, C., and Hiley, B. H.,** A quantum potential approach to the Wheeler delayed-choice experiment, *Nature (London),* 35, 294, 1985.
11. **Schulman, L. S.,** Definite measurements and deterministic quantum evolution, *Phys. Lett.,* 102, 396, 1984.
12. **Peres, A.,** On quantum-mechanical automata, *Phys. Lett. A,* 101, 249, 1984.
13. **Bohm, D. J.,** *Quantum Mechanics,* Prentice-Hall, Englewood Cliffs, N.J., 1951.
14. **Jauch, J. M.,** *Foundation of Quantum Mechanics,* Addison-Wesley, Menlo Park, Calif., 1968.
15. **Pagels, H. R.,** *The Cosmic Code: Quantum Physics as the Language of Nature,* Simon & Schuster, New York, 1982.
16. **Aspect, A., Grangier, P., and Riger, G.,** Experimental tests of realistic local theories via Bell's theorem, *Phys. Rev. Lett.,* 47, 460, 1981.
17. **Aspect, A. and Grangier, P.,** Experiment on Einstein-Podolsky-Rosen type correlations with pairs of visible photons, in *Proc. Int. Symp. on Foundations of Quantum Mechanics,* Kamefuchi, S. et al., Eds., Physical Society of Japan, Tokyo, 1983, 214.
18. **Feynman, R. P.,** Simulating physics with computers, *Int. J. Theor. Phys.,* 21, 467, 1982.
19. **Deutsch, D.,** Quantum theory, the Church-Turing principle and the universal quantum computer *Proc. R. Soc. London Ser. A,* 400, 97, 1985.

20. **Jaffe, C., Kanfer, S., and Brumer, P.,** Classical analog of pure-state quantum dynamics, *Phys. Rev. Lett.*, 54, 8, 1985.
21. **Malin, S.,** The meaning and significance of quantum states, *Found. Phys.*, 14, 1083, 1984.
22. **Jammer, M.,** *The Philosophy of Quantum Mechanics*, John Wiley & Sons, New York, 1974.
23. **Atkins, P. W.,** Quantum theory: certainty in uncertainty—a book review on *The Cosmic Code: Quantum Physics as the Language of Nature* by Heinz R. Pagels, *Nature (London)*, 296, 503, 1982.
24. **Cochran, A. A.,** Relationships between quantum physics and biology, *Found. Phys.*, 1, 235, 1971.
25. **Bennett, H. S.,** Muscle contraction and enzymic action: challenge to quantum chemistry, *Int. J. Quant. Chem.*, QBS8, 317, 1981.
26. **Pattee, H. H.,** Physical problems of decision-making constraints, in *The Physical Principles of Neuronal and Organismic Behavior*, Conrad, M. and Magar, M., Eds., Gordon and Breach, New York, 1973.
27. **Prigogine, I. and George, Cl.,** The second law as a selection rule: the microscopic theory of dissipative processes in quantum systems, *Proc. Natl. Acad. Sci. U.S.A.*, 80, 4590, 1983.
28. **Slack, J. M. W.,** *From Egg to Embryo: Determinative Events in Early Development*, Cambridge University Press, Cambridge, 1983.

Chapter 3

EQUILIBRATION

I. INTRODUCTION

Internal measurement exhibits a unique role as an endogenous agent for forming boundary conditions internally, while an exhaustive and complete control of boundary conditions is not feasible externally. Internal measurement is dynamic in generating and varying boundary conditions through the process of equilibration, eliminating disequilibrium between internal identification and realization of conservation laws. The process of equilibration refers to internal measurement in the making, whereas conservation laws are what internal measurement has accomplished. Internal measurement that endogenously examines how the null-conservedness of acting and reacting forces is fulfilled *a posteriori* is in force equilibration that underlies the materialistic process of actualizing the third law of mechanics. Also underlying the actualization of material flow continuity as a differential expression of the conservation of matter is internal measurement in the form of material flow equilibration. The materialistic actualization of the conservation of energy is the process of energy equilibration that eliminates disequilibria between internal identification and realization during the communication of the conservedness of energy.

A consequence of equilibration is infallible conservation laws in the record. More than that, equilibration constantly serves as a cause of the subsequent equilibration of similar nature. It is the on-going process of equilibration that can keep varying boundary conditions internally. In fact, the emergence of a boundary condition that makes nonequilibrium thermodynamics accompanied by fluctuations possible is the process of equilibration.

II. FORCE EQUILIBRATION

The third law of mechanics stating no net contribution from those forces of action and reaction admits two possibilities in its implementation. One is that the process underlying the implementation of the third law proceeds at an infinite velocity. That is to say, the null conservation of forces is accomplished instantaneously and is simultaneous with mechanical displacement of the medium. The other is that the process materializing null-conservedness of forces proceeds at a finite velocity. In fact, we have already observed that biological motility involves force equilibration eliminating disequilibrium between internal identification and realization of the null-conservedness of forces. It is force equilibration, in reality, that is responsible for implementing the third law.

The law of motion of the one-to-one mapping type that is compatible with completely fixed boundary conditions necessarily makes the third law of mechanics an attribute of boundary conditions, because the one-to-one temporal mapping guarantees the preservation of the third law ever since it has been set and identified as such at an initial moment. The preservation of the third law at any moment during development would thus make the process materializing null-conservedness of forces proceed at an infinite velocity. However, no physical processes are allowed to proceed beyond light velocity. The law of motion of the one-to-one mapping type would thus become inappropriate with regard to its implicit acceptance of some processes proceeding at an infinite velocity. An alternative must be either to regard the law of motion of the one-to-one mapping type supplemented by completely fixed boundary conditions simply as an approximation to the real law of motion or to seek the real law directly, instead.

The law of motion of one-to-one mapping type as experienced in classical and quantum

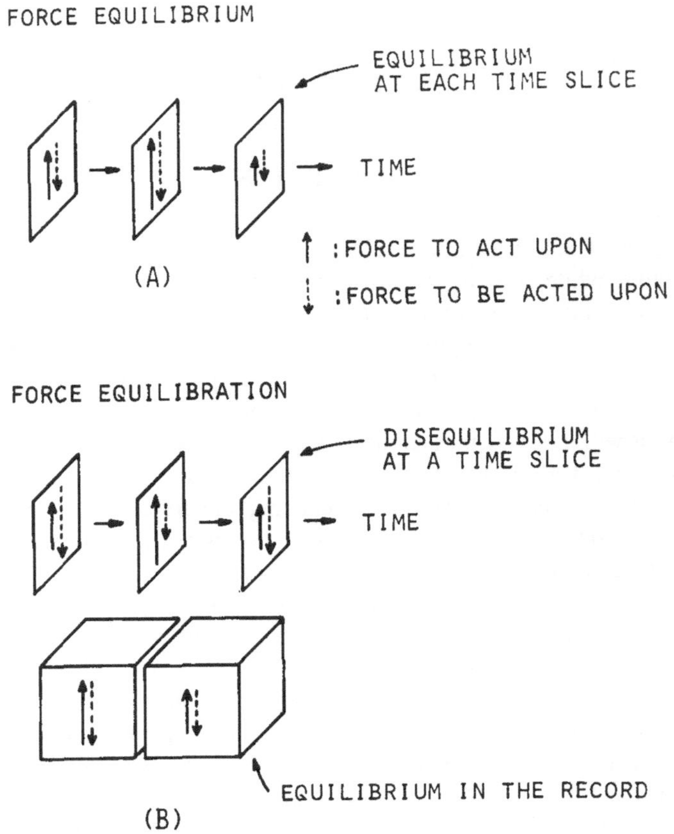

FIGURE 1. A conceptual difference between (A) force equilibrium and (B) force equilibration. The third law of mechanics gives force equilibrium such that the force of action is simultaneously counterbalanced by the force of reaction, if the law of motion of one-to-one mapping is the case. Force equilibration due to one-to-many temporal mapping, on the other hand, is involved in the process of internally selecting and actualizing both forces of action and reaction in a manner such that these two may be counterbalanced *a posteriori*. A causative factor for force equilibration is the hypothetical disequilibrium between those forces of action and reaction at each time slice. Force disequilibrium is not measurable as such. Measurement of force equilibration unfailingly yields the third law of mechanics in the record.

mechanics has been extremely successful. One cannot overemphasize its significance. Nevertheless, the price one has had to pay for adopting the law of motion is its complete separation from the associated boundary condition. The law of motion of the one-to-one mapping type could survive only under the circumstance that the separability between the law of motion and its boundary condition is guaranteed for some reason. Force equilibration proceeding at a finite velocity undoubtedly shows that the law of motion not of the one-to-one mapping type is the case. Internal measurement underlying the process materializing null-conservedness of forces makes the law of motion of one-to-one mapping type untenable.

Force equilibration assumes an internal process that can induce changes in force so as to actualize the third law of mechanics internally, by letting their repercussions continue to survive indefinitely, as depicted in Figure 1. There are in fact two sources of inducing changes in force. One is through mechanical displacement according to the second law of mechanics. If force is a property of field as with action-at-a-distance, mechanical displacement of any material unit would simultaneously induce the associated change in the force field in a unique manner. One more source is through the internal process actualizing the third law of mechanics. A mediating process of force is involved. What is characteristic to the mediating process of force is that mechanical displacement of the medium does not by itself uniquely determine its associated force, this being contrary to action-at-a-distance. There could be a one-to-many correspondence between mechanical displacement of the medium and the force acting in it. It is the mediating process of force proceeding internally which is responsible for singling out one from the possible many in the real time. The force whose change is mediated by other than forced mechanical displacement of the medium turns out to be an instance of action-through-medium.

Basic to the action-through-medium that underlies force equilibration is that both the force acting in the medium and the displacement of the medium that is acted upon cannot be uniquely identified externally due to the inseparability between the law of motion and its boundary condition. Mechanical displacement of the medium alone could not serve as a boundary condition to uniquely determine the force acting in the medium. Intrinsic partial fixedness of boundary conditions is consonant with the lack of uniqueness. Even if both the force acting in the medium and its displacement are measured externally, there could be many different values of the measured force acting in the medium against the same measured value of its displacement. Only internal measurement is responsible for supplementing intrinsic partial fixedness of boundary conditions upon the force acting in the medium.

At first sight, it might seem paradoxical to conceive of the law of motion underlying internal measurement because internal measurement originates in the impossibility of complete separation between the law of motion and its boundary conditions. What one can do instead is to describe the lack of complete separation between the law of motion and its boundary condition in less stringent terms. The law of motion of one-to-many mapping type supplemented by intrisically, partially fixed boundary conditions thus turns out to be an alternative specification of internal measurement.

The origin of one-to-many temporal mapping is of course within quantum mechanics the most fundamental basis of materialistic processes, but only with the reservations that unlimited controllability of boundary conditions is abandoned in theory and that complete separation between the quantum-mechanical law of motion and its boundary condition is untenable in reality. The probabilistic character of the quantum-mechanical wave function is not the direct origin of the one-to-many temporal mapping. Although quantum mechanics remains indefinite about the consequence of a single-shot measurement, it does yield a unique development of the probability amplitude insofar as the quantum-mechanical law of motion is supplemented by completely fixed boundary condition, at least in theory.[1] The relative frequency measured externally in repeated experiments is associated with the probability amplitude of the wave function. In contrast, the indefiniteness about the single-shot measurement is traced back not to quantum mechanics per se, but to the indefiniteness of the boundary conditions that has been actualized by the once and for all measurement provided externally. The quantum-mechanical origin of one-to-many temporal mapping is found in the indefiniteness of internal measurement.

Action-through-medium underlying force equilibration that proceeds at a finite velocity prohibits the establishment of a unique one-to-one relationship between the force acting in the medium and its displacement. If the force acting in the medium is uniquely related to the displacement of the medium at any moment, no matter how many microscopic variables may be involved in specifying the displacement, the third law of mechanics is unfailingly fulfilled for those forces acting in the whole medium. The third law of mechanics is infallible whenever forces are definable. In fact, insofar as the force is defined uniquely in relation to the displacement of the medium at every moment, force equilibration must proceed at an infinite velocity in the entire medium and force equilibrium immediately follows as exhibiting no net contribution from those forces of action and reaction.

The unique one-to-one relationship between the force acting in the medium and its displacement leads to the law of motion of the one-to-one mapping type via the second law of mechanics. In contrast, force equilibration proceeding at a finite velocity necessarily requires and justifies the lack of a unique relationship between the force acting in the medium and its displacement.

When the force acting in the medium is not uniquely definable in theory in relation to the displacement of medium, force equilibration due to internal measurement is now responsible for fixing the force that can be identifiable *a posteriori*. The lack of uniqueness in force necessitates the internal process that comes to realize only those forces that could satisfy the third law of mechanics *a posteriori*. The process concatenating the lack of *a priori* uniqueness and the *a posteriori* definiteness of force is fundamental to the one-to-many temporal mapping that lets force equilibration materialize, whereas such a concatenation has nothing to do with the law of motion of one-to-one mapping type because in the latter the *a posteriori* definiteness is necessarily equated with the *a priori* uniqueness. Force equilibration proceeding at a finite velocity points to the fact that it takes time to transform what lacks an *a priori* uniqueness into what has been actualized in accordance with the third law of mechanics, the latter of which again lacks an *a priori* uniqueness for its subsequent development. The lack of *a priori* uniqueness is inherent in the inseparability of the two roles the medium assumes as the mediator of communicating the null-conservedness of forces and as the material carrier of force.

Force varies its intensity through the displacement of medium, while the medium is displaced by the force acting there. Action-through-medium allows the displacement of medium to serve two roles. One is the actualization of force, and the other is the response to the force acting on the medium. These two roles are not clearly distinguished except for the fact that the third law of mechanics is unfailingly fulfilled *a posteriori*. Given the displacement of medium, there is some arbitrariness of force actualized there. It is the third law of mechanics that constrains the extent of arbitrariness *a posteriori*. In particular, being different from what is implied by the law of motion of the one-to-one mapping type in which the displacement of medium uniquely determines the force influencing the medium simultaneously, the lack of *a priori* uniqueness in the force for a given displacement of medium leaves the medium an endogenous capacity of actualizing the force that can be identified *a posteriori*. The actualization of force and the displacement of medium do not proceed concurrently. The absence of concurrency between the actualization of force and the displacement of medium is attributed to the finiteness of the propagation velocity of interaction changes in the medium, since it is the propagation of interaction changes which concatenates the displacement of medium with the actualization of force.

Action-through-medium observed in the acto-myosin complex of actin filaments[2] and in the dynein-tubulin complex of flagella[3] is associated with the hydrolysis of energy-rich ATP molecules. The consumption rate of ATP molecules per unit time is a factor involved in the actualization of force because the energy carried by force equilibration proceeding at a finite velocity originates in the energy released from ATP. The relationship between the actualized force and the ATP consumption rate is, however, not unique. Even if the same amount of ATP molecules is hydrolyzed, the intensity of force actualized accordingly depends upon the propagation velocity of changes in force. The force is inversely proportional to the propagation velocity of its change for the same amount of energy released per unit time. The ATP consumption rate cannot be the property uniquely related to the displacement of the medium carrying ATP hydrolysis. If the consumption rate were the unique property of the medium, the force would also have to be a unique property of the medium and the genesis of force due to ATP hydrolysis would have to be uniquely determined in the entire medium so as to fulfill the third law of mechanics. But, instantaneous and simultaneous coordination of forces in the whole medium for the third law would imply that changes in interaction are propagated at an infinite velocity. This hypothetical uniqueness contradicts the fact that action-through-medium being propagated at a finite velocity in reality.

More specifically, action-through-medium that underlies force equilibration is indefinite about the separation between the intensity of force to be propagated and its propagation. The propagation velocity of force equilibration as a boundary condition cannot be totally separated from force equilibration itself. Force equilibraton is associated with the process materializing conservedness of energy, that is to say, energy equilibration. The conservation of energy assumes energy equilibration that makes the law of motion inseparable from its boundary condition.

The conservation of energy and the third law of mechanics both join in constituting definitive boundary conditions if the law of motion of the one-to-one mapping type is the case. In contrast, the law of motion of the one-to-many mapping type that lacks a unique relationship between the force acting upon the medium and its displacement lets the coordination of energy in the entire medium vary in a manner that is not uniquely related to the displacement of the medium. The hydrolysis of ATP molecules is just such a particular instance exhibiting the coordination of energy that is not uniquely related to the displacement of the medium, because the uniquely related coordination would require that the propagation of changes in force proceeds at an infinite velocity and would let those changes in force follow the displacement instantaneously. Even if the displacement of the medium remains invariant, the force originating in the hydrolysis of ATP molecules could change. Any deviation from the unique relationship between the force acting in the medium and its displacement is deemed a variation applied to the boundary condition. Action-through-medium by means of ATP hydrolysis varies the boundary condition on the movement of the medium as much as it serves as an impetus to move the medium.

Action-at-a-distance as exemplified in the Coulomb force in classical mechanics, on the other hand, maintains a unique relationship between the force acting in the medium and its displacement. The boundary condition on the bodies interacting through action-at-a-distance is thus exogenous. The law of motion due to action-at-a-distance is completely separable from its boundary condition. The conservation of energy and the third law of mechanics are within the boundary conditions, and these two are transformed through the law of motion of the one-to-one mapping type. But, if one examines an action-at-a-distance acting in a continuous medium as in continuum mechanics,[4] it can be seen that variations in the force are accompanied by the propagation of the displacement of medium. Those superficial changes in the force are to be propagated in the medium at a finite velocity. This propagation of changes in the force at a finite velocity does, however, not originate in the indefiniteness about the separation between the law of motion and its boundary conditions.

Unique to the continuous approximation of action-at-a-distance is that the way in which the force is varied and propagated in the medium is already within the law of motion of the one-to-one mapping type and its completely fixed boundary conditions. A unique relationship persists between the force being propagated in the medium and its displacement. Because of this uniqueness, the action-at-a-distance lacks its capability of influencing the boundary condition endogenously. Force equilibrium as a consequence of the third law of mechanics is always associated with those superfical changes in force being propagated at a finite velocity in a continuous approximation of action-at-a-distance. Force equilibration for the third law is foreign to action-at-a-distance irrespective of whether an approximation of continuous medium may be taken.

Action-through-medium underlying force equilibration, being contrary to action-at-a-distance, distinguishes itself by exhibiting that the force acting in the medium carries the internal capacity to determine its own intensity in a manner not uniquely related to the displacement of medium. The functional origin of such an internal capacity to regulate the genesis of force in the medium is in the process materializing null-conservedness of forces that proceeds at a finite velocity. In this regard, action-through-medium underlying force equilibration is more fundamental than action-at-a-distance because no physical processes are allowed to proceed beyond light velocity. It is of course a common understanding that action-at-a-distance is a good approximation of forces acting in the realm of both classical and non-relativistic quantum mechanics. Nevertheless, in biological motility, one sees the presence of action-through-medium that cannot be approximated by action-at-a-distance. A more particular problem with action-through-medium will be how it develops and comes to materialize as distinguished from action-at-a-distance.

III. MATERIAL FLOW EQUILIBRATION

Force equilibration proceeding at a finite velocity has been confirmed with ATP-activated actin filaments in the presence of myosin and with dynein-tubulin complexes of flagella that are both biologically full-blown from an evolutionary perspective. The extent of being full-blown biologically is seen in the extreme smallness of the propagation velocity of force equilibraton that has been identified to be more than ten orders of magnitude less than light velocity. In fact, the process materializing conservedness, that includes force equilibration as its representative example, serves as a thread running through all the successive stages of evolving material aggregates. Because of its ubiquity, the process survives unless prohibitive boundary conditions intervene externally. The internal operation of communicating and materializing conservedness is not restricted to force equilibration. In addition to force equilibration, the process materializing conservedness also acts upon the conservation of matter, or equivalently, the continuity of material flow. One reason we pay a special attention to material flow continuity is that those aggregates open to material flow are ubiquitous at every phase of evolution of matter from subatomic to supramolecular aggregates. Cellular aggregates, both protobiological and biological, are open to material flow. We thus recognize that the process materializing conservedness, especially the one underlying material flow continuity, deserves special attention for its own sake.[5]

Material aggregates that are most significant protobiologically are those that could emerge in the primordial soup,[6] either natural or simulated. Atoms and abiotic molecules in the primordial soup can aggregate and dissociate. The combined process of aggregation and dissociation can give rise to material units which are open to material flow provided that the material aggregates once formed, can participate in further association with other elements during their inevitable partial degradation.

The maintenance of material aggregates which are open to material flow depends very much upon the three characteristic energies of association, dissociation, and thermal agitation. Material aggregation is unlikely to occur if the energy needed for material association is much greater than the thermal energy of the primordial soup. Also, if the energy needed for material dissociation is much greater than the thermal energy, a material aggregate, once formed, could avoid degradation. On the other hand, if the dissociation energy is much smaller than the thermal energy, the aggregates, whenever they are formed, will soon disintegrate due to thermal agitation. Only if the dissociation energy is of the order of thermal energy and if the association energy does not greatly exceed thermal energy, could one expect the maintenance of material aggregates which are open to material flow.

When the three energies of material association, dissociation, and thermal agitation are found within the appropriate range, open material aggregates appear as the result of interactions internal to the constituent material elements and not as the result of external action. The boundaries separating open systems from their surroundings are self-generative so long as the energy window for the appearance of open material aggregates is realized by some external means. It should be emphasized that the external means for providing the energy window for the genesis of open systems does not specify how each open system is phase-separated from the primordial soup. It provides no more than a weak macroscopic constraint and leaves enough room for internal determination. It is a necessary, though not sufficient,

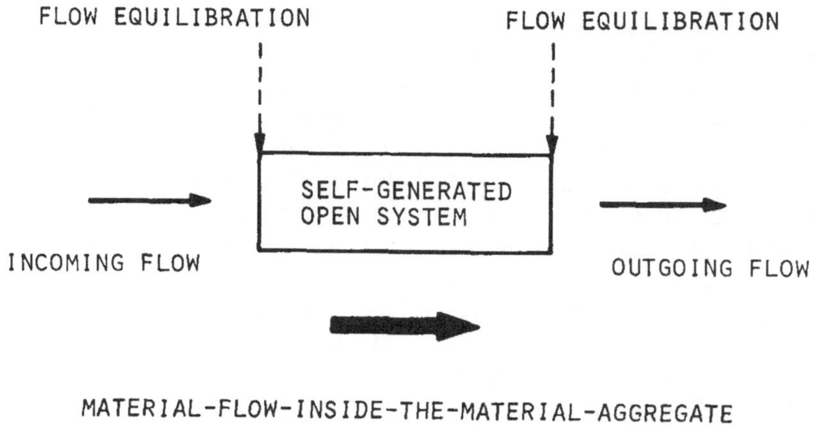

FIGURE 2. A schematic representation of material flow equilibration.

condition for the emergence of open systems in the primordial soup. Nevertheless, numerous experiments have already shown that a simulated primordial soup including various amino acids and thermal polyamino acids can give rise to microsystems open to material flow.[7] This demonstrates that the emergence of open systems in a self-generative way is not a mere theoretical surmise, but a repeatedly observable fact.

Open material aggregates with self-generative boundaries are unique in the sense that on the one hand their macroscopic contours must be shaped in a consistent manner with the underlying microscopic configurations of constituent atoms and molecules and on the other the microscopic configurations are subject to influences from the macroscopic entity which they constitute. This interplay between the microscopic configurations of constituent elements and their macroscopic contours is the key to the genesis of open systems. The microscopic interactions among atoms and molecules are always constrained by what they have accomplished.

In order to visualize the dynamic process underlying the interplay between the microscopic and the macroscopic, let us take a material aggregate open to material flow that is phase-separated in the primordial soup. It is in fact in a state of material-flow-inside-the-material-aggregate, as depicted in Figure 2, since material elements within the aggregate can be interchanged with others from the exterior. The process materializing conservedness giving rise to material flow continuity is at work here.

The material flow entering the aggregate equilibrates with the flow-inside-the-material-aggregate, which in turn equilibrates with the flow out of the aggregate. The continuity of material flow must be satisfied at both interfaces. The interaction between the material aggregate and the exterior is responsible for the maintenance of the continuity of material flow. However, it is important to notice that the material aggregate maintains the continuity by itself by adjusting its interaction with the exterior in an endogenous manner.

Internal regulation of the interactions in an open material aggregate is self-generative. Suppose that the continuity of material flow is first satisfied at both interfaces, and then the flow entering the aggregate happens to vary due to some change in the interaction with its exterior. Unless a proper adjustment of interactions is made, continuity between the flow into the aggregate and the flow-inside-the-material-aggregate would be destroyed. The process materializing conservedness for material flow continuity now emerges and internally induces equilibration between those material flows to be continued. In particular, material flow equilibration at the interface between the incoming flow and the flow-inside-the-material-aggregate forces the latter flow to change so as to preserve continuity with the former incoming flow. The change in the flow-inside-the-material-aggregate propagates toward the interface with the outgoing flow. Material flow equilibration at the interface now induces a change in the outgoing flow so as to recover the continuity by adjusting the interaction with the surroundings. The change in the interaction between the material aggregate and its surroundings accompanying equilibration at the interface with the outgoing flow again induces a variation in the incoming flow as the interaction change is propagated in the medium.

Most fundamental to the maintenance of the flow continuity through a material aggregate is that equilibration at the interface between the incoming flow and the flow-inside-the-material-aggregate does not occur concurrently with the similar equilibration at the interface between the flow-inside-the-material-aggregate and the outgoing flow. This is because the change in interaction is propagated in the medium at a finite velocity, not exceeding light velocity. An immediate consequence is that material flow equilibration at one interface always becomes a cause for the similar equilibration at other interfaces, in a successive manner. Material flow equilibration in open material aggregates continues to survive indefinitely once it has begun. The cause of material flow equilibration in an open material aggregate is always the preceding equilibrations.

The *a posteriori* principle of conservation as a boundary condition renders the law of motion inseparable from its boundary condition because the conservation assumes the process materializing conservedness due to the finiteness of the propagation velocity of interaction changes in the medium. Material flow continuity as a boundary condition is inseparable from the process identified as material flow equilibration. The *a posteriori* result of material flow equilibration is the continuity of material flow. However, material flow equilibration as a process implies more than the simple flow continuity that a boundary condition does. It includes the internal action for the continuity of material flow simply as a consequence of the fact that the change in interaction is propagated in the medium with a finite velocity. If the velocity were infinite, the difference between material flow equilibration and material flow continuity would vanish. And, the difference between the principle of conservation and the process materializing conservedness would also vanish.

We can summarize the process of material flow equilibration in a mathematical form. We denote the incoming material flow by $f^{(in)}$ and the flow-inside-the-material-aggregate at the interface with $f^{(in)}$ by $f_1^{(int)}$. The flow disequilibrium[8] at this interface is

$$\Delta f_1 \equiv f^{(in)} - f_1^{(int)}$$

Similarly, the flow disequiilibrium between the outgoing material flow $f^{(out)}$ and the flow-inside-the-material-aggregate $f_2^{(int)}$ at the interface with $f^{(out)}$ is

$$\Delta f_2 \equiv f_2^{(int)} - f^{(out)}$$

Flow continuity simply asserts the equalities $\Delta f_1 = \Delta f_2 = 0$. However, the internal processes realizing that the flow continuity differ according to whether the propagation velocity of interaction change in the medium is finite.

If the propagation velocity of interaction change remains finite as it should, the removal of flow disequilibrium $\Delta f_1 \neq 0$ successively causes a new flow disequilibrium $\Delta f_2 \neq 0$ to be removed subsequently. Again, the removal of the flow disequilibrium $\Delta f_2 \neq 0$ turns out to be a new cause for the flow disequilibrium $\Delta f_1 \neq 0$. A schematic representation of the process underlying the removal of the flow disequilibrium is shown in Figure 3. This is a demonstration of the internal action for the maintenance of continuity of material flow.

On the other hand, if the propagation velocity of interaction change were considered infinite, the removal of flow disequilibria $\Delta f_1 \neq 0$ and $\Delta f_2 \neq 0$ would take place simultaneously. The internal cause for generating the flow disequilibrium would vanish, and material flow equilibration then becomes indistinguishable from the simple continuity of material flow.

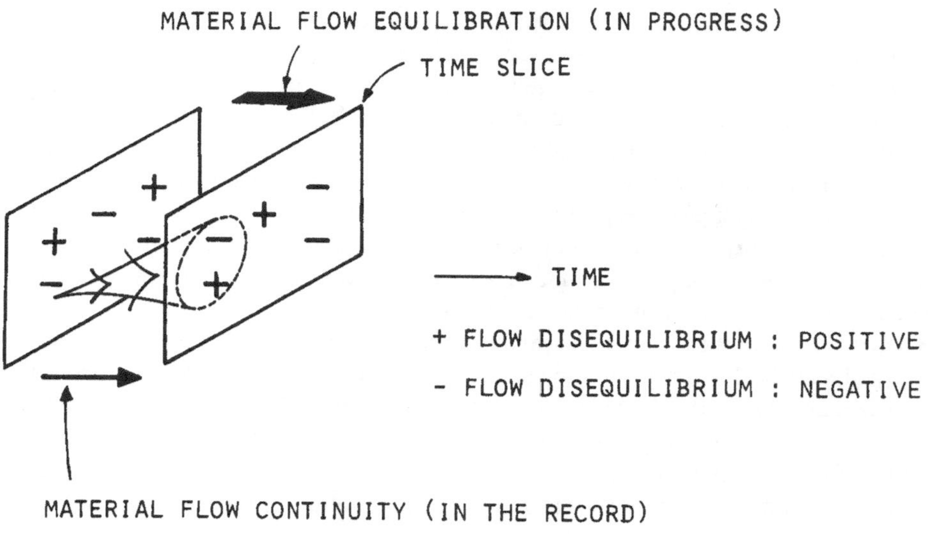

FIGURE 3. Material flow equilibration, flow disequilibrium, and flow continuity. A causative factor for material flow equilibration is flow disequilibrium at a time slice, and a consequence of the equilibration is flow disequilibrium at the succeeding time slice. Material flow continuity is guaranteed between an arbitrary pair of time slices.

Material aggregates under the influence of material flow equilibration induce adjustments of interaction with the surroundings. The preceding adjustments constantly supply endogenous causes for the subsequent ones in a successive manner, thus manifesting themselves as a form of variation imposed on the configuration of interaction among material elements constituting an open material aggregate. The variations are not arbitrary or random as only those fulfilling the conditions for material flow equilibration are realizable. Those hypothetical variations that would violate material flow continuity *a posteriori* are eliminated through the process of material flow equilibration. The source of hypothetical variations is within the indefiniteness about the inseparability between the law of motion and its boundary condition. An open material aggregate as a carrier of the inseparable complex of the law of motion and its boundary condition constrains with time the extent of indefiniteness about the inseparability by eliminating those potential complexes that would violate material flow continuity *a posteriori*. The preceding complex of the law of motion and its boundary condition is degenerated with the subsequent complexes, and the extent of being degenerate is successively dissolved with time. The separability of the law of motion and its boundary conditions becomes sharper and sharper with time, though the intrinsic inseparability persists.

If one examines the successive dissolution of degeneracy associated with the complex of the law of motion and its boundary condition in terms of the number of degrees of internal freedom in motion, the existence of unconstrained degrees of internal freedom must be admitted in order to appreciate the successive enhancement of constraint due to material flow equilibration. Only in an idealistic case in which the law of motion is separable from its boundary condition, the latter of which is completely fixable and identifiable, could the whole degrees of internal freedom involved in motion be controlled without leaving any unconstrained degrees of internal freedom. Material flow equilibration acts so as to decrease the number of unconstrained degrees of internal freedom in motion.

The presence of unconstrained degrees of internal freedom allows quantum statistical mechanics to be workable if their number remains invariant. Quantum statistical mechanics that abandons unlimited theoretical controllability of boundary conditions now admits only partially fixed boundary conditions that can be deemed as a statistical ensemble of local boundary conditions. Although it sets aside the problem of how each local boundary condition is generated and varied, quantum statistical mechanics shows that the statistical ensemble of local boundary conditions has been quite successful in a wide variety of interacting many-molecule systems insofar as the number of unconstrained degrees of internal freedom remains unchanged. Each local boundary condition indicates the extent to which the available constraints maintain themselves. Beyond this local limit, an arbitrary pair of local boundary conditions are independent irrespective of how they move with respect to each other. The intrinsic inseparability between the law of motion and its boundary condition lets each local boundary condition conceived within the framework of quantum statistical mechanics move independently of other local boundary conditions.

What is unique to the statistical ensemble of local boundary conditions is the symmetry property of statistical-mechanical movement that remains invariant against arbitrary relative motion among those local boundary conditions. However, the theoretical claim that the number of unconstrained degrees of internal freedom remains unchanged is no more than an artifact for approximating what is really going on. The number of unconstrained degrees of internal freedom decreases with material flow equilibration. If one approximates the later complex of the law of motion and its boundary condition by a statistical ensemble of local boundary conditions subject to statistical-mechanical movement, the symmetry property of the movement observed in the preceding complex will be lost there because of the enhancement of internal constraints that constantly limits the extent of arbitrary relative motion among local boundary conditions.

Material flow equilibration induces a certain symmetry breaking upon the statistical-mechanical movement which the carrier of the very equilibration maintains. The process of symmetry breaking points to the genesis of variation due to an endogenous application of constraints. The generated variation is nonrandom because it is unidirectional in breaking, instead of enhancing, the symmetry property observed at the preceding stage. If variation were random, a variation that could enhance the extent of symmetry with regard to statistical-mechanical movement would also be conceivable. Nonrandom variation associated with material flow equilibration is a consequence of eliminating those variations that could enhance the extent of symmetry among the available statistical-mechanical movements.

Material flow equilibration operates insofar as open material aggregates are available. Consequently, it can serve as a regulative principle acting upon those material aggregates that could evolve protobiologically and biologically, because of the ubiquity of their openness to material flow. Material flow equilibration is regulative internally[9] in the sense that it constantly constrains its carrier by endogenously breaking the symmetry property which the underlying material aggregate has had with regard to its statistical-mechanical movement. Unless prohibitive conditions intervene externally, the lesser symmetry, if any at all, will be approached in the course of development.

One outcome of the symmetry breaking that accompanies the decrease in the number of unconstrained degrees of internal freedom is an enhancement of local material accumulation at those aggregates appearing in the primordial soup. This is due to the fact that those molecules joining in a material aggregate lose the degrees of kinetic freedom they had previously in their statistical-mechanical movement. Molecules in the aggregate have to move in a somewhat correlated manner compared with the case otherwise. An evolutionary implication of material flow equilibration is the enhancement of local material accumulation at open material aggregates.

The enhancement of material accumulation can endogenously induce changes in the propagation velocity of interaction change in the medium. In order to see this, suppose that there are two material aggregates maintaining an identical molecular configuration of interaction that can enhance their material accumulation with time due to material flow equilibration, but that they differ only in the propagation velocity of interaction change. The difference in the propagation velocity now equivalently induces a change in the time scale. The smaller the propagation velocity of interaction change, the longer time it takes for material flow equilibration. Furthermore, the longer time for eliminating flow disequilibrium, the greater the material accumulation under a given positive flow disequilibrium. Equilibration eliminating negative flow disequilibrium would finally extinguish material aggregates altogether.

Material flow equilibration internally regulates itself so as to actualize the greater material accumulation and, accordingly, the smaller propagation velocity of interaction change. The internal cause for the smaller propagation velocity is in the nature of material flow equilibration that internally strives for the lesser symmetry with regard to the statistical-mechanical movement. As a result, the evolutionary decrease in the propagation velocity of interaction change accompanies material flow equilibration, since the greater material accumulation is an evolutionary character intrinsic to open material aggregates.

The effect of the evolutionary decrease in the propagation velocity of interaction change is, however, not restricted to material flow equilibration. Molecular flow also assumes both types of force, action and reaction, in molecules. In fact, those changes in the force are carried by the corresponding changes in the molecular interaction. Change in the molecular force go along with the changes in the molecular interaction. What is more, changes in the molecular force acting in an open material aggregate are subject to force equilibration as a matter of principle. For the third law of mechanics has to be materialized and no change in the force can be propagated beyond light velocity. Consequently, the evolutionary decrease in the propagation velocity of interaction change yields the evolutionary lowering of the propagation velocity of force equilibration. The upper limit of the propagation velocity is light velocity. In contrast, contemporary biosystems witness the propagation velocity of force equilibration being lowered even down to the order of $10^3 \mu m/s$ or less as evidenced in ATP-activated actin filaments and in the dynein-tubulin complex of flagella.

Force equilibration and material flow equilibration do not proceed independently of each other. Force equilibration accompanies material flow equilibration and vice versa, because both the conservation of energy and the continuity of material flow have to be fulfilled when measured under whatever circumstances. Equilibration is physical in its origin, but unique in its operation of constraining the possible when transformed into the actual. Equilibration serves as the regulative principle that acts with and within atoms and molecules.

The constraint acting in the process of transforming the possible into the actual may be viewed as a mode of selection. Equilibration in fact provides a materialistic ground for the nonanthropocentric selection to proceed. Nevertheless, if one takes a view that selection presumes an exhaustive list of items whose characters are completely identified even before the actual selection, this would not be the case with the process of equilibration. Equilibration can be viewed as being selective only *a posteriori* in the sense that when the possible is transformed into the actual, other alternatives, though not identifiable in an exhaustive manner, have been eliminated. Even the transformed actual cannot be exhaustively identified with regard to its later development because it constantly serves as a source of the possible to be subsequently transformed into the actual. If the objects upon which selection acts were completely identified, an external selective agent would have to be called on in order to secure selection in reality. For there is no room for a selective agent in the set of objects solely to be selected. Equilibration, quite to the contrary, maintains the internal indefiniteness that defies its exhaustive identification. It is in this internal indefiniteness that makes it possible to provide equilibration with the internally selective capability of constraining its own extent. The internal indefiniteness is inherent in the inseparability between the law of motion and its boundary condition.

Internal regulation inherent to equilibration exhibits a distinct contrast compared with the law of motion of the one-to-one mapping type in classical mechanics and in quantum mechanics, because in the latter the agent that regulates the motion is in the exogenous boundary condition. In spite of this apparent discrepancy, equilibration is materialistic as much as Newton's equations of motion in classical mechanics and Schrödinger's in quantum mechanics are. The only difference is whether the law of motion is of the one-to-many mapping type or the one-to-one. Equilibration is physical in providing the boundary condition under which the law of motion of the one-to-one mapping type could approximately work. The emergence of particular boundary conditions as experienced in nonequilibrium thermodynamics can be attributed to the process of equilibration.

IV. EQUILIBRATION AND NONEQUILIBRIUM THERMODYNAMICS

Material flow equilibration in process keeps constraining the indefiniteness it carries, and the same equilibration in the product is a manifestation of variations that are nonrandom. The genesis of variations and the constraints acting upon them are not separable in open material aggregates. The boundary condition on generating variations is under the influence of those variations that have already been generated. This convoluted mechanism of generating nonrandom variations is partially accounted for within the scheme of nonequilibrium thermodynamics.[10] There, the genesis of variations and the constraint acting upon them are hypothesized to be separate. This separation is dependent on a particular viewpoint, being specific only to the thermodynamic postulation.

Nonequilibrium thermodynamics assumes two types of exogenous boundary conditions, macroscopic and microscopic. The macroscopic boundary condition refers to the externally applied constraint that keeps maintaining the bounded system in a thermodynamically nonequilibrium state. On the other hand, the microscopic one refers to the external source of fluctuations or variations acting upon the system bounded by the macroscopic constraint. Nonequilibrium thermodynamics addresses itself to the interplay between the external macroscopic constraint and those microscopic fluctuations originating in the external source, in contrast to material flow equilibration that makes both the macroscopic constraint and the microscopic variation interwoven endogenously. The externalization of the source of fluctuations has been assumed for the benefit of thermodynamic theorizing at the expense of reality.

In the modelling of the primordial soup, for instance, representative thermodynamic variables include the concentrations of the reacting chemical species. Once the reacting chemical species and their reaction scheme are identified, one can write down a set of reaction kinetic equations of motion with fixed reaction rate coefficients as a macroscopic constraint. The thermodynamic variables and the fixed kinetic equations result from the theoretical constraints imposed arbitrarily upon the system.[11] Within the scheme of nonequilibrium thermodynamics, the evolutionary process cannot proceed without generating variations, the genesis of which is separated from the thermodynamic constraints, which are macroscopic. The source of primary variations or microscopic fluctuations has to be sought in the environment external to the macroscopic constraints that are currently in operation. Even if it is legitimate to say on a purely theoretical ground that a set of kinetic equations of motion can raise a nonperiodic and chaotic trajectory that could seemingly be associated with endogenous microscopic fluctuations,[12] the origin of those superficial fluctuations is still within the invariant macroscopic constraint that makes the kinetic equations possible. Unless microscopic fluctuations originating in the external source are present, the thermodynamic motion subjected to the invariant macroscopic constraint cannot exhibit an evolutionary character.

A postulation of nonequilibrium thermodynamics makes it tenable that thermodynamic variations or fluctuations of exogenous origin are applied to the macroscopic constraint identified in terms of the kinetic equations of thermodynamic variables. Since the kinetic equations, except for pathological cases, are nonlinear with regard to the thermodynamic variables, fluctuations can sometimes destabilize the current thermodynamic constraint. When the equations destabilize, a set of modified kinetic equations with a new set of the thermodynamic variables can be written down. The emergence of new thermodynamic variables is due to a macroscopic enhancement of microscopic fluctuations. A new chemical species can then establish itself in the modified kinetic equations. A particular part of microscopic fluctuations of exogenous origin is now to be incorporated into and internalized in the modified macroscopic constraint.

The structure of the kinetic equations can change with time only through fluctuation-induced instability.[13] The fluctuations themselves originate outside the thermodynamic constraints which have governed the form of the kinetic equations. This observation tells us that the thermodynamic constraints aided by microscopic fluctuations of exogenous origin only partially recognize material flow equilibration as the regulative principle working in open material aggregates. The thermodynamic analysis leaves the external source of microscopic fluctuations intact even if fluctuation-induced instability can induce a change in the thermodynamically macroscopic constraint, not to mention the fact that in between successive fluctuation-induced instabilities no structural changes in the constraint take place. Material flow equilibration goes beyond the scheme of fluctuation-induced instability upon thermodynamically macroscopic constraints in that the process of equilibration keeps changing the way of generating variations.

The difference between material flow equilibration and the thermodynamic theory of the constraint-fluctuation complex is not limited to how microscopic fluctuations are generated, in which the constraint-fluctuation complex implies the thermodynamic scheme of nonlinear kinetic equations supplemented by fluctuations applied externally. The difference is also manifested in how the macroscopic works. In order to examine the difference, let us once again take an open material aggregate likely to appear in the primordial soup as an illustrative example. If one tries to specify the thermodynamic boundary conditions upon the material aggregate, there would be at least two distinct ways. One is to fix the level of a generalized force that controls the amount of material accumulation in the aggregate. Even if one does not know what the generalized force looks like, it can serve as a useful theoretical device at the least because the material accumulation is measurable. The other is to control the amount of material flow going through the aggregate.

We thus observe there are thermodynamically at least two different types[14] of boundary conditions applicable to open material aggregates, force-controlled and flow-controlled.[11,15] One might be inclined to think that an appropriate combination of these two types of boundary conditions can properly specify the actual open material aggregate since it maintains a certain amount of material accumulation in the stream of a certain amount of material throughput. However, this is not the case. The force-controlled boundary condition does assume the presence of an external agent to monitor and then to control the level of force so as to maintain the amount of material accumulation as it is, and the flow-controlled one also assumes a similar external agent to control the throughput of material flow.

Although it is of course legitimate to imagine and then to set up either the force-controlled or the flow-controlled boundary condition or their arbitrary juxtaposition in experiments, a mere preparation of the primordial soup does not presume an introduction of such an external agent to control the material accumulation or the material throughput in the aggregate that emerges there. What looks like an internal boundary condition that would emerge in the primordial soup is a particular manifestation of molecular interaction acting there. In fact, material flow equilibraton that can adjust both the microscopic molecular interaction and the macroscopic contour of molecular configuration in a mutually consistent manner underlies the emergence of what would look like boundary conditions otherwise. If the search for an external agent that makes a nonequilibrium thermodynamic boundary condition realizable ends in vain, the necessary alternative will be an internal agent that has to be sought within the microscopic molecular interaction that could extend over to the macroscopic dimension via the process materializing conservedness.

Both the force-controlled and the flow-controlled boundary conditions are regulative only externally, whereas material flow equilibration is regulative internally, being inherent to atoms and molecules. Open material aggregates which are predominant during material evolution are self-generative in both their structure and function. Unless interference from the surroundings is so overwhelming as to rule out internal determination, material flow equilibration as the regulative principle provides the internal impetus for the evolution of interaction configurations in open material aggregates.

Experiments intended to demonstrate how the internally regulative principle could work in open material aggregates are paradoxical in the sense that they must not be so restrictive in designing their boundary conditions as to nullify the autonomy of the open aggregates but must be specific enough to be repeatable. In fact, we already have examples of experiments which verify the working of the regulative principle in open material aggregates. These examples include the self-generation of proteinoid microspheres from thermal proteins or what one calls proteinoids on contact with water,[7] where the formation of boundaries is internally regulated.

Formation of proteinoid microspheres can be predicted on thermodynamic grounds, also. If one compares the thermodynamic potentials in the presence of and in the absence of phase-separated microparticles in the aqueous suspension of thermal proteins, it will be found that the emergence of phase-separated microparticles can lower the thermodynamic potential due to the spatial isolation of hydrophobic parts of thermal proteins by those hydrophilic ones in the aqueous milieu. The lower the thermodynamic potential, the more stable the thermodynamic system becomes with a consequence of the emergence of proteinoid microspheres. Material flow equilibration, on the other hand, deals with the process of how the microscopic molecular interaction varies the macroscopic contour of molecular configuration and vice versa. Even if the end result coincides, material flow equilibration gives more than what nonequilibrium thermodynamics does. The internally regulative principle concerns itself even with the process of how boundary conditions of nonequilibrium thermodynamics could be constructed.

The difference between material flow equilibration and the thermodynamic postulation of the constraint-fluctuation complex has been identified by referring to what the relevant experiments would give in the end. But, the difference also consists of how they participate in the actual experimentation. One decisive factor in any experimentation is that experimental controllability of boundary conditions is intrinsically limited. There always remains room for internal autonomy on the part of material aggregates. Experiments on forming proteinoid microspheres point to the process of and the capacity for internal determination beyond what the experimental boundary condition controls on a macroscopic scale. Material flow equilibration potentially enhances this materialistic autonomy no matter how much the intervention from the exogenous control of boundary conditions may be.

In contrast, the thermodynamic postulation of the constraint-fluctuation complex is extremely theoretical, especially in the demarcation between the macroscopic constraints and the microscopic fluctuations. There is no definitive argument for distinguishing between the constraints and the fluctuations in reality. The demarcation is theoretical in its origin. Only when the macroscopic experimental boundary conditions controlled externally happen to coincide with the macroscopic constraints that are theorized thermodynamically could the thermodynamic theorization survive and be compatible with those experiments. The thermodynamic analysis alone does not tell us where the demarcation exists between the macroscopic constraints which could be experimentally controlled and the microscopic fluctuations that defy any exogenous control.

Fluctuation-induced instability certainly generates changes in the thermodynamic demarcation between macroscopic constraints and microscopic fluctuations. But it is quite another problem to examine how this scheme fits into the endogenous variation of macroscopic constraints observed in nature, especially from an evolutionary perspective, because in the thermodynamic theory, the generator of microscopic fluctuations supposedly remains indifferent to what would happen to the macroscopic constraints. Furthermore, the thermodynamically macroscopic constraints are sometimes controllable externally, for instance through varying control parameters. Variations of macroscopic constraint in the scheme of nonequilibrium thermodynamics therefore occur only through either fluctuation-induced instability or changes in so-called controllable bifurcation parameters.[13,16]

In self-generated open material aggregates such as proteinoid microspheres, thermodynamically macroscopic constraints cannot be imposed arbitrarily, though thermodynamics works there. Any self-generated open material aggregate produces nonrandom variations constrained by itself due to material flow equilibration. What is unique about material flow equilibration is that although it provides a constraint acting upon variations, the constraint itself never ceases to change its own content with time. Constraint and variation are not separable. Constraint generates variation, and variation generates constraint. Consequently, the imposition of externalized thermodynamic constraints interferes with the working of the internally regulative principle.

The thermodynamic variables within the thermodynamic constraints are aggregated macroscopic variables lacking any role as active agents. They vary their own values only by responding to the impetus originating outside. The microscopic elements involved in an aggregated macroscopic variable are forced to be under a common macroscopic constraint. The imposition of a common macroscopic constraint upon every microscopic element involved is instantaneous at the moment the aggregated macroscopic variable is defined. This fact necessarily leads to the assumption that the propagation of interaction change in the aggregate of microscopic elements constituting a macroscopic variable proceeds at an infinite velocity, thus depriving the thermodynamic variables of their capability for generating variations internally. What is to be propagated at an infinite velocity is the condition that every microscopic element involved in the macroscopic variable takes part in maintaining its value at a fixed level at every moment.

The regulative principle in self-generated open material aggregates, on the other hand, is compatible with the propagation of interaction change proceeding at a finite velocity. Consequently, an aggregated macroscopic variable such as the molecular concentration of self-generated open material aggregate in the primordial soup lets every molecular element take part in maintaining the concentration as it is and, at the same time, lets it vary the level of the concentration through the sequential repercussions among those molecular elements. The macroscopic variable can be seen as an active agent changing its own value with time. The regulative principle provides us with macroscopic variables whose values can change autonomously in distinct contrast to the thermodynamic variables whose motion always assumes impetus from something else.

Whenever macroscopic variables are defined under the premise that the propagation of interaction change proceeds in the medium at a finite velocity, as it actually does, the internal changes in the interaction configuration are maintained and in turn vary the values of macroscopic variables indefinitely. This is due to the simple fact that the microscopic molecular interaction influences as much as it is subject to the macroscopic contour of molecular configuration. Only when some external means of breaking the dynamic sequence between the microscopic molecular interaction and its covering macroscopic contour is available, could the active agents perceived in the macroscopic variables be phased out.

A common device for preventing macroscopic variables from becoming active agents is the so-called adiabatic approximation.[17] Any material system involves a wide spectrum of kinetic processes ranging from the very slow to the very fast. One can separate the entire spectrum into the two subgroups, the slower processes and the faster ones. The adiabatic approximation claims that the characteristic velocity of the faster processes could be put at infinity if only the slower processes are of interest. What is more, the faster processes are supposed to follow the slower ones instantaneously. This way of thinking, in one form or another, is found in all branches of physics,[5] not just in equilibrium and nonequilibrium thermodynamics.

Physics equipped with the adiabatic approximation conspires to make matter inanimate. If we free ourselves from the adiabatic approximation, matter can be seen as an active agent simply by virtue of the fact that interaction changes are propagated in the medium at a finite velocity. If it is to be applied to material evolution, nonequilibrium thermodynamics must be extended so that it can get rid of the fetters imposed by the adiabatic approximation. Molecules and molecular aggregates are not inanimate.

V. BOUNDARY CONDITIONS IN NONEQUILIBRIUM THERMODYNAMICS

Formation of open material aggregates is admittedly a consequence of quantum-mechanical processes proceeding in an interacting many-molecule system, whereas the quantum-mechanical law of motion and its boundary conditions are inseparably interwoven. Every molecule constitutes a part of those boundary conditions acting upon others interacting with it, because the process materializing conservedness influences the boundary conditions under which the quantum-mechanical law of motion is operative. The process of equilibration thus constantly induces variation in the ensemble of local boundary conditions generated endogenously. Furthermore, equilibration has an evolutionary tendency to cause those local boundary conditions to be more interdependent with time because of its intrinsic self-constraining capability. The emergence of open material aggregates in the primordial soup is in fact a morphological expression of the interdependence of those local boundary conditions on molecules there.

The extent to which the ensemble of local boundary conditions are interdependent or independent among themselves depends on the character of equilibration proceeding there. Each local boundary condition constitutes a spatio-temporal contour of the microscopic molecular interactions therein. Although equilibration makes it tenable for the macroscopic contour to be varied as responding to changes in the microscopic interactions inside, the contour can also vary in response to the impetus originating outside. It is the type of available molecular interactions that determines whether variations in the macroscopic contour are due primarily to molecules inside or to those outside. The extent of local boundary conditions being interdependent thus coincides with the extent to which the process of equilibration operates. Beyond this limit, local boundary conditions remain rather independent of each other because of the lack of a direct link between the macroscopic contour and the microscopic molecular interaction. The inevitable localization of equilibration yields an ensemble of independent local boundary conditions out of an interacting many-molcule system.

The origin of independent local boundary conditions is quantum-mechanical in the sense that the type of available interaction determines the linear decomposability of the local wave function and each local boundary condition is constructed by the surrounding local wave functions. The process of equilibration and the associated local wave function are conditioned by the surrounding local wave functions. Variations in the local wave function are imputed to those of the surrounding wave functions. What is more, those variations in the local wave function are sequential because of the involvement of equilibration proceeding at a finite velocity. The ensemble of independent local boundary conditions thus keeps varying each boundary condition as it responds to variations in the others. Accordingly, quantum mechanics yields quantum statistical mechanics, allowing an ensemble of independent local boundary conditions when the extent of equilibration is localized.

Quantum mechanics supplemented by the process materializing conservedness incorporates into itself a way of forming its own boundary conditions. Quantum statistical mechanics is just one mode of practicing quantum mechanics in such a manner that its boundary conditions are localized both spatially and temporally. The resulting ensemble of independent localized boundary conditions is statistical because of the spatio-temporal limitation of each condition but can be nonstationary because of the underlying process of equilibration.

The emergence of open material aggregates in the primordial soup and the conceptual origin of quantum statistical mechanics share the same physical underpinning, identified as the process of equilibration. The ensemble of locally independent boundary conditions can be transformed into localy more interdependent boundary conditions through this very process of equilibration, due to its endogenous capability of self-constraining. Unless prohibitive conditions intervene, the ensemble of local boundary conditions envisaged in quantum statistical mechanics can become more interdependent among themselves. Mixed states in quantum statistical mechanics attest to the presence of independent local boundary conditions that interfere with quantum mechanical coherence over those local wave functions belonging to different boundary conditions, whereas pure quantum states assume a quantum coherence extended over the entire system.[18] Such pure states, however, could not be guaranteed unless all of the microscopic elements are identified as subject to one and the same fixed boundary condition. Equilibration is in fact an agent of enhancing the quantum coherence internal to quantum-statistical mixed states.

Boundary conditions encountered in nonequilibrium thermodynamics are quantum-mechanical in their origin, unless controlled otherwise. The precursors of these boundary conditions are mixed states in quantum statistical mechanics, and the latter are due to internal measurement or the working of interaction that constitutes local boundary conditions internally. A macroscopic enhancement of local boundary conditions can be seen as an emergence of boundary conditions in the realm of nonequilibrium thermodynamics. In fact, it is the process of equilibration that brings about such a macroscopic enhancement by increasing the extent of interdependence among the previous mixed states.

VI. THE ORIGIN OF FLUCTUATIONS

The thermodynamic analysis of the constraint-fluctuation complexes also requires fluctuations in addition to those boundary conditions that make nonequilibrium thermodynamic constraints feasible. Fluctuations thus have to be sought in quantum mechanics as much as nonequilibrium thermodynamic constraints have to be. Classical mechanics based upon the law of motion of the one-to-one mapping type, supplemented by completely fixed boundary conditions, can introduce local boundary conditions and fluctuations only by artificially discarding the completely fixed boundary conditions. Quantum mechanics supplemented by the process materializing conservedness, on the contrary, intrinsically admit the partially fixed boundary conditions that enable internal measurement to generate and vary boundary conditions endogenously. The resulting ensemble of local boundary conditions can be either stationary or nonstationary depending on how the process of equilibration is involved.

Fluctuations are visible even if the ensemble of independent local boundary conditions is stationary. The ensemble can be stationary only statistically. Each local boundary condition constituting the ensemble delimits the unit of equilibration taking place there. Since each unit of equilibration subsequently constitutes a part of the boundary conditions on the neighboring units, local boundary conditions can vary individually. Fluctuations are thus within the variability of these local boundary conditions. If the units of equilibration are weakly interacting atoms and molecules, the standard interpretation of statistical fluctuations is recovered with regard to their stationary ensemble.

A unit of equilibration generates and varies its local boundary condition through the process materializing conservedness, though the neighboring units of equilibration also constitute a part of the boundary condition upon the former. Statistical fluctuations refer to the temporal variations in the exogenous contribution from the neighboring local boundary conditions. However, fluctuations cannot be exclusively of exogenous or of statistical origin. The process of equilibration constantly generates variations in the molecular interaction configuration within its carrier. Equilibration-induced fluctuations are endogenous in and inseparable from constraints acting upon the material unit that carries equilibration.

We thus observe two different types of fluctuations, statistical and equilibration induced. Of these two, equilibration-induced fluctuations are more fundamental in that they can vary the constraint acting upon each local boundary condition, whereas statistical fluctuations presume a stationary ensemble of those local boundary conditions that have already been generated. If equilibration-induced fluctuations were quiescent, only statistical ones would remain and no changes in the macroscopic constraints could be expected. An agent for varying what looks like a stationary ensemble of statistical fluctuations is equilibration-induced fluctuations.

VII. SUMMARY

The process materializing conservedness yields force equilibration if the null-conservation of forces is invoked, and similarly yields material flow equilibration if material flow continuity is referred to. Both the equilibrations proceed concurrently in reality. The concurrent operation of both force equilibration and material flow equilibration gives rise to the evolutionary implication that an open material aggregate, if any, enhances its material accumulaton while decreasing the propagation velocity of force equilibration there.

The emergence of self-generated aggregates open to material flow is certainly a phenomenon in nonequilibrium thermodynamics. However, self-generated open-material aggregates cannot be modeled by either force-controlled or flow-controlled boundary conditions, because their boundaries are formed through the interplay between the microscopic molecular interaction and the macroscopic contour of molecular configuration. Endogenous formation of local boundary conditions is due to internal measurement in quantum mechanics that delimits each unit of equilibration. The resulting ensemble of local boundary conditions can be dealt with in quantum statistical mechanics. What is more, if local boundary conditions become sufficiently macroscopic, nonequilibrium thermodynamics could be invoked.

Nevertheless, quantum statistical mechanics can be applicable to the process of equilibration only to the limited extent that the ensemble of local boundary conditions could be deemed to lack an endogenous capacity of varying the ensemble. Also, the thermodynamic postulation of the constraint-fluctuation complex is limited compared with the process of equilibration in that both the boundary conditions identified as constraints and those identified as fluctuations are theoretically imposed instead of being generated internally. The material carrier of equilibration, on the contrary, generates both its boundary or constraint and its inevitable variations or fluctuations. Equilibration-induced fluctuations generate variations in the ensemble of those material units carrying equilibration and thus go beyond those fluctuations associated with a stationary statistical ensemble.

Equilibration and equilibration-induced fluctuations are evolutionary and can be unidirectional in retrospect. The presence of unidirectionality needs to be examined in the light of the second law of thermodynamics.

REFERENCES

1. **Schulman, L. S.,** Definite measurements and deterministic quantum evolution, *Phys. Lett. A,* 102, 396, 1984.
2. **Yanagida, T., Nakase, M., Nishiyama, K., and Oosawa, F.,** Direct observation of motion of single F-actin filaments in the presence of myosin, *Nature (London),* 307, 58, 1984.
3. **Hiramoto, Y. and Baba, S. A.,** A quantitative analysis of flagellar movement in echinoderm spermatozoa, *J. Exp. Biol.,* 76, 85, 1978.
4. **Landau, L. D. and Lifshitz, E. M.,** *Fluid Mechanics,* Addison-Wesley, Reading, Mass., 1959.
5. **Matsuno, K.,** Open systems and the origin of protoreproductive units, in *Beyond Neo-Darwinism,* Ho, M. W. and Saunders, P. T., Eds., Academic Press, London, 1984, 61.
6. **Florkin, M., 1975,** Ideas and experiments in the field of prebiological chemical, in *Comprehensive Biochemistry,* Vol. 29B, Florkin, M. and Stotz, E. H., Eds., Elsevier, Amsterdam, 1975, 231.
7. **Fox, S. W.,** Proteinoid experiments and evolutionary theory, in *Beyond Neo-Darwinism,* Ho, M. W. and Saunders, P. T., Eds., Academic Press, London, 1984, 15.
8. **Matsuno, K.,** Disequilibrium dynamics of autonomous systems and their structural transformation, *Int. J. Gen. Syst.,* 6, 75, 1980.
9. **Webster, G. and Goodwin, B. C.,** The origin of species: a structuralist approach, *J. Soc. Biol. Struct.,* 5, 15, 1982.
10. **Kreuzer, H. J.,** *Nonequilibrium Thermodynamics and Its Statistical Foundations,* Clarendon Press, Oxford, 1981.
11. **Epstein, I. R.,** Complex dynamical behavior in "simple" chemical systems, *J. Phys. Chem.,* 88, 187, 1984.
12. **Eckmann, J. P. and Ruelle, D.,** Ergodic theory of chaos and strange attractor, *Rev. Mod. Phys.,* 57, 617, 1985.
13. **Nicolis, G. and Prigogine, I.,** *Self-Organization in Nonequilibrium Systems: From Dissipative Structures to Order Fluctuations,* John Wiley & Sons, New York, 1977.
14. **Prigogine, I.,** *Introduction to Thermodynamics of Irreversible Processes,* 3rd ed., Interscience, New York, 1969.
15. **Eigen, M. and Schuster, P.,** Stages of emerging life — five principles of early organization, *J. Mol. Evol.,* 19, 47, 1982.
16. **Hunding, A.,** Bifurcations of nonlinear reaction-diffusion systems in prolate spheroids, *J. Math. Biol.,* 17, 223, 1983.
17. **Landsberg, P. T.,** *Thermodynamics and Statistical Mechanics,* Oxford University Press, Oxford, 1978.
18. **Landau, L. D. and Lifshitz, E. M.,** *Statistical Physics,* Pergamon Press, Oxford, 1969.

Chapter 4

INTRINSIC IRREVERSIBILITY

I. INTRODUCTION

Equilibration due to the process materializing conservedness is intrinsically irreversible because of its self-constraining capability in time. The conservation of energy, or equivalently the first law of thermodynamics, assumes the process materializing conservedness of energy. Thus, already implicit in the first law is its irreversible mode of operation. The second law of thermodynamics points to an irreversible character unique to the way the first law is implemented in reality. Only if the microscopic law of motion is of the one-to-one mapping type supplemented by completely identifiable boundary conditions, would the first law become independent of the second law, which could establish its irreversible character simply by abandoning the once-assumed unlimited theoretical controllability of boundary conditions.

Entropy as an indicator of irreversibility differs in its implication depending upon the context under which it appears. Entropy in thermodynamics will require an artificial lessening of the controllability of boundary conditions in order to secure its irreversibility if the underlying microscopic law of motion is of the one-to-one mapping type. Entropy in quantum statistical mechanics can exhibit an irreversible character if the ensemble of local boundary conditions remains stationary. However, the process of equilibration implemented in quantum mechanics can also induce irreversible changes in the ensemble of local boundary conditions. Irreversibility found in physical processes is not necessarily associated with an increase of entropy in quantum statistical mechanics. Time's arrow is inherent in every process of equilibration.

II. FIRST AND SECOND LAWS OF THERMODYNAMICS

Thermodynamics of an interacting many-molecule system defines the state of equilibrium toward which an internal adjustment of interaction among those molecules moves in an irreversible manner. The second law of thermodynamics is identified as the underlying principle of irreversibility. Although there have been many different versions of what it is all about, the second law is based upon the incontrovertible empirical fact, according to Lord Kelvin, that it is impossible to convert an amount of heat completely into work by a cyclic process without leaving any other changes behind or, equivalently that no system can return to its earlier state without disturbing its surroundings.[1]

The equilibrium state in thermodynamics is so defined that it lacks any further tendency toward internal change in contact with the surroundings. Once an irreversible movement toward the state of equilibrium is guaranteed, one can naturally observe Le Chatelier's principle in action, namely, that any system in thermodynamic equilibrium, under the influence of a change of every single one of the factors determining the equilibrium, undergoes a transformation in such a direction that, if this transformation took place alone, it would produce a change in the opposite direction of the factor in question.[2,3] A more dynamic formulation of irreversibility comes from Spencer who observes that whatever quantity of motion any member constituting a thermodynamic system has in any direction, is from moment to moment counterbalanced by an equivalent motion in some other part of the same system in an opposite direction.[3] These discussions are derivatives from the statement claiming that there is an equilibrium state. The relationship between the mere presence of the equilibrium state and the second law of thermodynamics is, however, circular because either one of the two requires the other for its existence. Therefore, further scrutiny of the state of equilibrium will be required for elucidating the nature of the second law more explicitly.

One attempt to justify the state of equilibrium and accordingly the second law of thermodynamics is to start from the microscopic law of motion of the one-to-one mapping type. However, the law of motion of the one-to-one mapping type does remain reversible insofar as it is supplemented by completely identifiable boundary conditions. Only when unlimited theoretical controllability of boundary conditions is abandoned,[4,5] can the resulting motion become irreversible. The problem at stake concerns who or what should be responsible for abandoning the once-assumed unlimited theoretical controllability of boundary conditions on the law of motion of the one-to-one mapping type.

The tricky nature of the present problem is due to the fact that if unlimited theoretical controllability of boundary conditions is claimed upon materialistic grounds in the first place, it will be hard to justify abandoning the very same unlimited theoretical controllability on the same materialistic ground as well. If an observer who can measure only in a coarse-grained manner is invoked, one can recognize an inevitable limitation of experimental controllability of boundary conditions. Nevertheless, limited experimental controllability by no means provides the materialistic ground for abandoning unlimited theoretical controllability. Both limited experimental and unlimited theoretical controllabilities of boundary conditions can coexist in a queer manner. The difficulty in finding a materialistic reason for abandoning unlimited theoretical controllability stems from the initial theoretical premise concerning the law of motion supplemented by completely identifiable boundary conditions.

An alternative to unlimited theoretical controllability of boundary conditions is a proper appreciation of the process materializing conservedness that makes the law of motion and its boundary condition inseparable. The inseparability forces the law of motion to be of a one-to-many mapping type and the associated boundary condition to be only partially identifiable as a matter of fact. The law of motion of the one-to-many mapping type generates the process materializing conservedness that is intrinsically irreversible. The first law of thermodynamics identified as the principle of the conservation of energy in the product now gives rise to the process of equilibration in the making. Intrinsic irreversibility is found to be latent in the first law.

Equilibration associated with the process materializing conservedness refers to the principle of conservation. Both an impetus to and an endogenous goal of equilibration are the principle of conservation. Equilibration however does not immediately yield the state of equilibrium because equilibration never fails to generate the subsequent equilibrations. Equilibration is irreversible without invoking the state of equilibrium. Insofar as one admits that any change in interaction is propagated in the medium at a finite velocity, the first and second laws of thermodynamics cannot be independent of each other, due to the fact that the process materializing conservedness continues to survive through its successive repercussions. In fact, it is conceptually difficult to imagine at the same time both the state of equilibrium that lacks an endogenous capacity of generating further changes as it interacts with its surroundings and intrinsic irreversibility that keeps generating a unidirectional change. The state of equilibrium and intrinsic irreversibility simply cannot coexist due to a lack of internal consistency.

There are two conceptual choices in coping with the presence of the equilibrium state and intrinsic irreversibility. One is to emphasize the state of equilibrium at the expense of saving irreversibility in its fullest implication. This view is found in the standard formulation of thermodynamics. The other is to emphasize the process of equilibration or the process materializing conservedness at the expense of the state of equilibrium. Equilibration is ubiquitous at least to the extent that open material aggregates are available, as in the form of material flow equilibration. If one understands that the state of equilibrium conceived in thermodynamics is a specific state subject to the principle of the conservation of energy, that is, the first law of thermodynamics, equilibration might seem to be capable of leading to the state of equilibrium. However, the state of equilibrium is not a necessary consequence of the process of equilibration. Unless the endogenous goal of equilibration is forcibly set to be the state of equilibrium by some external means, equilibration can imply more than what the state of equilibrium could. Intrinsic irreversibility is a character unique to equilibration, and not to the state of equilibrium.

Intrinsic irreversibility associated with equilibration also goes beyond the irreversibility observed with a stationary state in nonequilibrium thermodynamics. The nonequilibrium stationary state, if stable, can maintain its stability in the face of arbitrary disturbances from outside. The stability is viewed as an indication of irreversibility in the sense that the stable stationary state will be approached in the end even in the presence of disturbances. There are in fact several examples of nonequilibrium stationary states that have Lyaponov functions exhibiting an irreversible monotonicity with the elapse of time, as in the case of Boltzmann's H-theorem.[6,7] It is however important to notice that the irreversibility associated with a nonequilibrium stationary state is no more than its stability against disturbances. The way a nonequilibrium stationary state comes into existence is not found within its own irreversibility.

The second law of thermodynamics can be applicable even to nonequilibrium thermodynamics. Its premise is the existence of a stationary state toward which internal adjustments of interaction in a given thermodynamically nonequilibrium system move. This movement has to be irreversible because of the presence of the goal which is eventually approached. However, the existence of a nonequilibrium stationary state poses a formidable problem of how one can justify the nonequilibrium state as in the case of the state of equilibrium in equilibrium thermodynamics. The problem in its essence is a matter of boundary conditions. If one could prepare a boundary condition that guarantees the existence of a nonequilibrium stationary state, the problem would become superficial. The real problem, on the other hand, is about how the boundary condition of the sort that can sustain a nonequilibrium stationary state, conceived only theoretically, could come into existence in the first place. The situation is similar to the case of the equilibrium state in thermodynamics. Thermodynamics guarantees the presence of the state of equilibrium in theory. But, it is quite another problem to justify how the boundary condition that renders the state of equilibrium possible could come into existence.

It has been a tradition in physics to understand the irreversibility expressed in the second law of thermodynamics in reference to the state of equilibrium or the stable stationary state.[8] Nevertheless, we still encounter the problem of whether irreversibility can be derived without referring to the state of equilibrium. Equilibration just happens to serve as an endogenous agent for irreversibility without identifying the goal state at which the unidirectional movement may aim. Equilibration points to the second law of thermodynamics without referring to the goal state, if any. However, equilibration is not the second law itself. If the goal state is not identifiable, the second law cannot stand alone because it lacks a reference against which the irreversibility that is empirically infallible is measured. The rescue comes from the first law of thermodynamics. The principle of the conservation of energy unfailingly serves as a reference against which irreversibility is identified, only with the physically understandable and justifiable reservation that any change in interaction is propagated in the medium at a finite velocity.

III. ENTROPY

A unique characteristic in thermodynamic analysis is its claim for the presence of various thermodynamic variables or state functions. More than that, all of the thermodynamic variables establish a set of closure relationships among themselves such that every one of them can be derived from the others. A most significant thermodynamic variable is entropy, which

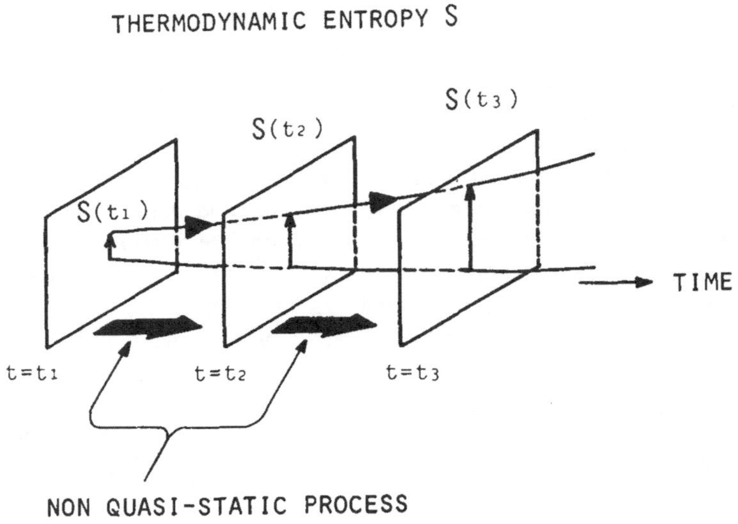

FIGURE 1. An irreversible increase of thermodynamic entropy. Responsible for the irreversible increase is nonquasi-stationary process in thermodynamics.

is basically defined as the ratio of heat energy relative to temperature. This does not imply that heat and temperature are more fundamental primitives than thermodynamic entropy. Heat energy and temperature will finally require thermodynamic entropy for their definitions.

These thermodynamic variables vary their values as they move along the quasi-static trajectory that is thermodynamically in equilibrium with the surroundings from moment to moment. However, thermodynamic entropy is exceptional in that it can specify its own behavior even away from the quasi-static trajectory in thermodynamic equilibrium. The thermodynamic entropy of a system in contact with a heat reservoir can be defined even off the quasi-static trajectory and can increase its value as the state of thermodynamic equilibrium is inevitably approached. The recovery to the quasi-static track is not the quasi-static process by itself, but is irreversible.

The irreversible character of thermodynamic entropy as depicted in Figure 1 requires its own dynamics that is different from the one on the quasi-static track. Classical mechanics allowing for the law of motion of the one-to-one mapping type supplemented by completely identifiable boundary conditions does not fulfill the stated purpose because of its intrinsic reversibility. It is of course legitimate to say that if unlimited controllability of boundary conditions that is conceived only theoretically is weakened so as to admit an inclusion of some indefiniteness in their identification, a superficial irreversibility would follow. But the difficulty with the present conceptual scheme is that classical mechanics by itself does not provide any reason that would justify abandoning the once-assumed unlimited theoretical controllability of boundary conditions. Even if limited experimental controllability of boundary conditions witnesses a functioning of irreversibility in classical mechanics, it has nothing to do with irreversibility in theory. Our problem is to see just how irreversibility comes into existence, especially in terms of entropy. An alternative to classical mechanics has to be sought, and quantum mechanics would come up as its candidate.

Quantum mechanics does not take it for granted that the boundary condition on an arbitrary quantum mechanical system be completely identifiable and fixable in theory. Every molecule provides quantum-mechanically a part of the boundary conditions on all of the others interacting with it through internal measurement. A natural consequence of internal measurement in quantum mechanics is the appearance of the ensemble of local boundary conditions. Those endogenous boundary conditions are the products of as well as the constraints upon

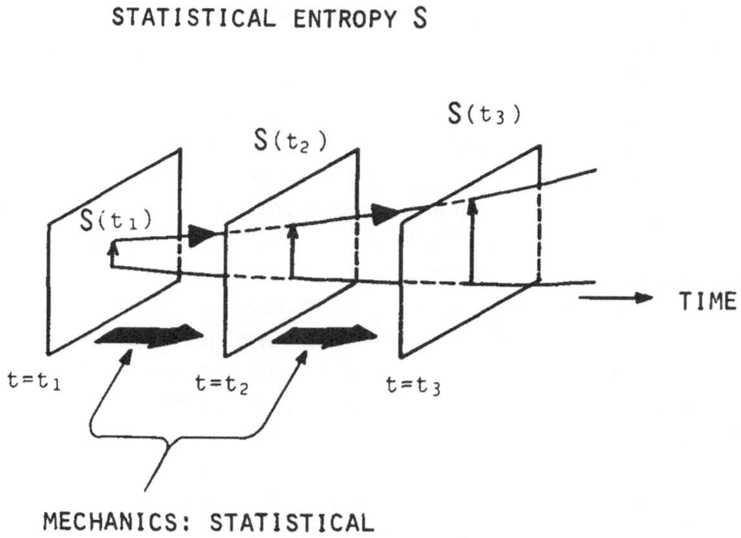

FIGURE 2. An irreversible increase of statistical entropy. Responsible for the irreversible increase is mechanics in the statistical sense, in which unlimited theoretical controllability of boundary conditions is abandoned.

the quantum-mechanical interaction taking place there. Each member constituting the ensemble of local boundary conditions can specify a local quantum state because of the unique one-to-one relationship between the two. Every quantum state presumes its unique boundary condition. Nevertheless, each local boundary condition in the ensemble cannot be totally independent of the others since its very origin is due to the interaction with the others. The quantum-mechanical ensemble of local boundary conditions can vary itself through the interaction inherent in the ensemble. It is only when the ensemble of local boundary conditions remains stationary that a local quantum state belonging to each local boundary condition attains its long-standing status.

As a result, the first requirement of approaching thermodynamics starting from quantum mechanics is to recognize the ensemble of local boundary conditions. The ensemble can be stationary or nonstationary.

If the ensemble of local boundary conditions is stationary, the standard formulation of quantum statistical mechanics can be invoked. The statistical interpretation of the quantum-mechanical ensemble of local boundary conditions makes it tenable that relatively long-standing local quantum states are equally populated all over the available local boundary conditions. In order to see its further implication, let us suppose that a quantum-mechanical system S could allow two nonoverlapping stationary ensembles of local boundary conditions, E_1 and E_2. What the present statistical interpretation suggests to us is that the realizable ensemble of S is a combined one of both E_1 and E_2, instead of being restricted to either E_1 or E_2. The stationary ensemble incorporates into itself all of the possible local boundary conditions and does not leave any possible ones behind. The reasoning is purely on statistical grounds, and the same reasoning can yield Boltzmann's statistical entropy in terms of the number of available local boundary conditions which has been referred to as the number of microscopic states.

If the stationary ensemble of local boundary conditions is stable against any of the slightest disturbances from outside, the recovery process is irreversible in the sense that the stationary ensemble will finally be approached. Boltzmann's statistical entropy increases its value in the recovery process toward the stationary ensemble as demonstrated in Figure 2. The potential of Boltzmann's entropy to increase is entailed by the fact that some possible local boundary conditions inherent to the stationary ensemble have not yet been actualized in the ensemble. However, the recovery to the stationary ensemble is no more than a form of generating variations in the ensemble of local boundary conditions. Only when the presence of the stationary ensemble is guaranteed for some reason can the irreversibility be argued for on statistical grounds. Otherwise, the presence of the stationary ensemble and the associated irreversibility cannot be reconciled in a consistent manner. Intrinsic irreversibility with the ensemble of local boundary conditions, on the other hand, has to be examined without invoking it hypothetical stationariness.

Irreversible variations in the ensemble of local boundary conditions require a quantum-mechanical underpinning. A common denominator among all quantum-mechanical processes is the process materializing conservedness. Unique to the process materializing conservedness identified as the process of equilibration is its endogenous capacity of enhancing the extent of being constrained. Given an interacting many-molecule system, only those molecular configurations that satisfy the principle of conservation *a posteriori* are realizable in such a sequentially multiplicative manner that the preceding configurations successively provide raw materials to be constrained. A necessary consequence is an enhancement of interdependence in the ensemble of local boundary conditions with the elapse of time.

Take for instance a system N of interacting molecules. If the quantum-mechanical interaction among N molecules is sufficiently weak so that every molecule can be deemed to be locally bounded by all of the other (N − 1) molecules, the local boundary condition varies depending upon such a quantity as the amount of energy imparted to the molecule. The ensemble of local boundary conditions reduces to that of microscopic states in which all of the molecules move almost independently of each other. At the opposite extreme, on the other hand, if the interaction is such that all of the molecules are interdependent in a correlated manner, every molecule can be deemed to be subject to one and the same boundary conditions in the construction of which all of the molecules are involved. The ensemble of local boundary conditions is to have only one member. The process of equilibration can be responsible for decreasing the members involved in the ensemble of local boundary conditions.

At first sight, it might seem that the diminishing ensemble of local boundary conditions contradicts the irreversible increase of Boltzmann's statistical entropy based upon the presence of the stationary ensemble of microscopic states. The difference is however superficial. Energy equilibration in the interacting many-molecule system does not presume that it is energetically isolated from outside. It can take in and throw out energy in the process of energy equilibration. If the energy thrown out in the form of a radiation field is properly discarded as in our present expanding universe,[9] the increase in the number of those local boundary conditions or microscopic states available to the emitted radiation field can overcompensate the decrease in the number of local boundary conditions for molecules.[10,11] So far as the radiation field keeps increasing the number of the constituent members involved in its ensemble of local boundary conditions, the diminishing ensemble of molecules can be compatible with the irreversible increase of these constituent members involved in the combined ensemble of local boundary conditions for both the molecules and the radiation field. The process of equilibration can underlie the irreversibility of Boltzmann's statistical entropy if the boundary conditions on the radiation field are so specific that the self-interaction within the radiation field can safely be disregarded. However, the reverse is not true. The irreversible character of Boltzmann's entropy in terms of the number of available microscopic states is a consequence of statistical inference at best, and does not provide a materialistic temporal nexus that can eventually yield irreversibility.

The process of energy equilibration in the form of interaction changes being propogated at a finite velocity is followed by energy flow. If the energy flow velocity happens to be equal to the propagation velocity of interaction changes as in the case of the radiation field in which both the energy and the change in interaction are carried by photons, the process of energy equilibration ceases to function. For the propagation velocity of interaction changes relative to the energy flow vanishes. Irreversibility due to the process of energy equilibration is limited to the case in which changes in interaction precede the associated spatial and temporal changes in the amount of energy. However, there also exists another theoretical possibility for irreversibility. Even if the energy flow precedes the associated change in interaction as in the almost interaction-free, many-molecule system, one more type of irreversibility is expected such that the phase volume or the number of microscopic states available to each almost-interaction-free molecule can attain its possible maximum value unless limited otherwise. This is in fact statistical irreversibility conceived in formulating Boltzmann's entropy that tends to maximize the number of available microscopic states, each of which is thought of as almost interaction free among themselves.

Interaction-dominant irreversibility, due to the process of equilibration, apparently differs from interaction-recessive irreversibility based upon statistical inference. But, they are not totally independent of each other. Interaction-recessive irreversibility simply presumes a set of almost interaction-free units, each of which is an aggregate of their subunits interacting strongly among themselves. On the other hand, interaction-dominant irreversibility can provide a materialistic temporal nexus of how such an almost interaction-free unit can come into existence out of these subunits interacting strongly with each other. Interaction-dominant irreversibility is more fundamental than interaction-recessive irreversibility in providing those material units to which the latter irreversibility addresses itself. In contrast, interaction-recessive irreversibility is conceptually more useful than interaction-dominant irreversibility in providing Boltzmann's statistical entropy.

The second law of thermodynamics as an empirical law[12] can imply both interaction-dominant and interaction-recessive irreversibilities. However, statistical entropy is limited to interaction-recessive irreversibility. Those units involved in interaction-recessive irreversibility nonetheless do not totally lack interaction among themselves. Even during a relatively limited time interval over which more than one unit interacts or collides with another the process of energy equilibration that is fundamentally irreversible can work. The law of motion of one-to-many mapping type is involved simply by virtue of the fact that the propagation velocity of interaction changes remains finite even over such a short distance.

It has been argued that interaction-recessive irreversibility is due to the statistical nature of interacting units. If the statistical inference did not require an alternative to the law of motion of the one-to-one mapping type, the source of irreversibility would have to be sought not in a materialistic process, but in the observer or the way of observing. As far as statistical inference goes, an irrevocable approach to the goal identified as the one with the statistically greatest likelihood is inevitable. Statistical reasoning is based on a logic that can necessarily yield irreversibility. We are thus asked to identify a materialistic cause that makes the statistical inference of irreversibility possible. A candidate for the materialistic cause is the process of energy equilibration, no mattter how limited the time interval may be over which changes in interaction take place. Sequential accumulation of energy equilibration can bring about irreversibility over a long period.

The increase in the number of available microscopic states is a consequence of those interactions that work at most intermittently. In between successive interactions, every microscopic state available to each unit of interaction-recessive irreversibility has an equal probability of actualization because of the lack of interaction. The irreversibility during interaction prevents those microscopic states from repeating their cyclic changes and increases with time the number of available microscopic states. Intermittent interaction and the relatively long periods of interaction-free flight underlie the inference of irreversibility for statistical entropy. It is during the time of free flight when Boltzmann's statistical entropy is definable, since the entropy is a numerical measure of the number of available microscopic

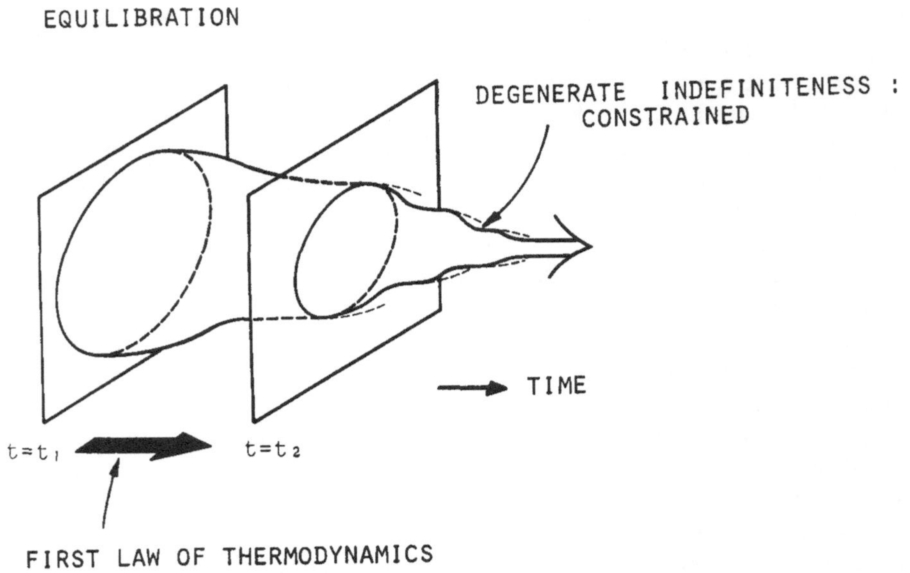

FIGURE 3. Intrinsic irreversibility latent in the process of equilibration. Indefinite degeneracy is dissolved with time. Responsible for the irreversibility is the first law of thermodynamics, or simply the conservation of energy, in the scheme of the law of motion of the one-to-many mapping type.

states. In contrast, the materialistic origin of irreversibility is within the intermittent interaction that makes energy equilibration inevitable.

Boltzmann's statistical entropy cannot decrease with time by definition. This is because the entropy refers to the number of available microscopic states that are presumed to be interaction free among themselves. If it decreased, the microscopic states that specified the entropy at a previous time would not have been interaction free among themselves and would decrease their number with the lapse of time. Henceforth, statistical entropy would be convenient if an interaction-free reference is available. The usefulness of statistical entropy for measuring irreversibility does not however guarantee the availability of such an interaction-free reference. The problem of whether an interaction-free reference would be available at all is independent of the matter of irreversibility. Intrinsic irreversibility originates within interaction. Statistical entropy is by itself an expression of a consequence of materialistic irreversibility in the interaction-free coordinate space, whereas the process of equilibration expresses its irreversibility in terms of interaction. Statistical entropy is indirect compared with equilibration in addressing irreversibility.

The process of equilibration can decrease the number of local boundary conditions found within an interacting many-molecule system by enhancing the extent of being constrained as depicted in Figure 3. Since each local boundary condition is associated with a corresponding quantum state, the number of microscopic states could decrease accordingly. Nevertheless, the decrease in the number of local boundary conditions does not contradict the statistical increase in the number of available microscopic states that are almost interaction-free. The ensemble of local boundary conditions is intrinsically interaction-dominant within itself at least over a long period of time. An interaction-free reference is not available within an interacting many-molecule system.

Irreversibility based upon a unidirectional increase of statistical entropy remains legitimate if an interaction-free reference is available and guaranteed by whatever means. Irreversibility due to equilibration, on the other hand, is independent of whether or not an interaction-free reference is available. The process of equilibration that provides a materialistic cause for the statistical irreversibility can cope with a type of irreversibility that goes beyond the realm of statistical entropy.

The microscopic origin of materialistic, that is to say, intrinsic, irreversibility is within the process of interaction inevitably accompanied by equilibration. It thus happens that a microscopic entropy,[13] instead of a statistical one, could be defined for the process of equilibration by referring to a quantity which exhibits a monotonic change with time on a microscopic scale. The monotonically varying quantity on a microscopic scale can be measured if it is feasible to conceive of an ensemble of quantum states on the same microscopic scale for the corresponding interacting many-molecule system. The distribution of quantum states, if available, does guarantee a measure of irreversibility by referring to the distribution function. Still, microscopic entropy assumes that interaction be at most intermittent. Otherwise the quantum states that are supposed to be almost interaction-free would not be definable. Both statistical and microscopic entropies point to interaction-recessive irreversibility, in which each interaction of molecules is followed by a relatively long period of free flight.

Entropy, whether statistical or microscopic, basically requires an interaction-free reference for its own underpinning. It is however paradoxical to think of an interaction-free reference in an interacting many-molecule system. On the other hand, though it is not a quantity, equilibration refers to materialistic irreversibility even if interaction is dominant and lasting.

IV. TIME'S ARROW

Irreversibility is sometimes associated with time's arrow. One recognizes that there are several different types of time's arrow depending upon whether the concept is used in physics, biology, or psychology.[14] Nevertheless, time's arrow is unitary once it is conceived as such. Time's arrow in biology has to be of a piece with the similar arrow in physics, otherwise one could not envision how the physics of becoming could be responsible for encompassing protobiology and biology. Likewise, time's arrow in psychology has to be based on the one in biology, because psychology cannot be thinkable without the underpinning of biology. It thus becomes a major issue of intrinsic irreversibility to answer the question of how the unitarity of time's arrow could be worked out.

A principal time's arrow in physics is the one in thermodynamics. Thermodynamic time's arrow, conceived in terms of thermodynamic entropy, first requires that the entropy is definable as such. Thermodynamic entropy presumes for its own sake, among other things a thermodynamically isolated system, thermodynamic equilibrium and the quasi-static process preserving thermodynamic equilibrium. Thermodynamic time's arrow is associated with the process of approaching thermodynamic equilibrium. This consideration now raises a formidable problem of how the initial state that deviates from thermodynamic equilibrium could be prepared at all and how thermodynamic entropy can be defined away from thermodynamic equilibrium. One way out is to require a statistical mechanical extension and interpretation of thermodynamic entropy.

Statistical entropy can be defined in terms of the distribution of quantum states even if away from thermodynamic equilibrium, with the reservation that those quantum states are *a priori* assumed to be almost interaction-free among themselves. Once it is defined, statistical entropy increases with time toward its possible maximum value. It is however important to notice that the possible maximum value of statistical entropy depends entirely upon how the interaction-free reference could be provided and guaranteed. Unless such an interaction-free reference is available, the distribution of almost interaction-free quantum states and the associated statistical entropy would not be conceivable. In this regard, the Big Bang cosmology[15] gives us a sound clue to thermodynamic time's arrow in the sense that the ever-expanding universe does not fail to provide an interaction-free reference for statistical entropy to remain legitimate.[10]

If we put $S_{max}(t)$ as the possible maximum value of and $S(t)$ as the real value of the statistical entropy of our universe at time t thermodynamic time's arrow denotes an irreversible increase of S(t) with time. In spite of this, if the possible maximum value $S_{max}(t)$ increases far faster than the real value S(t) as we see in the present universe, the relative entropy $s(t) = S/S_{max}$ can decrease with time. This consequence is consonant with the present nonhomogeneous universe we observe. Thermodynamic time's arrow does not necessarily lead to an inevitable heat death in a homogeneous universe. The Big Bang scenario is so peculiar in that an interaction-free reference under which statistical entropy is definable is rapidly moving. Thermodynamic time's arrow by itself does not specify how the condition that lets statistical entropy be definable comes into existence. If the interaction-free reference under which statistical entropy is definable is stationary as in a theoretically conceived isolated system, the difference between the real and the possible maximum values of entropy would disappear shortly in the course of time development. Thermodynamic time's arrow in terms of statistical entropy is a derivative of the condition under which that very entropy is defined. Irreversibility associated with thermodynamic time's arrow in the real materialistic world is thus traced back to the way an interaction-free reference for statistical entropy comes into existence and develops with time.

Thermodynamic time's arrow is totally conditioned upon how an interaction-free reference for evaluating statistical entropy could be provided, not vice versa. It is one thing that one could imagine that our present universe might happen to be followed by a contracting one on the hypothetical basis of nonzero neutrino's mass,[16] but it is quite another whether there could be an interaction-free reference that enables one to define entropy in such a scheme. Even if a moving interaction-free reference for defining statistical entropy were available, it would not follow that the then-defined statistical entropy would be a monotonically increasing function of time. For the possible maximum entropy $S_{max}(t)$ might decrease in the contracting universe.[9] If a theoretically conceived isolated system is contracting for some reason (and allowing for a moving interaction-free reference), thermodynamic time's arrow would point in the direction along which the statistical entropy decreases. On the other hand, thermodynamic time's arrow would point toward just the opposite direction along which the statistical entropy increases if the stationary or expanding isolated system is available and if the entropy there is definable as such.

Thermodynamic time's arrow in terms of statistical entropy depends upon the boundary condition to which the entropy is subject. If one envisions the statistical entropy from the perspective of the law of motion of the one-to-one mapping type supplemented by completely identifiable boundary conditions as in the case of classical mechanics, only the externally controlled boundary conditions could determine the direction of thermodynamic time's arrow. A mere evaluation of statistical entropy cannot tell us the direction toward which thermodynamic time's arrow points. On the other hand, however, if the law of motion of the one-to-many mapping type is taken for granted as it should be, time's arrow points toward the direction along which the extent of being constrained is enhanced with the process of equilibration. Time's arrow in equilibration is unidirectional in the sense of enhancing the constraint upon the material carrier of equilibration and remains indifferent to the nature of exogenous boundary conditions, if any.[17]

The law of motion of the one-to-many mapping type is inherent to the process of equilibration in which interaction is dominant. In contrast, an interaction-free reference upon which statistical entropy is defined can be well approximated by the presence of interaction-free motion being subject to the law of motion of the one-to-one mapping type, because of the lack of interaction. Consequently, the process that makes statistical entropy definable is the one governed by the law of motion of the one-to-one mapping type for a relatively long period and then intermittently intercepted by that of the one-to-many mapping type. The irreversibility in regard to the law of motion of the one-to-many mapping type over relatively short periods, whereas the value of statistical entropy depends upon the boundary conditions on the law of motion of the one-to-one mapping type that survives over relatively long periods. Even if the relative period over which the law of motion of one-to-many mapping type survives is short, every process of interaction of one-to-many temporal mapping provides an irreversibly constrained initial condition on the succeeding process of one-to-one temporal mapping.

A specific constraint upon or selection of initial conditions is envisioned as a source of irreversibility in the dynamics that follows the law of motion of the one-to-one mapping type. Sequential multiplication of the process of one-to-many temporal mapping can thus bring about an overall irreversibility which could be expressed in terms of statistical entropy. It is however important to recognize that the irreversibility associated with statistical entropy is derivative from microscopic entropy due to the law of motion of the one-to-many mapping type, or equivalently to the process of equilibration. Thermodynamic time's arrow in the sense of increasing statistical entropy coincides with time's arrow of microscopic irreversibility originating in equilibration only when the boundary condition on a macroscopic scale is prepared so as to guarantee the coincidence. Of course, nondecreasing entropy is duly expected insofar as a stationary or expanding isolated system is conceivable. Nevertheless, the feasibility of a stationary or an expanding isolated system cannot be equated with the microscopic irreversibility due to equilibration. Time's arrow in equilibration remains unchanged even if a conceived isolated system is contracting, whereas thermodynamic time's arrow in the sense of increasing statistical entropy is circumstantial.

Thermodynamic time's arrow is an outcome of pairing a strange couple, statistical entropy and microscopic irreversibility. Whether statistical entropy increases, decreases, or remains invariant with time in being subject to microscopic irreversibility depends upon the boundary condition that defines the entropy. Even if the boundary condition that lets statistical entropy increase is quite conceivable as in the present expanding universe, ever-increasing statistical entropy is not a cause of thermodynamic time's arrow but a consequence of the more fundamental process that is intrinsically irreversible. Microscopic irreversibility due to equilibration and an appropriate boundary condition on a macroscopic scale, put together, yield a nondecreasing statistical entropy that can be identified as thermodynamic time's arrow in the usual sense.

A time's arrow that is dependent upon the macroscopic boundary condition is not restricted only to the one in thermodynamics. Time's arrow in wave mechanics is also dependent upon what sort of initial and boundary conditions are prepared. If Sommerfeld's condition prohibiting advanced potentials is met or if waves coming into a material system from the future are prohibited, irreversible energy dissipation from the system takes place by emitting waves or radiation only into the future. The lack of incoming waves from the future indicates time's arrow in wave mechanics. However, the law of motion in wave mechanics, whether in electromagnetics or in general relativity, remains invariant against time reversal. The time-reversal law of motion is admittedly of the one-to-one mapping type. The law of motion of the one-to-one mapping type by itself allows a wave propagating toward the past as well as one propagating toward the future. As a result, evidence of empirical time's arrow in wave mechanics that eliminates those waves propagating toward the past has to be sought totally in the external boundary conditions upon the law of motion in wave mechanics.

The law of motion of the one-to-one mapping type and time's arrow are simply incompatible. If one tries to derive time's arrow in reality, two choices would seem to be available. One is to seek the origin of time's arrow in the uniqueness of boundary condition on the law of motion of the one-to-one mapping type. The other is to look for a law of motion that incorporates time's arrow intrinsically. These two approaches, however, cannot be independent of each other. Quantum mechanics of interaction fundamentally takes it for granted that a complete separation between the law of motion and its boundary conditions is not attainable because of the involvement of equilibration, with the inevitable consequence that the law of motion has to be of the one-to-many mapping type unless otherwise controlled externally. Local boundary conditions generated endogenously in a quantum mechanical system have been subject to at least the preceding process of equilibration that is self-constraining in the sense that only those interactions fulfilling the principle of conservation *a posteriori* are realizable. The genesis of local boundary conditions and their temporal variation do not originate in the law of motion of the one-to-one mapping type.

Time's arrow in quantum mechanics viewed from the perspective of the law of motion of the one-to-many mapping type is endogenous within the mere existence of quantum mechanics of interaction. Consider, for instance, two metals A and B, hot and cold, with temperatures T_A and T_B ($<T_A$) put together. Thermodynamic time's arrow suggests that the difference between temperatures T_A and T_B will finally disappear. There is no empirical evidence indicating that the hot becomes hotter while the cold becomes colder. If one relies upon the law of motion of the one-to-one mapping type in order to justify the exclusion of the case that the hot becomes hotter and all that, only the initial condition would serve to do this. The law of motion of the one-to-one mapping type, if accepted, would be able to push the initial condition back into the remote past, even further back to the very beginning of everything, no matter how ridiculous this line of thinking may seem. On the other hand, if one examines the same problem within the framework of quantum mechanics, which duly admits an endogenous generation and variation of local boundary conditions, the law of motion will be seen to be of the one-to-many mapping type. The hot metal becomes colder because the initial condition to be obeyed at every moment is chosen so as to eliminate those that could make the hot hotter, instead of preserving a constant initial condition during development.

The direction toward which time's arrow in wave mechanics points depends upon the boundary conditions on a macroscopic scale. Nevertheless, time's arrow in wave mechanics, like the one in thermodynamics, can gain its microscopic irreversibility from the quantum mechanics of interaction. As far as irreversibility that is microscopic in scope compared with the cosmological counterpart in space and time is concerned, one can refer to microscopic irreversibility due to quantum mechanics of interaction without invoking the cosmological boundary condition that determines the direction of time's arrow in thermodynamics and in wave mechanics. But the distinction between the microscopic and the macroscopic would have disappeared in the very vicinity of the Big Bang of our expanding universe. According to the grand unified therory, the decay of X-particles, which was supposed to exist at the very early stage of our universe when the temperature was above 10^{27} K, breaks time-reversal symmetry and accordingly the conservation of baryon number.[18,19]

The breaking of time reversal symmetry implies that the decay of X-particles accompanied by the increment of baryon number ΔB does not occur at the same rate as with the decay of its antiparticle $\overline{X}$ accompanied by the decrease of baryon number by ΔB. As the universe was expanding, its temperature decreased and a nonvanishing net contribution from baryons, that is to say, matter, could have been squeezed out due to the difference of decay rate between X and $\overline{X}$. Although the disparity of decay rate between X and $\overline{X}$ is a matter of theory, we are already acquainted with experimental evidence of the analogous breaking of time-reversal symmetry for the K-meson.

Whether it may be for X-particle in the grand unified theory or for K-meson in the accelerator machine, the breaking of time-reversal symmetry on a microscopic scale could occur only under the extraordinary conditions that the distinction between the microscopic and the macroscopic disappear. In particular, microscopic irreversibility due to the disparity of the decay rate between particle and antiparticle does not have direct influence over irreversibility in protobiology and in biology, because in the latter there is no room for antiparticles which are imaginable only at the very early stage of our expanding universe or in a huge high-energy accelerator machine.

Microscopic irreversibility inherent to quantum mechanics of interaction continues to survive as long as the process of interaction is guaranteed. Quantum mechanical decay of a particle into others is an instance of interaction, but does not last long. Instead, equilibration of quantum mechanical origin is ubiquitous to the extent that interaction among material units is not totally prohibited. Even if microscopic irreversibility originating in quantum mechanical decay as exemplified by K-mesons has been exhausted, other types of quantum mechanical interaction can give rise to microscopic irreversibility as is the case with equilibration.

Time's arrow originating in the asymmetry in particle-antiparticle decay as in the decay of the K-meson is a particular case of time's arrow in quantum mechanics. Irreversibility intrinsic to equilibration that allows the inseparability between the law of motion and its boundary conditions is one more instance of time's arrow in quantum mechanics. It is however important to recognize that time's arrow in quantum mechanics is not associated with the time coordinate in the quantum-mechanical equation of motion under completely identifiable boundary conditions, but rather with the inseparable dichotomy between the equation of motion and its boundary conditions.

Protobiology and biology are characterized by a particular time's arrow that is not thermodynamic, wave-mechanic, nor K-mesonic.[20,21] The problem we now face is what type of time's arrow is the predecessor of time's arrow in protobiology. A unique characteristic of time's arrow in biology that should be connected with the one in protobiology is its multiplicative irreversibility as evidenced in both phylogeny and ontogeny.[22] A candidate for this would be time's arrow in quantum mechanics especially in the form of equilibration because the latter is also multiplicative in its irreversibility with time.

It may sometimes be argued that phenomenological irreversibility can be decomposed into a particular choice of initial conditions and the law of motion of the one-to-one mapping type. The irreversibility is not within the law of motion, but within the initial condition in the sense that its time reversal such as the inversion of velocity coordinates is unthinkable and inadmissible. In order to make the point clearer, let us examine the dynamics of gases in the two boxes A and B, hot and cold, put together within the framework of classical mechanics that admits the law of motion of only the one-to-one mapping type. We know on an empirical basis that it could never happen that the hot gas becomes hotter and the cold one becomes colder, even though classical mechanics alone would not totally eliminate such a possibility depending upon its choice of initial conditions. The classical mechanical law of motion of the one-to-one mapping type would thus force us to accept the view that the observed irreversible approach of the two temperatures to the same value is due to the choice of a unique initial condition that prohibits the hot becoming hotter and the cold becoming colder. On the other hand, however, if one examines the same problem within quantum mechanics of interaction, a completely different perspective would arise. The whole system is now visualized as an ensemble of endogenously generated local boundary conditions. The law of motion and its boundary conditions are inseparable, thus yielding the process of equilibration. The present inseparability makes the law of motion to be of one-to-many mapping type. The origin of irreversibility is to be sought in the temporally asymmetric law of motion of one-to-many mapping type, instead of in a particular choice of initial condition.

Irreversibility can be classified into two groups: one is phenomenological due to the law of motion of the one-to-one mapping type supplemented by a very special initial and boundary condition of exogenous origin. The other is intrinsic due to the law of motion of the one-to-many mapping type. Intrinsic irreversibility yielding the process of equilibration refers to the law of motion of the one-to-many mapping type that is duly recognized within the framework of quantum mechanics. One more simple metaphor may help in this regard. The classical law of motion of the one-to-one mapping type may provide a reason why two boxes of the same gases, hot and cold, approach the same temperature when put together, instead of widening the temperature gap, if it is supplemented by a special initial condition. In contrast, the law of motion of the one-to-one mapping type is not sure about a rationale for why developmental process in biology cannot be reversed, e.g., why an adult cannot grow into a baby. If a biological developmental process is of a one-to-many temporal mapping or if an adult is the actualization of a particular out of the possible envisioned at the stage of baby, its irreversibility is to be within the temporally asymmetric law of motion, not solely within a particular choice of initial condition. In fact, quantum mechanics of interaction provides a materialistic carrier of the law of motion of the one-to-many mapping type.

V. SUMMARY

The conservation of energy as stated in the first law of thermodynamics intrinsically implies irreversibility if the process of interaction is properly taken into account. The process materializing conservedness refers to an on-going adjustment of interaction. When one examines the conservation of energy in terms of the state of an interacting many-particle system, it is imperative to claim the availability of the state fulfilling the conservation of energy at every moment irrespective of what is actually going on there. In contrast, if the conservation of energy is examined in terms of the process of interaction, what is kept going there turns out to be responsible for fulfilling the conservation of energy. We thus notice an inevitable discrepancy between the state-description and the process-description, especially with regard to the conservation of energy. If it were asserted that the state preserving conservedness of energy is definable at every instant, the process underlying a realization of the conservation of energy would proceed at an infinite velocity. However, no physical process is allowed to proceed at a velocity greater than light velocity.

The state-description overreaches itself when it comes to the process materializing conservedness. On the other hand, the process-description remains legitimate even if it is incomplete in identifying the state involved at every moment. The first law of thermodynamics is intrinsically irreversible within the framework of the process-description whereas it would be no more than a tautology in the state-description. The state preserving conservedness of energy that is definable at every moment and the process materializing conservedness simply cannot coexist. Irreversibility intrinsic to the first law of thermodynamics has its origin in the process materializing conservedness or in the process of equilibration.

The second law of thermodynamics formulated in terms of entropy requires a framework in which the entropy is definable, in addition to a process that generates irreversibility. Insofar as an interaction-free reference is available, statistical entropy as a state function can be defined. Statistical entropy in quantum mechanics assumes an ensemble of endogenously generated local boundary conditions. Boltzmann's statistical entropy measured in terms of the number of available microscopic states can increase in an irreversible manner only when an appropriate macroscopic boundary condition, such as our present expanding universe, is provided externally.

Thermodynamic time's arrow in terms of statistical entropy depends upon both the definability of entropy on a macroscopic scale and the process generating irreversibility on a microscopic scale. Time's arrow in the quantum mechanics of interaction, on the other hand, is solely microscopic in the genesis of irreversibility and can provide statistical entropy with its irreversibility. Since it is multiplicative in time as far as interaction is guaranteed, time's arrow in quantum mechanics can generate and support time's arrow in protobiology and in biology.

REFERENCES

1. **Landsberg, P. T.,** *Thermodynamics and Statistical Mechanics,* Oxford University Press, Oxford, 1978, 34.
2. **Lotka, A. J.,** *Elements of Mathematical Biology,* Dover, New York, 1956, 281.
3. **Baily, K. D.,** Equilibrium, entropy and homeostatis: a multidisciplinary legacy, *Syst. Res.,* 1, 25, 1984.
4. **Ehrenfest, P. and Ehrenfest, T.,** *The Conceptual Foundations of the Statistical Approach in Mechanics* (transl.), Moravcsik, M. J., Cornell University Press, Ithaca, N.Y., 1959.
5. **Kirkwood, J. G.,** The statistical mechanical theory of transport processes, *J. Chem. Phys.,* 14, 1211, 1946.
6. **Lasalle, J. P. and Lefshetz, S.,** *Stability by Liapounov's Direct Method,* Academic Press, New York, 1961.
7. **Glansdorff, P. and Prigogine, I.,** *Thermodynamic Theory of Structure, Stability and Fluctuations,* John Wiley & Sons, New York, 1971.
8. **Kreuzer, H. J.,** *Nonequilibrium Thermodynamics and Its Statistical Foundations,* Clarendon Press, Oxford, 1981.
9. **Hawking, S. W.,** The edge of spacetime, *Am. Sci.,* 72, 355, 1984.
10. **Frautschi, S.,** Entropy in an expanding universe, *Science,* 217, 593, 1982.
11. **Wicken, J. S.,** A thermodynamic theory of evolution, *J. Theor. Biol.,* 87, 9, 1980.
12. **Misra, B. and Prigogine, I.,** Irreversibility and nonlocality, *Lett. Math. Phys.,* 7, 421, 1983.
13. **Prigogine, I. and George, Cl.,** The second law as a selection principle: the microscopic theory of dissipative processes in quantum systems, *Proc. Natl. Acad. Sci. U.S.A.,* 80, 4590, 1983.
14. **Denbigh, K. G.,** *Three Concepts of Time,* Springer-Verlag, Berlin, 1981.
15. **Weisskopf, V. F.,** The origin of the universe, *Am. Sci.,* 71, 473, 1983.
16. **Ellis, J.,** The very large and the very small, in *The Nature of Matter,* Mulvey, J. H., Ed., Clarendon Press, Oxford, 1981, 126.
17. **Matsuno, K.,** Open systems and the origin of protoreproductive units, in *Beyond Neo-Darwinism,* Ho, M. W. and Saunders, P. T., Eds., Academic Press, London, 1984, 61.
18. **Wilkinson, D.,** The organization of the universe, in *The Nature of Matter,* Mulvey, J. H., Ed., Clarendon Press, Oxford, 1981.
19. **Gaillard, M. K.,** Toward a unified picture of elementary particle interactions, *Am. Sci.,* 70, 506, 1982.
20. **Wagner, G. P.,** On the necessity of a system theory of evolution and its population biologic foundation: comments on Dr. Regelmann's article, *Acta Biotheor.,* 32, 223, 1983.
21. **Regelmann, J. P.,** On the necessity of a system theory of evolution and its population biologic foundation: a reply to Gunter Wagner's commentary, *Acta Biotheor.,* 33, 133, 1984.
22. **Slobodkin, L. B.,** Listening to a symposium — a summary and prospectus, in *Biotic Crises in Ecological and Evolutionary Time,* Nitecki, M. H., Ed., Academic Press, New York, 1981, 269.

Chapter 5

SYMMETRY-BREAKING

I. INTRODUCTION

Intrinsic irreversibility is observed in material self-assembly processes, in which the preceding products successively provide the endogenous boundary condition for the subsequent production process. Material self-assembly leading to protobiological and biological evolution is the process of breaking the symmetry property observed within the preceding products. The symmetry-breaking process induces a dissolution of the degeneracy latent within the preceding products, by decreasing the number of indistinguishable individuals while increasing the number of distinguishable ones. The process materializing conservedness or the process of equilibration, when viewed as a form of one-to-many temporal mapping, underlies the dissolution of the degeneracy of the future possibilities being latent within the actual at the present moment. Accordingly, the symmetry property found within the endogenous boundary condition is successively broken with time. Empirical evidence of symmetry-breaking in material self-assembly processes is available from the laboratory synthesis of amino acids in a simulated reduced atmosphere and from thermal polymerization of amino acids. Similar evidence of symmetry-breaking is also available from the nonrandomness of point mutation in pseudogenes that are full-blown biologically.

The symmetry-breaking has its functional underpinning in the law of motion of the one-to-many mapping type. Thermodynamic stability criteria suggesting that the most likely products minimize their thermodynamic potentials are of course applicable to chemical synthesis of monomers and polymers from their constituent elements. However, the stability criteria are subject to moving boundary conditions forced by the process materializing conservedness. The origins of observed successive symmetry-breaking are in the endogenous variability of thermodynamic boundary conditions.

II. PRODUCTION PROCESS

Irreversibility leading to the emergence and evolution of protobiological substances is accompanied by the synthesis of monomeric and polymeric molecules from the much smaller molecules and atoms.[1] If one examines the molecular synthesis within a framework of thermodynamics, the thermodynamic boundary condition upon an arbitrary reacting molecule is found to be imposed by its nearby molecules. Each reacting molecule constitutes a part of the boundary condition upon others as well as being subject to the boundary condition made by others. If the propagation of influencing nearby reactants proceeded at an infinite velocity, the internal adjustment between what is influencing and what is being influenced would be completed instantaneously. The ensemble of local thermodynamic boundary conditions would remain stationary unless disturbed externally. Chemical reactants that minimize their thermodynamic potentials would most likely appear because the stationary boundary conditions guarantee the availability of thermodynamic stability.

On the other hand, if those influences from reacting chemical are propagated at a finite velocity as they should be, each local boundary condition is still in the process of varying itself by being subject to variations imputed to the nearby chemical reactants. Thermodynamic stability criteria cannot be applicable in the fullest sense. The relaxation process whose relaxation time is much greater than the characteristic time over which local thermodynamic boundary conditions could remain stationary does not belong to the thermodynamic fluctuation pertinent to those boundary conditions. Instead, such a slow process with a large relaxation time is now responsible for varying local thermodynamic boundary conditions. The relaxation process with the greater relaxation time participates in sequence in forming these local boundary conditions.

Quantum mechanics in fact guarantees that thermodynamic boundary conditions constitute an ensemble of endogenously generated local boundary conditions viewed from the standpoint of the law of motion of the one-to-many mapping type. In the context of reacting chemicals, the preceding products constitute the boundary condition for the subsequent production process. The preceding products, while inevitably degraded through their relaxation into their constituent elements, further participate in forming the boundary condition for the subsequent production process, and new products result. The sequence from product to production can continue indefinitely. The surviving products are those with the greater relaxation time of their degradation, while those with the shorter one are naturally sieved out. The indefinite sequence from product to production is thermodynamically a process of endogenously generating those local boundary conditions with the greater relaxation time, while it is quantum-mechanically a process of one-to-many temporal mapping that makes the law of motion inseparable from its boundary condition. It thus turns out that the emergence of the local boundary conditions with the greater relaxation times is a consequence of the quantum-mechanical law of motion of the one-to-many mapping type supplemented by intrinsically partially fixed boundary conditions.

The successive sequences from product to production in reacting chemicals is self-constraining due to the underlying law of motion of the one-to-many mapping type, implying that an actualization constantly eliminates other possibilities. The products with the smaller relaxation times are eliminated in the process of one-to-many temporal mapping. Accordingly, one particular operational mode of the one-to-many temporal mapping is seen in the successive generation of a relaxation process alternating with the one with the greater relaxation time. Those with the smaller relaxation times turn out to fail to fulfill the condition for equilibration that any surviving products have to satisfy.

The relaxation process under fixed boundary conditions, that is common in thermodynamic analysis, differs from the similar process under endogenously generated local boundary conditions. The difference is due to the manner of generating various relaxation processes. If boundary conditions fixed externally are accepted, the corresponding external agent, whatever it may be,[2] would be responsible for raising a wide variety of processes to be relaxed within the bounded region. The boundary condition, once fixed, would remain invulnerable whatever relaxation processes might be raised and excited externally. Thermodynamic relaxation processes envisaged from the microscopic viewpoint of one-to-one temporal mapping cannot vary their own boundary condition.

By contrast, those relaxation processes in the scheme of one-to-many temporal mapping vary the boundary condition that is responsible for generating the subsequent processes that keep the ones with the greater relaxation time. The law of motion of the one-to-many mapping type, when examined from the perspective of relaxation process, is responsible for both generating and limiting those relaxation processes. Most of the generated relaxation processes die away in the course of development. They are passive in being subject to a temporal sieve, but at the same time active though only in the sense of participating in the generation of subsequent relaxation processes. The law of motion of the one-to-many mapping type maintains the relaxation process as an irreducibly fundamental attribute of dynamics, whereas the law of motion of the one-to-one mapping type lets relaxation process be a derivative of an externally controlled boundary condition.

The sequence from product to production within the law of motion of the one-to-many mapping type is a process of dissolving the degeneracy observed within the preceding products. For the prior products serving as an endogenous boundary condition for the subsequent production process can enhance those relaxation processes that could hold only temporarily, not to mention the ones that would survive for a relatively long period. The symmetry property observed within the initial condition that is constituted by prior products is broken in the real process of one-to-many temporal mapping, contrary to the case of one-to-one mapping that preserves the symmetry property once it is set initially. The symmetry-breaking is associated with the dissolution of the degeneracy within the prior products.[3] The symmetry property perceived in the realm of the possible is broken in the actual, the latter of which constantly serves as a source of the possible for the subsequent actualization.

A quantitative figure related to the dissolution of degeneracy and to the symmetry-breaking is the number of unconstrained degrees of internal freedom involved in the motion of one-to-many temporal mapping. The existence of unconstrained degrees of internal freedom is imperative for the law of motion of the one-to-many mapping type in order to appreciate the presence of the possible that could include those possible candidates to be eventually eliminated in the subsequent process of actualization. The situation is, however, completely different from the case of the law of motion of the one-to-one mapping type; in the latter, every degree of freedom in motion is constrained in principle by the boundary condition applied externally.

The law of motion of the one-to-many mapping type necessarily admits the presence of an indefinite number of degrees of freedom on the part of its boundary condition due to the inseparability between the law of motion and its boundary condition. An earlier boundary condition is made more constrained with the elapse of time because of the self-limiting transformation from the possible to the actual inherent to the one-to-many temporal mapping. The number of unconstrained degrees of freedom cannot remain finite, otherwise the law of motion of the one-to-many mapping type would cease to function beyond the stage at which all of the degrees of freedom in motion would have been constrained. The unconstrained degrees of freedom remain indefinite. But, the extent of indefiniteness is constrained with time.

The indefinite freedom in motion that is carried by the endogenous boundary condition at an arbitrary time point is now identified as the degeneracy within the boundary condition itself at every moment. Each of the unconstrained degrees of freedom cannot be distinguished from the others that are unconstrained because of its indefiniteness in motion. For instance, if there is only one constraint upon ten degrees of freedom in motion, one cannot distinguish one from the others because of the lack of any means for the purpose. Consequently, one expression of the unconstrained indefinite degrees of freedom is made possible by invoking the presence of indistinguishable individuals in motion. The degeneracy of the endogenous boundary condition is within the indistinguishable individuals involved. By the same token, the symmetry property of the boundary condition is within a set of operations, either real or hypothetical, that preserve the degeneracy with regard to the indistinguishable individuals as it is.

The law of motion of the one-to-one mapping type leaves the degeneracy associated with the indistinguishable individuals invariant and preserves the symmetry property once observed with those individuals. By contrast, the law of motion of the one-to-many mapping type dissolves the degeneracy and accordingly generates the distinguishable individuals out of the former indistinguishable ones. Even if those individuals are enumerable, the number of degrees of freedom to identify each individual remains indefinite in the scheme of the one-to-many temporal mapping. Those individuals, either indistinguishable or distinguishable, are simply units for measuring the degeneracy and the symmetry property. The relaxation process associated with the sequence from product to production points to how the indistinguishable individuals are combined and how the distinguishable ones are squeezed out therefrom. The transition from being indistinguishable to distinguishable originates in the law of motion of the one-to-many mapping type, particularly through the sieve of the relaxation process under endogenously generated boundary conditions.

The law of motion of the one-to-many mapping type can increase the number of distinguishable individuals at the expense of the number of indistinguishable ones. The enhancement of distinguishable individuals is an indication of the dissolution of degeneracy and also of the breaking of the symmetry property observed among the former indistinguishable individuals. Although the dissolution of degeneracy and the symmetry-breaking are associated with the indefinite number of unconstrained degrees of freedom that is inherent to the one-to-many temporal mapping, those unconstrained degrees of freedom cannot be directly referred to because of the lack of enumerability. Instead, indistinguishable individuals have been introduced. They can raise new distinguishable products when concatenated together due to their mutual interaction. The degeneracy is within a multitude of hypothetical products conceived from the preexisting indistinguishable individuals, and the set of operations that leave the multitude of the hypothetical ones invariant refers to the underlying symmetry property.

The hypothetical genesis of products refers to the process of one-to-one temporal mapping that could preserve the identifiability of the boundary condition with arbitrary precision at any moment. Those hypothetical products viewed as aggregated individuals are no more than a one-to-one transformation of the externalized boundary condition. In fact, the real products that can survive through one-to-many temporal mapping are always limited and constrained compared with the hypothetical ones conceived previously. The dissolution of degeneracy and the symmetry-breaking follow as a consequence of enhancing the number of distinguishable individuals, which is however far less than the number of hypothetically concatenated ones. The transition from those hypothetical products to the real ones materializes in the relaxation process under endogenously generated boundary conditions. The relaxation process is fundamental in the law of motion of the one-to-many mapping type, though only secondary in the one-to-one temporal mapping. The increase in the number of distinguishable individuals is accompanied by the symmetry-breaking in the relaxation process that keeps only those products with the greater relaxation time.

In order to visualize the dissolution of degeneracy and the symmetry-breaking, let us consider a mixture of monomers of the three different types, A, B, and C, and further imagine a process of forming dimers and trimers as demonstrated in Figure 1. If we pay attention to trimers, we could hypothetically expect 27 different species, depending upon the internal permutation of monomers. The origin of those hypothetical trimers is within the thermal motion of the one-to-one mapping type that preserves the symmetry property initially observed with the mixture. Furthermore, if already-made trimers can influence further polymerization of other dimers and trimers in any way, only those trimers which are mutually most supportive of their own formation could survive if resources are plentiful and if dissociation reactions are not prohibited. Underlying this is the relaxation process under endogenously generated local boundary conditions that preserve those products with the greater relaxation time. This mutually supportive formation that allows prior products to be the boundary conditions for the subsequent production is due, more specifically, to the positive feedback of polymerization in which the previous polymeric products positively influence the polymerization to come.

The positive feedback of polymerization leads to an exponential growth of products with time. Because of the present exponential growth, only those processes with the largest production rate would finally dominate all the other processes with the smaller rates, and the number of different species of surviving products would thus be constrained. The production processes with the largest production rate coincide in retrospect with those products with the largest relaxation time. Others with the smaller relaxation times are eventually phased out. If the number of different types of trimers which can survive turns out to be less than 27, one can say that the number of types of available species or of distinguishable individuals is constrained compared with that of hypothetically possible ones. The boundary

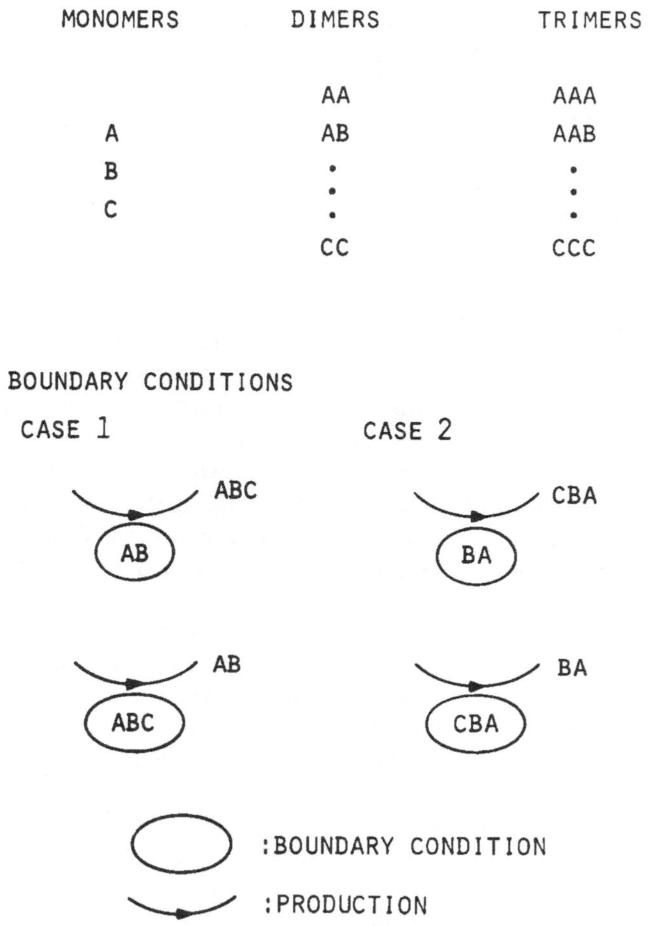

FIGURE 1. A schematic representation of a relationship between products and their production process. When the preceding products come to serve as a template, or as a boundary condition upon the succeeding production process, the boundary condition turns out to be endogenously variable. When two autocatalytic reactions, Case 1 and Case 2, are available, the one with the greater multiplication rate will dominate the boundary condition upon the production process to come.

condition that specifies the process of making trimers changes endogenously with time. Accordingly, the symmetry property observed with the hypothetical products is broken as the real production proceeds. The number of types of distinguishable trimers increases at the expense of the number of indistinguishable monomers.

What this hypothetical example suggests is that if an already made polymer can influence a polymer to be made next in any way, the boundary condition for polymerization can change in an endogenous manner. The underlying law of motion is necessarily of the one-to-many mapping type, instead of being of the one-to-one mapping type, because in the latter the boundary condition is simply of exogenous origin.

Furthermore, the sequence from product to production in the scheme of the one-to-many temporal mapping leaves the surviving products open to material flow. The production implies material flow into prior products to make them bigger, whereas the inevitable relaxation process necessitates material flow out of those products already made. The law of motion of the one-to-many mapping type with regard to open material aggregates functions in the form of material flow equilibration. The process materializing conservedness or, equivalently the process of equilibration, is visualized in the relaxation process under endogenously generated local boundary conditions if the production process is not prohibited. The symmetry-breaking associated with the endogenous sequence from product to production is thus synonymous with material flow equilibration.

III. SYMMETRY-BREAKING: PROTOBIOLOGICAL

Material aggregates under the influence of material flow equilibration induce adjustments of interaction with their surroundings. The preceding adjustments constantly supply endogenous causes for the later ones in a successive manner. Successive adjustments of interaction for material flow equilibration are manifest as a form of variation imposed upon the configuration of interaction among material elements constituting the open material aggregate through the continual sequence from product to production. Although one could imagine arbitrary variations in the configuration of interaction, which in turn influence the energies of material association and dissociation, only those fulfilling the condition for material flow equilibration are realizable in a material aggregate open to material flow. Random variation means at least that every conceivable configuration of interaction between atoms and molecules could appear with equal likelihood, otherwise a nonrandom bias of physical origins would intervene. Yet the real variation is by no means random. Configurations of interaction which violate material flow equilibration are internally eliminated. Endogenously generated variations in open material aggregates are already constrained compared with hypothetically random ones, and constrained variations provide an impetus to the genesis of the succeeding constrained variations. The symmetry property that would maintain the invariance of hypothetically random variations is broken in the real ones.

Material flow equilibration which always generates the subsequent equilibration of the similar character is an agent for causing constrained nonrandom variations. A material aggregate open to material flow equips itself with the regulative capability of limiting allowed variations only to those which are nonrandom and constrained by material flow equilibration. Both the successive genesis of variations and the constraints acting upon them originate in material flow equilibration. What would appear as vitalistic in the successive genesis of constrained variations is firmly based upon the physical process of equilibration, which is identified as the process materializing conservedness that is propagated in the medium at a finite velocity.

Experimental evidence will support the successive genesis of constrained variations due to material flow equilibration. Evidence for this is available from a simulated primordial soup. In particular, a mixture of free amino acids serves this purpose because of its conceptual clearness.[4] Although it might be argued that a mixture including monomers other than amino acids could also be considered, this would cause a further complication of the simpler one. The choice of the mixture of free amino acids should be judged according to what kind of results it can provide. If no results of protobiological significance were available, the previous choice of the simulated primordial soup would have to be abandoned and an alternative tried. On the other hand, if the result turns out to be relevant to the protobiological process, as we shall actually recognize, further attempts to simulate a more complex and more realistic primordial soup are feasible by continuously transforming the mixture of free amino acids into a more complex one through, for instance, adding monomers other than amino acids.

Open material aggregates expected during thermal polymerization in the mixture of free amino acids are thermal proteins, i.e., proteinoids. The endogenous constraint due to material flow equilibration applies to the amino acid sequence of synthesized polypeptides. Consider, for instance, the tripeptide synthesis[5] due to thermal polymerization of the three different amino acids, glutamic acid, glycine, and tyrosine, in a simulated primordial soup. It has been observed that the identifiable sequences are restricted only to pyroglutamyl-tyrosylglycine (<glu-tyr-gly) and pyroglutamyl-glycyltyrosine (<glu-gly-tyr).[5,6] No other theoretical tripeptide isomers have been identified. The hypothetical tripeptide tyr-glu-gly resulting from glu-tyr-gly by permutational change between a glutamic acid and a tyrosine, for instance, has not been identified in the products of thermal polymerization. On the other hand, the mixture of free amino acids of the three different kinds, especially before the stage of thermal polymerization, does not change the character of its thermal molecular motions even if any pair of, let us say, a glutamic acid and a tyrosine interchange their mutual positions. These artificial positional interchanges are not distinguishable from similar interchanges originating in the thermal motion inherent in the mixture of free amino acids. This indicates a change in the symmetry property during thermal polymerization.

A permutation symmetry between an arbitrary pair of different types of amino acids, that was preserved before thermal polymerization, is broken after the polymerization. The symmetry property before thermal polymerization is entailed by the hypothetical law of motion of the one-to-one mapping type that would preserve the initial condition that has been set for the mixture of free amino acids. The real symmetry-breaking originates in the law of motion of the one-to-many mapping type which functions in the form of material flow equilibration in open material aggregates.

The symmetry-breaking is a result of applying an extra constraint to a system which would otherwise not lose the symmetry property. Consequently, material flow equilibration as a generator of successive constraints applied to open material aggregates constantly induces symmetry-breakings. The decrease in the number of unconstrained degrees of internal freedom is associated with the symmetry-breaking because of the increase in the number of additional constraints that the material system must comply with.

One more aspect of the symmetry-breaking is the dissolution of the previous degeneracy in the endogenous boundary condition observed at a higher symmetry stage. The degeneracy means the multitude of indistinguishable individuals constituting the boundary condition, as we have already recognized. For the sake of concreteness, let us again consider the mixture of free amino acids comprising the three different amino acids of glutamic acid, glycine, and tyrosine. The mixture of free amino acids could preserve the initial-boundary condition even if a single glycine and tyrosine interchange their positions. This is because the two hypothetical mixtures, one of which differs from the other only with regard to the interchange between a glycine and a tyrosine, are indistinguishable with respect to their boundary conditions insofar as the thermal motion of one-to-one temporal mapping could survive. Thus, the two mixtures would be degenerated. On the other hand, the mixture of free amino acids changes its symmetry property after thermal polymerization. The tripeptide <glu-tyr-gly certainly differs from <glu-gly-tyr. This marked difference indicates that thermal proteins as products from the mixture of free amino acids are less degenerate compared with prior free amino acids.

When the preceding products constitute the boundary condition for the subsequent production, the dissolution of degeneracy and the symmetry-breaking accompany it. The dissolution of degeneracy brings about an increase in the number of distinguishable individuals as evidenced in the thermal polymerization of amino acids.[7] The increase in the number of distinguishable individuals is associated with a decrease in the number of indistinguishable individuals, as is evident in the previous example of the mixture of free amino acids. The mixture carries those indistinguishable individuals whose number is on the order of the total number of amino acid molecules. In contrast, the number of discernible amino acids is only three of glutamic acid, glycine, and tyrosine, in the present case. This imbalance between the number of indistinguishable individuals and the number of types of distinguishable ones is a cause of the degeneracy. If one can imagine or prepare such a system in which the number of indistinguishable individuals far exceeds the number of types of distinguishable individuals, the dissolution of degeneracy and the symmetry-breaking can proceed through the sequence from product to production.

The symmetry-breaking due to the decrease in the number of unconstrained degrees of internal freedom and the accompanying dissolution of degeneracy with boundary conditions are the two fundamental characteristics of evolving open material aggregates. The evolutionary capability of matter is found in the higher symmetry with the greater degeneracy of its boundary conditions.[8] The arrow of evolutionary time is in the direction along which the symmetry breaks and at the same time the degeneracy is dissolved. Unless there are the symmetry to be broken and the degeneracy to be dissolved, the evolution of matter would not be implied. In particular, the primordial soup could have made itself highly degenerate with regard to its boundary conditions because of the extremely small ratio of the number of distinguishable elements to the number of total elements inherent in the soup. Symmetry-breaking and the dissolution of degeneracy are common to the evolution of any open material aggregate.

The increase in the number of types of distinguishable individuals in the course of material evolution sheds some light on the problem of how the diversity in the different types of protein molecules has been promulgated with evolutionary time. The earlier material system comprising a smaller number of different types of protein is the source matrix for the later ones having a greater number of different types. The precursor proteins are limited in their types. Still, they are highly degenerated in the sense that the number of total proteins far exceeds the number of their different types. It is this degeneracy that can yield the wider variety of different types of protein in the later stage of evolution. Unless there is such a degeneracy in an earlier boundary condition, the symmetry-breaking due to material flow equilibration cannot bring in further new types. The evolutionary process would meet a dead end there. However, the degeneracy in the endogenous boundary conditions, specifically with regard to terrestrial proteins, is not exhausted so far as the number of the total protein molecules is concerned, since they are far greater in number than the number of different types of protein molecules. The multitude of precursor proteins of the same type is the source of the evolutionary diversification in the types of successor proteins. The evolutionary potential of protein molecules is thus found in the degeneracy of, that is to say, the multitude of, the precursor proteins of the same type.

Thermal polymerization of amino acids involves a symmetry-breaking at the protobiological level. Similar symmetry-breaking is also identified in the synthesis of simple monomers such as amino acids and nitrogen bases from much smaller molecules including methane, ammonia, hydrogen cyanide, formaldehyde, and the like.[9] The Miller-Urey experiment on synthesizing amino acid molecules from their reducing ingredients,[10,11] for instance, breaks the symmetry property that the ingredient molecules retained during their thermal motion prior to joining in the amino acid synthesis. When the energized ingredient molecules move almost freely in the gaseous state, every such molecule can be seen to follow the process of one-to-one temporal mapping in an approximate sense because of the relative insignificance of interaction.

The symmetry property preserved in the approximate one-to-one temporal mapping is broken after the ingredient molecules form an amino acid. The particular conformation of an amino acid molecule is a result of eliminating other possible spatial configurations that the involved smaller molecules could have realized during their collisions even temporarily while they move mostly as following the law of motion of the one-to-one mapping type in the form of thermal motion.

It is also understandable that the conformation which an amino acid molecule would assume is in accord with what the thermodynamic stability criterion requiring the minimization of thermodynamic potential suggests. The synthesis of amino acid molecules is thermodynamically a relaxation process, preserving the products having the greater relaxation time. The synthesized amino acid molecule again serves as a material unit carrying the law of motion of the one-to-many mapping type. Thermal polymerization of amino acid molecules is a mode of relaxation in which the boundary conditions on amino acid molecules are endogenously varied with the consequence that the products with the greater relaxation time are kept. The symmetry-breaking process, however, does not stop here. It can continue indefinitely.

Thermal polymers of amino acid can form a spherical structure in an aqueous milieu through their noncovalent bondings.[12] This is another indication of symmetry-breaking, especially with regard to the spatial configuration of those thermal polymers. When the polymers move almost freely in the aqueous suspension, those intermittent collisions among themselves may yield a wide variety of spatial configurations. In reality, however, the underlying law of motion of the one-to-many mapping type restricts the range of realizable configurations compared with those that may be conceived in the approximate process of one-to-one temporal mapping. The emergence of spherical structures of diameter 1 to 5 μm in aqueous suspension of thermal polymers of amino acid is undoubtedly an instance of keeping the products with the greater relaxation time while varying their boundary conditions endogenously.

The symmetry-breaking at the protobiological level thus continues to survive insofar as the law of motion of the one-to-many mapping type defies approximation to that of the one-to-one mapping type, in the latter of which the process materializing conservedness or the process of equilibration is supposed to proceed at an infinite velocity. As a matter of fact, the law of motion of the one-to-one mapping type could stand as a good approximation if the propagation velocity of interaction changes underlying the process materializing conservedness is far greater than the other velocities of interest. One thus observes that the law of motion of the one-to-many mapping type and the associated symmetry-breaking would become significant only when the propagation velocity of interaction changes is sufficiently lowered.

We have already recognized an evolutionary lowering of the propagation velocity of interaction changes due to the fact that the smaller the propagation velocity the greater the accumulation of evolving matter. Furthermore, evolutionary material accumulation endogenously enhances the lowering of the propagation velocity of interaction changes in the medium. Force equilibration proceeding at velocity of order of 10 to 10^3 μm/s (as evidenced in the flagellar movement and in the ATP-activated actin filament) is evidence that biologically full-blown processes allow the propagation of interaction changes at a velocity far less than light velocity. Symmetry-breaking at the protobiological level is a point of departure from nonbiologically physical processes with respect to lowering the propagation velocity of interaction changes from its possible maximum value of light velocity.

IV. SYMMETRY-BREAKING: BIOLOGICAL

The symmetry-breaking under endogenously generated boundary conditions continues to survive through the sequence from product to production. Development as well as evolution in biology is also seen as an instance of a production process under endogenously generated boundary conditions. The biological production process generates variation either developmentally or evolutionarily. Accordingly, the process of generating variations breaks the symmetry property with time due to the underlying equilibration that constantly limits the extent of the possible when it is transferred to the actual. The most fundamental character of the symmetry-breaking is the lack of time-reversal symmetry in generating variations. This implies that the pattern of variations to be generated at an earlier time is less constrained compared with that to be generated at a later time. The pattern of variations constrains itself with time. It thus turns out that those variations are by no means random in the sense that

Table 1
RELATIVE SUBSTITUTION FREQUENCIES (%) IN PSEUDOGENES

	A	T	C	G	59.2
A	—	4.7 ± 1.3	5.0 ± 0.7	9.4 ± 1.3	19.1
T	4.4 ± 1.1	—	8.2 ± 1.3	3.3 ± 1.2	15.9
C	6.5 ± 1.1	18.2 ± 2.3	—	4.2 ± 0.5	31.7
G	20.7 ± 2.2	7.2 ± 1.1	5.3 ± 1.0	—	33.2
	31.6	**32.9**	**18.5**	**16.9**	

Note: Each matrix element (i, j) represents the relative frequency (%) with which one type of nucleotide i is replaced by another nucleotide j. The relative frequency with which one type of nucleotide is replaced by any of the other nucleotides appears in the last column, indicating that C and G are more likely to be replaced than A and T.

Modified from Li, W.-H., Wu, C.-I., and Luo, C.-C., *J. Mol. Evol.*, 21, 58, 1984. With permission.

the pattern of variations does not remain invariant. The nonrandomness points to a lack of temporal invariance with regard to the pattern of variations generated.

One particular process of generating variations in the realm of biology is found within point mutation as demonstrated in nucleotide substitutions in DNA molecules. Spontaneous point mutation means a substitution of one of the four nucleotides, adenine(A), cytosine(C), guanine(G), and thymine(T), by one of the remaining three others in a DNA sequence. Since A is complementary to T and G to C in the double-strand structure of DNA, one can identify the consequence of point mutation by referring only to the sense strand, i.e., the untranscribed strand. Thus, an A-G substitution denoting an A base in the sense strand being replaced by G naturally induces a substitution of T by C in the complementary strand because of the base complementarity. An A:T pair in the double-stranded DNA molecule is replaced by a G:C pair if point mutation A → G occurs on the sense strand. Likewise, mutation A → T induces a substitution of pair A:T by T:A, and so on.

A suitable candidate for measuring the effect of spontaneous point mutation is pseudogenes such as goat pseudogenes.[13] The use of pseudogene sequences is due to the fact that pseudogenes are apparently subject to no functional constraints in connection with proteins so that the pattern of nucleotide substitution preserves the pattern of spontaneous point mutation as it happened. Their selective neutrality guarantees that none of the mutations that could occur in these pseudogenes would have been subject to selection, whatever it may be.[14,15] The pattern of spontaneous point mutation is identified by measuring the difference between two closely related pseudogenes derived from a common ancestral pseudogene, neglecting the possible multiple substitutions of simultaneous nature compared with a single substitution because of their probabilistic unlikelihood.[16] A numerical result is found in Table 1.

Point mutations of twelve different members (A → C, A → G, . . . , T → G) are classified into the two groups, transitional and transversal. The transitional mutations include the four types of A → G, G → A, T → C, C → T, and the transversal ones include the remaining eight members. If spontaneous point mutation is random among the 12 different types, the relative frequency of transitional mutation would be 33% ($^4/_{12}$) because of the equal likelihood of each mutation. The pattern of those random mutations would preserve its symmetry property in the sense that the uniform pattern of random mutations remains invariant with time. The measured reality, however, does not conform with the hypothetical symmetry based upon random mutation. The study of two closely related pseudogenes evolved from a common ancestral one reveals that the relative frequency of transitional mutation is 59.2%, almost twice as high as the value of 33% that would be expected under the assumption of random mutation. Another independent study of pseudogenes also confirms the nonrandom character of spontaneous point mutation by giving the relative frequency 56% of transitional mutation.

Nonrandom point mutation breaks, with time, the symmetry property observed with the pattern of variations. Those generated at an earlier time constitute the endogenous boundary condition upon the generation of subsequent variations at a later time. The generated variations at an arbitrary time are degenerated with the possible transversal as well as transitional mutations to come. In reality, however, the possibility of transversal mutations is more likely to be eliminated with time as the transitional ones are actualized. The symmetry-breaking occurs in the process of eliminating the latent transversal mutations in greater proportion compared with the transitional ones with time. This elimination of transversal mutations relative to transitional ones is irreversible and, accordingly, asymmetric with time.

The source of nonrandom point mutation is in the whole biological machinery that serves as an endogenous boundary condition for synthesizing these pseudogenes, in which the machinery is a construct of the preceding products that include pseudogenes. The biological construct from these products is a manifestation of the generation of nonrandom variation as well as a nonrandom variant that has itself been generated. Nonrandom point mutation as evidenced in the pseudogenes is thus evidence of both a generator of nonrandom variation as well as an already generated nonrandom variant.

The relative dominance of transitional mutations over transversal ones in the pseudogenes requires biological machinery to accommodate such an unevenness. Symmetry-breaking in the nucleotide sustitutions is found in the biological developmental process that is responsible for constructing the machinery required for producing the biased point mutations. Furthermore, the biological developmental process incorporates into itself a unique symmetry-breaking, especially in morphology. Holoblastic cleavage formation of a fertilized egg is an evidence of the symmetry-breaking in morphology.[17] We thus come to recognize an interplay between the symmetry-breakings in the nucelotide substitutions in DNA and in morphological development.

If one takes the view that development is a readout process of the developmental program stored in DNA molecules, the symmetry-breaking production process in morphology would be seen to be subject to the boundary condition specified by DNA. Nevertheless, the development program stored in DNA cannot remain totally indifferent to the production process it instructs. The DNA molecules that are reshaped or replicated during development constitute the endogenous boundary conditions for the subsequent development process. The endogeneity of boundary condition keeps maintaining the self-constraining capability originating in the process materializing conservedness or the process of equilibration. Development is always more constrained than the developmental program that could be identified in the DNA at the time of fertilization.

Symmetry-breaking in morphology could be justified even if the developmental program stored in the DNA that is equated with the nucleotide sequence remains intact. This is because the symmetry-breaking in morphology proceeds as long as the self-assembly of morphological material survives, for material self-assembly of any kind exhibits symmetry-breaking in the coordinate space of the constituent molecules. At this point, it is important to recognize that development is not a one-to-one transformation of the developmental program stored in DNAs onto the morphological space. Development as a material self-assembly under endogenously generated boundary conditions is subject to the law of motion of the one-to-many mapping type, instead of one-to-one. What is transformed during development is the developmental program itself which enhances its specificity with time due to the character of one-to-many temporal transformation.

We thus observe a convoluted mutual relationship between development and developmental program. Development is subject to the developmental program encoded in the nucleotide sequence of DNA when the program is unfolded in the morphological space, whereas variations in the developmental program in the program space are subject to the very developmental process in the morphological space. If there were only a one-way influence from developmental program to development, the program could not be varied unless an external agent intervenes. As a matter of fact, the law of motion of the one-to-many mapping type underlying material self-assembly in the developmental process induces self-constraining variations upon the accompanying developmental program. The impetus for inducing changes in the developmental program is found not within the carrier of the program, but within the very material self-assembly process of following the very program. The developmental program in the program space varies or, more specifically, constrains itself only through the materialistic developmental process in the morphological space unless an external agent intervenes.

Symmetry-breaking in the morphological space, though originating in the material self-assembly subject to the instruction from the developmental program, now induces another type of symmetry-breaking in the program space. The symmetry-breaking material self-assembly in the morphological space constantly constrains the extent of correlation among aggregated material units, in the latter of which the material carrier of the developmental program is embedded, because of the self-limiting character of the underlying one-to-many temporal mapping. An effect upon the program space from the one-to-many temporal mapping in the morphological space is the genesis of nonrandom variations in the former program space. The nonrandomness points to an unevenness of variants generated over a limited time interval. What is more, the unevenness is varied with time in the manner that the preceding variants that have been actualized condition the genesis of variants to come.

The nonrandomness of spontaneous point mutation observed in pseudogenes refers to an unevenness in the distribution of mutations accumulated over the period during which a common ancestral pseudogene has evolved into two closely related ones. Accordingly, the nonrandomness is varied with time and it cannot remain stationary. Nonrandom variation in the nucleotide substitutions or, more loosely, in the space of the developmental program originates in the symmetry-breaking in the morphological space. The symmetry-breaking in the nucleotide substitutions is at most a derivative from the similar breaking in the morphological space, for an examination of the nonrandomness presumes those variants are already made. What brings about the symmetry-breaking in the nucleotide substitutions is not their nonrandomness, but the morphological process. It is the morphological production process which gives rise to those products that exhibit an uneven distribution among themselves. Because of the successive feedback from product to production, nonrandom products lead to nonrandom production.

Symmetry-breaking in the development program is actualized in the symmetry-breaking in the morphological process. The developmental program is a kind of specification of how molecules are coordinated in the morphological space, which in turn determines the boundary condition for the program to be read out. The developmental program inherent to molecules is intrinsically degenerate. The dissolution of the degeneracy proceeds only through the production process in the morphological space. For instance, a nucleotide molecule in a DNA is less degenerate than a nucleotide of the same type in isolation. The emergence and evolution of DNA molecules are actualized through the process of lessening the extent of degeneracy associated with the constituent nucleotide molecules. The intrinsic degeneracy of developmental programs in DNA is dissolved only to the extent that the boundary condition in the morphological space can exert influence. Those products in the morphological space, while produced as following the instruction from the developmental program, also serve as a means of dissolving the degeneracy within the program itself.

The morphological space that dissolves the degeneracy in the developmental program is also available in the realm of protobiology. An amino acid molecule in a polyamino acid made thermally differs from the amino acid of the same type in contemporary proteins. The difference is within the morphological space available to polymerization of amino acids. Each amino acid can carry, in principle, an indefinite capacity for forming different polymers depending on the configuration of all of the other molecules in the morphological space. The boundary conditions actualized in the morphological space, though instructed by the developmental program intrinsic to the constituent molecules, further determine how the degenerate program within each molecule is to be differentiated and to be read out. The developmental program cannot be read out unless a reading machine is available in the morphological space. A unique characteristic of the reading process is that while reading, it generates the reading machine for the next generation. The successive generation of the reading machine thus makes it possible to read out the developmental program in a less degenerate way with time because of the operation of one-to-many temporal mapping. This is equivalent to saying that nonrandom production yields nonrandom products, which then enhance further nonrandom production.

Symmetry-breaking in the morphological space is common in both protobiology and biology. The only difference is that the boundary condition available in the morphological space to read out the developmental program is more specific and more particular in biology than in protobiology, because of the sequential accumulation of dissolving the degeneracy in the developmental program.

V. COMPLEXITY AND COMPLEXITY INCREASE

Symmetry-breakings at the protobiological as well as at the biological level enhance the degree of complexity in the molecular configuration. Complexity, although necessarily defined in an anthropomorphic perspective,[18] is a property that indicates the extent of how particular an object under examination may be. In order to identify how complexity varies with time, it is necessary to examine the underlying law of motion. If the law of motion is of the one-to-one mapping type as in classical mechanics, the sole origin of complexity is within the initial-boundary condition. The law of motion simply transforms the complexity, once fixed, solely through one-to-one temporal mapping. Unless controlled externally, the complexity remains invariant no matter how many degrees of freedom may be involved. The law of motion of the one-to-one mapping type can be particular and complex in its implication only to the extent which its boundary condition externally specifies. Room for complexity increase is ruled out there.

On the other hand, if the law of motion is of a one-to-many mapping type as it should be, a totally different picture of complexity and its increase will emerge. The extent of particularity increases as the possible is successively transformed into the actual, which is only a part of the possible. The process of complexity increase is inevitable in the one-to-many temporal mapping because the actual is within the persistent process of limiting what was formerly possible. The source of complexity is in every atom and molecule, and the potential for complexity to increase originates in the law of motion of the one-to-many type.

Complexity increase, if singled out, may sound teleological in the sense that the more complex state would be aimed at. However, there is no such agent that will aim at the more complex state. The superficial appearance of teleology is simply due to the anthropomorphic connotation inevitable to complexity and its increase. Complexity increase can be teleological as much as the law of motion of the one-to-many mapping type can be; in the latter, the process materializing conservedness is just the working of internal teleology without any anthropomorphic intervention. The internal teleology is entirely materialistic. Nonanthropomorphically expressed complexity increase amounts to symmetry-breaking in the molecular configuration.

One mode of complexity is increase in the number of distinguishable individuals at the expense of the number of indistinguishable ones. The degree of particularity is undoubtedly enhanced with the emergence of new distinguishable individuals. Nevertheless, the separation between indistinguishable and distinguishable individuals still may sound anthropomorphic, since this separation requires an agent to distinguish. This charge would have to be taken seriously if no materialistic process serving for the purpose of their distinction were available. In reality, however, the successive sequence from product to production serves as a materialistic agent for differentiating indistinguishable and distinguishable individuals with the aid of the symmetry-breaking which is endogenous in the process. The symmetry-breaking process in the space of material configuration is a cause for the emergence of new distinguishable individuals, for such a distinction can be reproduced if one follows the same operation of symmetry-breaking, though in retrospect. The materialistic process of symmetry-breaking is a prerequisite for the anthropomorphic recognition of distinguishable individuals. Complexity increase is a consequence of materialistic symmetry-breaking.

The degree of distinction between distinguishable individuals depends upon what sort of materialistic operations, both symmetry-breaking and -preserving, are available. Symmetry-breaking operations make those involved individuals distinguishable, whereas symmetry-preserving ones leave a set of individuals indistinguishable. Distinguishable individuals can be recognized as such only to the extent that a set of symmetry-breaking operations that are basically internal to a materialistic process is externalized by an external agent.

Complexity increase is intrinsically a quantum-mechanical phenomenon in maintaining the boundary conditions that are endogenous. Although it is possible to think of an externally fixed boundary condition upon Schrödinger's equation of motion, at least in theory, the externalization of boundary conditions would be admissible only when the device for the purpose is contrived either in theory or in experiment. Otherwise, the quantum mechanical process is in principle subject to endogenously generated boundary conditions. Symmetry-breaking in protobiology and in biology is a process of quantum-mechanically generating boundary conditions internally. Complexity increase is a materialistic consequence of generating quantum-mechanical boundary conditions without invoking external agents. Characteristic of protobiological and biological symmetry-breaking is complexity increase of endogenous origin.

VI. SUMMARY

Insofar as preceding products keep constructing the endogenous boundary condition for subsequent production, the process of symmetry-breaking continues. The symmetry-breaking in material self-assembly is observed in the genesis of nonrandom variants and their uneven distribution. Furthermore, the distribution varies its form with time because of the self-constraining capability of the process. Thermal polymerization of amino acids confirms the operation of symmetry-breaking at the protobiological level, indicating that only a limited kind of polymers is realizable out of the hypothetically conceivable alternatives. Nonrandom nucleotide substitutions in the pseudogenes also confirm the symmetry-breaking at the biological level. Symmetry-breaking in products indicated by their uneven distribution presumes the symmetry-breaking in the production space, that is to say, in the morphological space. Development is a symmetry-breaking process in the morphological space, whereas a nonrandom product is a carrier of the developmental program. Variations in the developmental program are subject to symmetry-breaking in the morphological space. While it is following instructions from the developmental program, the process of development in the morphological space varies the program through generating *de novo* nonrandom variations no matter how limited their scope may be. Complexity increase in material self-assembly is a consequence of symmetry-breaking in the morphological space.

REFERENCES

1. **Matsuno, K.,** Is matter inanimate: protobiological information from within, *Origins Life,* 14, 489, 1984.
2. **Kondepudi, D. K. and Nelson, G. W.,** Weak neutral currents and the origin of biomolecular chirality, *Nature (London),* 314, 438, 1985.
3. **Anderson, P. W.,** Suggested model for prebiotic evolution: the use of chaos, *Proc. Natl. Acad. Sci. U.S.A.,* 80, 3386, 1983.
4. **Fox, S. W. and Harada, K.,** Thermal copolymerization of amino acids to a product resembling protein, *Science,* 128, 1214, 1958.
5. **Nakashima, T., Jungck, J. R., Fox, S. W., Lederer, E., and Das, B. C.,** A test for randomness in peptides isolated from a thermal polyamino aicd, *Int. J. Quantum Chem.,* QBS4, 65, 1977.
6. **Hartmann, J., Brand, M. C., and Dose, K.,** Formation of specific amino acid sequences during thermal polymerization of amino acids, *BioSystems,* 13, 141, 1981.
7. **Matsuno, K.,** Determinism and freedom in early evolution, in *Individuality and Determinism: Chemical and Biological Bases,* Fox, S. W., Ed., Plenum Press, New York, 1984, 203.
8. **Conrad, M.,** Bootstrapping model of the origin of life, *BioSystems,* 15, 209, 1982.
9. **Miller, S. and Orgel, L. E.,** *The Origins of Life on the Earth,* Prentice-Hall, Englewood Cliffs, N.J., 1974.
10. **Miller, S. L.,** The production of amino acids under possible primitive earth conditions, *Science* , 117, 528, 1953.
11. **Miller, S.,** Production of some organic compounds under possible primitive earth conditions, *J. Am. Chem. Soc.,* 77, 2351, 1955.
12. **Fox, S. W., Harada, K., and Kendrick, J.,** Production of spherules from synthetic proteinoids and hot water, *Science,* 129, 1221, 1959.
13. **Gojobori, T., Li, W.-H., and Graur, D.,** Pattern of nucleotide substitution in pseudogenes and functional genes, *J. Mol. Evol.,* 18, 360, 1982.
14. **Kimura, M.,** Evolutionary rate at the molecular level, *Nature (London),* 217, 624, 1968.
15. **King, J. L. and Jukes, T. H.,** Non-Darwinian evolution, *Science,* 164, 788, 1969.
16. **Li, W.-H., Wu, C.-I., and Luo, C.-C.,** Nonrandomness of point mutation as reflected in nucleotide substitutions in pseudogenes and its evolutionary implications, *J. Mol. Evol.,* 21, 58, 1984.
17. **Slack, J. M. W.,** *From Egg to Embryo: Determinitive Events in Early Development,* Cambridge University Press, Cambridge, 1983.
18. **Saunders, P. T. and Ho, M. W.,** On the increase in complexity in evolution, *J. Theor. Biol.,* 63, 375, 1976.

Chapter 6

INFORMATION

I. INTRODUCTION

Symmetry-breaking and complexity increase in the morphological space are also specified in terms of information. Information presumes a message to be transmitted, a sender, and a receiver. If the message receiver is a fellow human being, information is anthropomorphic and it is taken for granted that the receiver is informed once he has received the message. On the other hand, if the message receiver is a material aggregate, information is nonanthropomorphic and it is necessary to identify how materialistic information processing can be made possible at all. Symmetry-breaking and complexity increase in the morphological space refer to nonanthropomorphic information, in which the triad of message, sender, and receiver has its materialistic underpinning.

Characteristic of nonanthropomorphic information is the inseparability between the message received and its sender. Transmission of materialistic information is actualized in the process materializing conservedness that is necessarily of one-to-many temporal mapping. The law of motion of the one-to-many mapping type is intrinsically informational because the material carrier of one-to-many temporal mapping is constantly constrained and informed with its endogenous operation. Nonanthropomorphic information is latent in atoms and molecules. Its actualization in the morphological space requires the law of motion of the one-to-many mapping type. Nonanthropomorphic information and the law of motion of the one-to-many mapping type are inseparable, whereas anthropomorphic information to a human observer is separable from the law of motion, if any, that might be applicable to the receiver. Once nonanthropomorphic information is taken as anthropomorphic information by a human observer, it would lose the further materialistic capability of information processing, for the human receiver cannot be a substitute for the material carrier of one-to-many temporal mapping. Nonanthropomorphic information associated with the law of motion of the one-to-many mapping type constantly enhances the degree of its particularity.

II. ANTHROPOMORPHIC INFORMATION

It is commonly taken for granted that biological products carry the specific information which distinguishes them from others. Take, for instance, contemporary proteins that are made from 20 different kinds of amino acid. One might be able to measure the information content of each amino acid residue if the amino acid sequence happens to be a particular one fixed through a random sampling out of the 20 different kinds of amino acid. A crude estimate of the information content of each amino acid residue based upon random polymerization would give $-\log_2(^1/_{20}) \simeq 4.33$ bits per amino acid residue. This amount of information would be required for identifying each amino acid residue in a protein molecule if it were synthesized through random coming together of amino acids. Then, questions arise about the implication of the information content thus derived.[1] Who or what generates information in the first place, and for what purpose? Who or what recognized the information as such? What does the information do for its own sake?

Shannon's information provides a unique answer to the stated questions.[2] An intelligent, necessarily anthropomorphic, receiver underlies Shannon's information. Anthropomorphic information presumes a source matrix of information that determines the probability measure of each message to be sent to the receiver. When the receiver anticipates a message to be received with a probability of occurrence p and when that message is actually received, the receiver will decrease the ambiguity with regard to the source matrix of information by the amount $-\log_2 p$ in terms of informational entropy and the message will be understood as conveying the information of the amount of $-\log_2 p$. The receiver has to be intelligent enough to be able to evaluate the source matrix of information in terms of the probability distribution of messages or events. Both the invariant source matrix of information and a particular message to be sent are prerequisite to anthropomorphic information.

If the source matrix of information for proteins is such that each amino acid residue would appear with equal likelihood among the 20 different members, one particular protein having 100 amino acid residues would be materialized with the probability of occurrence $(^1/_{20})^{100}$ among all of the hypothetically possible proteins of the same length. Accordingly, the actualized protein would carry the information content of 433 b (bits). The information thus evaluated is relevant only to the intelligent information receiver who views the source matrix of proteins as arising from random polymerization of the 20 different kinds of amino acid. The invariant source matrix of information maintains the distribution of all of the possible products invariant, and random polymerization has been taken as the agent that yields the invariant source matrix. Random polymerization of proteins from the 20 different kinds of amino acid is certainly a manifestation of a materialistic production process that can give rise to the source matrix of information for proteins which the intelligent observer identifies. Nevertheless, it is quite another problem to examine whether random polymerization can be legitimate as such in reality.

If random polymerization is questioned, on the other hand, the corresponding source matrix of information would have to be revised accordingly. The information content of proteins depends upon how the source matrix of information for proteins would be prepared in the first place, in which the source matrix is materialistic in its origin. Anthropomorphic information does require an invariant source matrix of information as a reference against which the information content of a given message can be measured. As a matter of fact, it is the materialistic production process, not anthropomorphic information itself, that determines the source matrix of information.[3]

The source matrix of information provides the set of all of the possible products. An actualized product represents a message sent from the source matrix. Consequently, we observe that two types of materialistic process are involved, one for preparing the source matrix of information and the other for generating and singling out the actual products from the source matrix. The generation of information, in fact, refers to the latter actualization process of products, since it is these actualized products that carry the relevant information to the receiver. Anthropomorphic information presumes that the intelligent receiver has some knowledge about how the source matrix of information is prepared and how the actual products are singled out. The knowledge the intelligent receiver carries is summarized in the form of the probability distribution of products.

The probability distribution is a resultant knowledge about both the genesis of the source matrix of information and the message to be sent therefrom. Anthropomorphic information thus does not distinguish between the source matrix of information and the message therefrom. When one says, for instance, that a protein having 100 amino acid residues carries its information content of 433 b, this anthropomorphic information is found to imply two things. One is that the source matrix of information for proteins is prepared by purely random polymerization of amino acids of the 20 different kinds, and the other is that an actualized protein is a consequence of random sampling from the source matrix without being subject to any biased selection. Randomness appears both in preparing the source matrix of information and in embodying the actual proteins. If the randomness is mitigated on the part of the intelligent receiver, a protein having 100 amino acid residues would appear to have an information content of less than 433 b, depending upon the extent to which the receiver is knowledgeable about the nonrandomness that is prevalent.

If the source matrix of information for proteins is provided by random polymerization of amino acid, the nonrandomness, if any, can intervene in the process of singling out the actual proteins from the source matrix. A biased selection from the source matrix is a mode of nonrandomness working in determining the actual proteins to be identified by the intelligent observer. Although it cannot be denied that the observer might be ignorant about the mechanism of the biased selection, this ignorance does not entail that the selection would be unbiased and random. If the selection assumed a uniquely definable selective value for each product in the source matrix and if the one that maximizes the selective value were realizable, the actual protein could be identified by the intelligent observer who is knowledgeable about the selective value even before its actual identification. The actual protein having 100 amino acid residues would carry its information content of 0 b, instead of 433 b, if it could maximize the selective value among all of the possible alternatives having 100 residues and if it could be selected accordingly. The process of actualization would not add to any further increase of information on the part of the intelligent observer.

The information content that the actual protein can carry depends very much upon how well informed the intelligent observer is about the selection mechanism. Still, the important problem is in the possible discrepancy between the selection mechanism per se and the knowledge that the intelligent observer may have about that mechanism. If the discrepancy exists as it should, the information content will be specific only to the information receiver and not to the materialistic character of the information carrier. It should not be appropriate to confuse the materialistic aspect of information with the capability of knowing about the materialistic character.

The selection mechanism that functions in protobiology and in biology is materialistic and natural in its operational mode. If the source matrix of information is prepared in a random manner, natural selection will be fully responsible for raising the actual out of the possible. There is no argument against natural selection, but it is quite another problem to examine to what extent the intelligent observer is knowledgeable about natural selection. Natural selection is still indefinite in its implication. One can identify the functioning of natural selection *a posteriori* by pointing to the products selected accordingly, but cannot tell the same result *a priori*. This lack of symmetry between *a posteriori* and *a priori* knowledge indicates an inevitable limitation of the intelligent observer with regard to the selection mechanism. Unless complete knowledge about natural selection is available, the information content of an arbitrary product that comes through the sieve of natural selection could not claim objectivity. For the information content depends upon the anthropomorphic condition of the observer, not upon the intrinsically materialistic character. Anthropomorphic information about natural products cannot claim its objectivity unless the intelligent observer boasts about one's complete knowledge about natural selection.

Natural selection is irrefutable, but refers to an intricate web of influences and interdependences among all of the possible entities in the protobiological and biological realm.[4] As far as one recognizes the inevitable discrepancy between *a posteriori* and *a priori* knowledge about natural selection, anthropomorphic information cannot attain its objectivity. The present incompleteness of anthropomorphic information originates in the anthropomorphic separation between the preparation of the source matrix of information and the actualization of natural products. Furthermore, the source matrix of information has been prepared through a random production process that is hypothetical. If the source matrix is prepared so as to include all of the possible products that could be conceivable even hypothetically, only the process of natural selection would be responsible for the actual. Nevertheless, one cannot claim an anthropomorphic information about natural selection that can be justified both objectively and materialistically.

The same charge can also apply to the hypothetical preparation of the source matrix of information in a random manner. The randomness refers either to the ignorance of the intelligent observer about how the source matrix should be prepared or to the real randomness on a materialistic ground. If the randomness were real, one would have to take it into account that natural selection that is selective and nonrandom in its operation would have to be responsible for raising random products. For natural selection provides the source matrix of products for the subsequent process. But, it is simply unacceptable both empirically and theoretically that something nonrandom be responsible for leading to something random, for natural selection accentuates the degree of nonrandomness in a cumulative manner with time. Once due attention is paid to natural selection, the random preparation of the source matrix of information cannot be justifiable in reality. Likewise, if the randomness in preparing the source matrix originates in the ignorance of the observer, the resulting anthropomorphic information will lose its objective and materialistic underpinning.

Anthropomorphic information about natural products, if legitimate, presumes an objectively justifiable knowledge about how the source matrix of natural products is prepared and how the actual products come into existence. However, the process of natural selection, which is real, defies the possibility of a complete knowledge about natural selection, because of the lack of symmetry between *a posteriori* and *a priori* knowledge about it. As far as it depends upon the contingent knowledge that the intelligent observer may have, anthropomorphic information will lack its materialistic underpinning. Of course it is conceivable to think of anthropomorphic information on a hypothetical basis. But the hypothesis also requires scrutiny. The real incompleteness of anthropomorphic information is within an arbitrary separation between the preparation of the source matrix of information and the actualization of real products that can be identified by the intelligent observer.

As a matter of principle, it is by no means evident whether the preparation of the source matrix and the actualization of real products are separable. What is more, anthropomorphic information is incomplete in the sense that the information is not clear by itself about what the intelligent observer would do once informed. The biological machinery of protein synthesis can make its own proteins by following the instruction or the information encoded in DNA sequences. The materialistic information receiver makes its proteins when it is informed. On the other hand, the intelligent observer like our fellow human being (let us say, a biochemist) cannot make proteins simply because he can decipher the corresponding DNA sequences.

Anthropomorphic information is useful on a hypothetical basis, but incomplete and even erroneous on a materialistic ground. In addition to the lack of objective standard about the separation between the preparation of the source matrix of information and the actualization of real products, anthropomorphic information necessarily introduces an anthropomorphic intervention in the materialistic process that has nothing to do with the intelligent observer. Still, information is useful both in protobiology and in biology as demonstrated in a frequently quoted cliche saying that the biosynthesis of proteins proceeds by following the information encoded in DNA. A way to save the notion of information is to make a clear distinction between anthropomorphic and nonanthropomorphic information. It is anthropomorphic information that causes a practical and also a conceptual difficulty in interpreting information in protobiology and in biology. A way out for protobiological and biological information is an appraisal of nonanthropomorphic information.

III. NONANTHROPOMORPHIC INFORMATION

One of the two difficulties with anthropomorphic information is in the arbitrariness with which the process of generating information is hypothesized. The other is in the lack of specification about what would occur to the information receiver once it is informed. These two difficulties with anthropomorphic information can be overcome if information is reinterpreted as being nonanthropomorphic and if the entire problem is rephrased accordingly.

Table 1
INFORMATION: ANTHROPOMORPHIC AND NONANTHROPOMORPHIC

	Informer	Informed	Message
Shannon's (anthropomorphic)	Source matrix of information	Intelligent receiver	Symbols
Protobiological & biological (nonanthropomorphic)	Prior products	Subsequent production	Boundary conditions (endogenous)

If one pays attention to the materialistically successive sequence from product to production, the ensemble of products due to the preceding production will be seen to constitute the endogenously generated boundary condition for the subsequent production. This sequence is informational in the sense that the ensemble of the preceding products serves as a source matrix for the subsequent actualization of the real products. At the same time, the actualization process is selective or, more precisely, self-constraining in limiting the extent of the subsequent production due to the irrevocable law of motion of the one-to-many mapping type. Nonanthropomorphic information as a measure indicating the extent of materialistic self-constraint is intrinsic to the successive sequence from product to production. Nonanthropomorphic information does not require either the invariant source matrix of information nor the actualization process of singling out the real products therefrom. The real ensemble of products is self-constraining and accordingly informational purely on a materialistic ground. This is equivalent to saying that the law of motion of the one-to-many mapping type is by itself intrinsically informational, though in a nonanthropomorphic sense. What comes to the foreground in nonanthropomorphic information is the clarification of what the materialistic information receiver does after it is informed. The materialistic information receiver points simply to the products synthesized from the preceding products and the role of the information receiver is to let these products thus generated be the endogenous boundary condition for the subsequent production. A comparison of anthropomorphic and nonanthropomorphic information is summarized in Table 1.

The distinctive characteristic of nonanthropomorphic information, in contrast to the anthropomorphic counterpart, is its explicitness about what would happen to an information receiver that is unfailingly materialistic. Its consequence is a disappearance of the clear separation between the preparation of the source matrix of information and the actualization of real products therefrom. In fact, the source matrix of information ceases to have its primary importance in nonanthropomorphic information, for the one that is destined to claim an exhaustive identification of the source matrix is absent in the nonanthropomorphic scheme. Contrary to anthropomorphic information that assumes an intelligent information receiver without asking how the presence of such a receiver could be possible at all, nonanthropomorphic information does not suppose the presence of any intelligent receiver. Fundamental to nonanthropomorphic information is the operation of one-to-many temporal mapping that is materialistic. The law of motion of the one-to-many mapping type is a materialistic explanation of how the materialistic information receiver could be made possible in the first place.

Nonanthropomorphic information is purely materialistic in its entire operation, whereas the anthropomorphic counterpart makes inevitable an interference from an intelligent information receiver into the materialistic process. Nevertheless, it is also undeniable that the anthropomorphic information receiver can be informed to a certain extent if he is exposed to the nonanthropomorphic processing of information. It is conceivable that the human intelligent observer will become knowledgeable about the DNA sequence, if deciphered, even though one cannot make the corresponding proteins simply on the basis of this operation of deciphering. Although the operation of nonanthropomorphic information is self-consistent in sustaining materialistic information processing indefinitely without allowing any intervention from an intelligent observer, it is still possible that a particular phase of the nonanthropomorphic operation can serve as a temporary reference against which anthropomorphic information is definable and conceivable. A candidate for such a temporary reference is the ensemble of real products available at an arbitrary time point in the course of the successive feedback from product to production.

The ensemble of real products, that are of course a consequence of nonanthropomorphic information processing, can define the relative frequency of each product that is again interpreted as the probability distribution. Once the probability distribution of products is identified, it is straightforward to estimate the information content of each product by following Shannon's scheme. The information content is, however, meaningful only to the intelligent information receiver who identifies the relative frequency of each product in the real ensemble that can be measured at an arbitrary time point. Measurement of these relative frequencies by the intelligent observer is prerequisite to anthropomorphic information.

It is important to recognize that the derivable anthropomorphic information is not available before measurement by an intelligent observer. An uneven distribution of real products is quite conceivable, and the association of the unevenness with anthropomorphic information would be natural if its measurement is possible. The measured distribution function is an expression of how each real product can be singled out from the source matrix of information. One can thus envision an appropriateness of anthropomorphic information even in the purely materialistic process. Nevertheless, the use of such anthropomorphic information is fundamentally limited. Anthropomorphic information requires the measurement of the probability distribution function of real products, and this procedure implies an anthropomorphic intervention. Furthermore, the anthropomorphic intervention cannot tell us how the ensemble of real products varies with time. Anthropomorphic information is intrinsically incapable of describing how the ensemble of real products comes into existence and evolves. It is the materialistic sequence from product to production that is responsible for varying the ensemble of real products.

It is possible to measure the information content of each amino acid residue in a protein if the ensemble of proteins is available.[5] The information content that is necessarily anthropomorphic depends upon what sort of ensemble of proteins is feasible in the first place. The information content of each amino acid residue in contemporary proteins can be measured if one prepares the source matrix of information, namely, the statistical ensemble of all possible contemporary proteins. The relative frequency with which one can find a particular amino acid at a particular site in the sequence determines the quantitative measure of the information content of the amino acid. Identification of the source matrix of information is prerequisite to the measurement of the information content. Similarly, one can define the information content of each amino acid residue in thermal proteins that are made from thermal polymerization of amino acids, as referring to its own source matrix of information that reduces to the statistical ensemble of all possible thermal proteins. We thus find that the same quantitative measure called information content of each amino acid residue is used for the two different source matrices of information, one for contemporary proteins and the other for thermal.

The information content with respect to the source matrix of information is anthropomorphic, because it is imperative there to claim a knowledge about the source matrix in advance. What is more, the source matrix of information represents a particular phase of nonanthropomorphic information processing when viewed from the perspective of materialistic process.

The information content is dependent upon the context fixed by the source matrix of information. The two different information contents of an amino acid residue belonging to

Table 2
C-TERMINAL AMINO ACID COMPOSITION OF THERMAL COPOLYMERS OF 18 DIFFERENT KINDS OF AMINO ACID

Amino acid	Case 1 composition (mol %)	C-terminal composition (mol %)	Case 2 composition (mol %)	C-terminal composition (mol %)
Lys	1.7	14.5	5.8	18.5
His	1.0	5.2	2.2	5.4
Asp	68.5	4.7	54.9	9.2
Glu	16.4	10.1	12.9	1.1
Gly	1.4	8.0	3.0	6.9
Ala	2.4	18.1	5.9	11.8
$^1/_2$Cys	1.4	7.6	3.7	11.0
Val	0.9	4.3	1.3	3.4
Met	1.0	6.0	1.8	6.0
Ile	0.9	5.1	1.0	4.4
Leu	0.9	5.0	1.2	3.8
Tyr	1.0	6.0	1.8	6.6
Phe	1.9	5.4	1.6	5.0

Note: Composition mol% was calculated without ammonia.

From Harada, K. and Fox, S. W., *BioSystems*, 7, 213, 1975. With permission.

the two different source matrices of information are not directly comparable quantitatively. By denoting the context in brackets ( ), we observe that the information content of each amino acid residue (thermal proteins) qualitatively differs from the information content of each amino acid residue (contemporary proteins), even if these two give the same numerical figures.

Anthropomorphic information can be quantified at the expense of evolutionary change, whereas the nonanthropomorphic counterpart can be evolutionary at the expense of quantification. If thermal proteins served as an evolutionary precursor of contemporary proteins by whatever means, the evolutionary nexus between the two source matrices of information would have to be worked out.

The information content of each amino acid residue in thermal proteins, H(thermal proteins), can be estimated from the C-terminus analysis of thermal proteins made from the mixture of the 18 different types of amino acids that are biologically relevant.[5] An experimental result is found in Table 2. If the amino acid sequences in thermal proteins were random, the compositional molar percent of each amino acid in thermal proteins would be proportional to the C-terminal compositional molar percent of the same amino acid. However, the experimental observation indicates nonrandomness in the sequences. The relative frequency with which each amino acid appears as a C-terminus of thermal protein determines the information content H(thermal proteins). The two independent measurements that have been tried are found here to give the same numerical figure:[6]

$$\text{H(thermal proteins)} \simeq 3.47 \pm 0.003 \text{ b per amino acid residue}$$

These figures have been evaluated without considering the contribution from three amino acids; arginine, half-cystine, and tryptophan. Nevertheless, this does not affect the significance of nonrandomness of the amino acid sequences in thermal proteins. If the amino acid sequences were totally random, the corresponding measurement would give

$$\text{H(thermal proteins)}_{\text{random}} = \log_2 18 \simeq 4.17 \text{ b per amino acid residue}$$

Contemporary proteins, on the other hand, are found to have the information content H(contemporary) $\simeq$ 4.153 b per amino acid residue, which is slightly less than $\log_2 20 \simeq$ 4.33 b based upon purely random polymerization.

If thermal proteins served as an evolutionary precursor of contemporary proteins, the source matrix of information for thermal proteins would have had to be alternated with the one for contemporary ones during evolution. The evolutionary alternation would also have had to be quantitatively responsible for how the information content of each amino acid residue in proteins has changed from H(thermal proteins) $\simeq$ 3.47 b to H(contemporary proteins) $\simeq$ 4.153 b. At this point, we recognize that the evolutionary alternation of the source matrix of information is purely materialistic, in which the successive sequence from product to production assumes a role of primary importance.

A simplified scheme of the successive sequence from product to production that is conceivable in polymerization of amino acids is given in the form of autocatalytic reaction:[7]

$$a_1 + \ldots + a_n + \nu E \rightarrow (\nu + 1)E$$

in which $a_1, \ldots, a_n$ represent energized amino acids, E is a polyamino acid, and ν is the order of autocatalysis. The usefulness of an autocatalytic reaction is in the fact that it can incorporate the preceding product, let us say polyamino acid E, in the above scheme, into the subsequent production process. The autocatalytic polymer that is certainly a product of the preceding production process serves as a new member, constituting the endogenous boundary condition for the immediately following production. It is, however, important to notice that the operation of the sequence from product to production is not limited exclusively to the autocatalytic reaction of a fixed order of autocatalysis. A possibility of endogenous variations in the order of autocatalysis may not be excluded from the operation of the successive sequence from product to production. In fact, if thermal proteins, whose synthesis is subject to thermal polymerization of amino acids of the zeroth order ($\nu = 0$) of autocatalysis, are supposed to be an evolutionary precursor of contemporary ones which are certainly autocatalytic in their making, the evolutionary change in the order of autocatalysis would have to occur.

The source matrix of information that can maintain a stationary ensemble of products as a particular phase of nonanthropomorphic information processing refers to the one with a fixed order of autocatalysis, for the invariant order of autocatalysis does not change the way that the autocatalytic reaction proceeds. Anthropomorphic information in terms of information bits is thus available to the source matrix of information under a fixed order of autocatalysis. On the other hand, variations in the order of autocatalysis go beyond and cannot be traced in the anthropomorphic information, because the variations induce changes in the manner of production and accordingly, in the ensemble of products. Nonanthropomorphic information processing is responsible for these variations in the order of autocatalysis. Nevertheless, one can still hope to appreciate the functioning of nonanthropomorphic information, though implicitly, by parameterizing the source matrix of information in terms of the order of autocatalysis and by figuring out a physical concatenation between the source matrices of different orders of autocatalysis.

Denoting the information content of each amino acid residue for the case of the ν-th order of autocatalysis by H(ν), one can expect to find a rule of transformation between H(ν = x) and H(ν = y) (x < y) on physical and chemical grounds if the transition of autocatalysis from order x to y really occurs in the course of nonanthropomorphic information processing. Most interesting is the case of the transition from the nonautocatalytic thermal polymerization of x = 0 to an autocatalytic one of y > 1, and it will be examined.

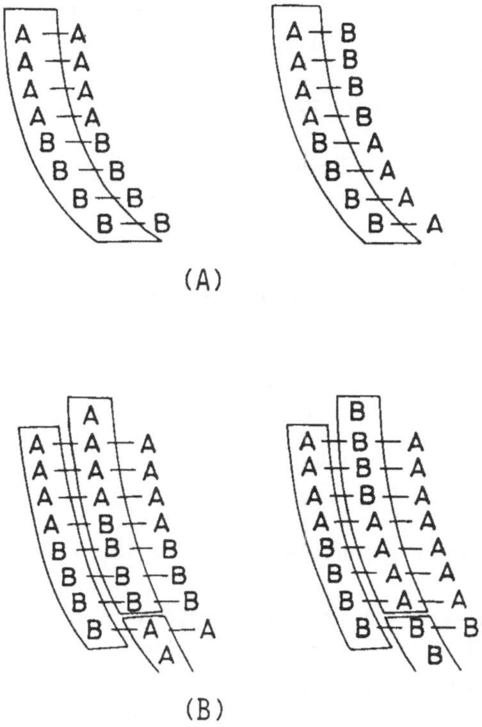

FIGURE 1. Information transcription from a polymer of monomeric sequence. (A) If a single polymer is used for instructing a sequence of monomers to be polymerized, the information content of each monomer in the product cannot increase beyond the similar content in the template. (B) On the other hand, if more than one template is employed for the instruction, the information content of each monomer in the product can become greater than the similar content in the template.

The number of autocatalytic polymers E necessary for instructing the making of one such polymer is ν in the reaction scheme $\ldots + \nu E \rightarrow (\nu + 1)E$ when $\nu < 1$. A schematic representation of such an autocatalytic instruction is demonstrated in Figure 1. Since the y-th order autocatalysis is supposed to spontaneously emerge from the matrix of thermal proteins, the information used for instructing the initial synthesis of autocatalytic polymer E of the y-th order is measured with reference to thermal proteins and the information content available for instructing the making of polymer E turns out to be

$$I_y = yH(\nu = 0) \qquad \text{(b per amino acid residue)}$$

since the number of thermal protein molecules used for the instruction of making one polymer E is y. The autocatalytic polymerization brings about a transformation of the source matrix of information, from the matrix of thermal proteins to that of autocatalytic proteins. The information of each amino acid residue in the new source matrix of information H ($\nu = y$), now satisfies

$$H(\nu = y) \leqslant I_y$$

since the product cannot exceed the instructor in its information content.

It is nonanthropomorphic information processing that spontaneously gives rise to autocatalytic proteins from the source matrix of thermal ones. The source matrix of autocatalytic proteins is latent within that of thermal proteins from the perspective of nonanthropomorphic information. As a matter of fact, the generation of and even the enhancement of information, when viewed from the anthropomorphic perspective, are in the spontaneous raising of autocatalytic polymerization.

Thermal proteins have been found to have the information content

$$H(\nu = 0) \left(\equiv H(\text{thermal proteins})\right) \simeq 3.47 \text{ b per amino acid residue}$$

If contemporary proteins are the evolutionary successor of thermal proteins as we shall visualize in the following, the order of autocatalysis of contemporary proteins ν_{cont} must satisfy the inequality

$$\nu_{cont} > \frac{4.153}{3.47} \simeq 1.2$$

because of[6]

$$H(\nu = \nu_{cont}) \left(\equiv H(\text{contemporary proteins})\right) \simeq 4.153 \text{ b per amino acid residue}$$

An evolutionary dynamics on the order of autocatalysis will further be needed.

The materialistic underpinning of nonanthropomorphic information is within the law of motion of the one-to-many mapping type, which underlies the process materializing conservedness. When the law of motion is applied to an open material aggregate, material flow equilibration results. The operation of material flow equilibration is intrinsically self-constraining because only those interaction configurations fulfilling material flow continuity *a posteriori* are realizable. The constraint due to material flow equilibration yields the decrease in the number of degrees of internal freedom. What is more, the decrease in the number of degrees of internal freedom brings about symmetry-breakings among the constituent material elements because of the increase in the number of conditions that all of these elements have to comply with in a correlated manner.

One particular form of the decrease in the number of degrees of internal freedom is the increase in material accumulation in the material aggregate. The material elements newly accumulated in the material aggregate have lost some of the former degrees of internal freedom they had, especially those degrees of freedom in the kinetic motion, by joining in the correlated macroscopic motion of the material aggregate. Material flow equilibration thus operates so as to enhance material accumulation at those aggregates open to material flow. The increase in the order of autocatalysis in protein synthesis is one process, among many others, for enhancing the decrease in the number of degrees of internal freedom through the increase in the material accumulation. This is because the production of proteins synthesized per unit time is proportional to the amount of the products of the same type present raised to the power of its order of autocatalysis, as in the form of the equation of motion $[\dot{E}] \propto [E]^{\nu}$ in which [ ] denotes the concentration of the product inside. Material flow equilibration as a mode of nonanthropomorphic information processing has an evolutionary directive factor to fix the autocatalytic reaction with a higher order of autocatalysis.

There have been several experiments suggesting an evolutionary enhancement of the order of autocatalysis. Simulated experiments of protobiogenesis have already revealed the fact that thermal proteins, especially lysine-rich ones,[8] have the catalytic capability for both peptide and internucleotide bond syntheses. There is also experimental evidence that the peptide synthesis catalyzed by lysine-rich thermal proteins is influenced by the presence of polynucleotide. Base on these experimental results, we expect a theoretical possibility of peptide-making-peptide of the first order of autocatalysis such as

$$\ldots + B \rightarrow 2B \tag{1}$$

to appear in an early evolutionary stage, where B is an autocatalytic polypeptide, and . . . represents the energized monomers to be polymerized.

One more theoretical possibility expected from these experimental findings is the autocatalytic reaction in which polynucleotide C also participates, as

$$\begin{aligned} \ldots + B &\rightarrow C + B \\ \ldots + BC &\rightarrow B + BC \end{aligned} \tag{2}$$

with

$$B + C \rightleftarrows BC$$

The cyclic reaction with respect to both polypeptide B and polynucleotide C is a theoretical extension of the results obtained in simulated protobiogenesis and serves as a prototype of hypercyclic catalytic reaction,[9,10] or a protohypercycle.[11] The first order autocatalytic polymerization modeled in Equation 1 is supposed to be an evolutionary precursor of the protohypercycle modeled in Equation 2, partly because of the enhancement of complexity in the latter. In order to justify the protohypercycle as the evolutionary successor of the simple autocatalytic cycle upon the ground applicable to open material aggregates, its order of autocatalysis ν_{proto} must be shown to be greater than unity. It is already implicit that all of the reactions proceed in open material aggregates such as protocells or cells.

We shall evaluate the order of autocatalysis ν_{proto} for the protohypercycle modeled in Equation 1. By denoting the concentration of arbitrary chemical substance A by [A], one can express the production kinetics of polypeptide B of the x-th order of autocatalysis as

$$[\dot{B}] = \alpha[B]^x$$

in the small concentration limit, where the dot on the left-hand side means the time derivative and α is the reaction rate coefficient. As far as only the order of autocatalysis is concerned, there is no difference in the analysis even if we consider only the small concentration limit. Similarly, the production of polynucleotide C can be expressed as

$$[\dot{C}] = \beta[C]^u$$

in the small concentration limit, where β is the reaction rate coefficient and u is a parameter. Expressions in Equation 2 also follow:

$$[\dot{C}] = a[B]$$

$$[\dot{B}] = b[BC]$$

$$[\dot{BC}] = c[B][C]$$

where a, b, and c represent the corresponding reaction coefficients. Dimension analysis now yields

$$x = \frac{4}{3} \quad \text{and} \quad u = \frac{3}{2}$$

This result reveals the fact that the protohypercycle has the order of autocatalysis in protein synthesis with its value

$$\nu_{\text{proto}} = \frac{4}{3}$$

The information content of each amino acid residue found in polypeptides synthesized through the protohypercycle turns out to be

$$H(\nu = \nu_{\text{proto}}) < \nu_{\text{proto}} H(\nu = 0) \simeq 4.63 \text{ b per amino acid residue}$$

The information content $H(\nu = \nu_{\text{proto}})$ of the protohypercycle can become greater than that of thermal proteins, though the protohypercycle is already latent in the source matrix of thermal proteins. Material flow equilibration is responsible for constraining the source matrix of thermal proteins to that of protohypercycles. An evolutionary enhancement of the order of autocatalysis, however, should not stop here.

A reaction kinetic model of contemporary protein synthesis is, in principle, well described by Eigen's hypercycle.[9,10] Its most simplified form is expressible in terms of polypeptide B′ and polynucleotide C′ as

$$\begin{aligned} \ldots + C' &\rightarrow B' + C' \\ \ldots + B'C' &\rightarrow C' + B'C' \end{aligned} \tag{3}$$

with

$$B' + C' \rightleftarrows B'C'$$

where . . . also denotes the necessary material resources in the form of energized monomers. Polypeptide B′ is synthesized upon polynucleotide C′ acting as a template. Polynucleotide C′ is replicated with the help of polypeptide B′ as a polymerase. The reaction scheme in Equation 3 is, however, undoubtedly insufficient because the polypeptide coded by a particular polynucleotide differs from the one serving as a polymerase of the same polynucleotide and because the interaction between anticodons and amino acids is indirect. Nevertheless, the following discussion will remain legitimate as far as only the order of autocatalysis is concerned.

Since we are interested in estimating the order of autocatalysis of protein synthesis in the hypercycle, ν_{hyp}, only the small concentration limit will be considered. The reaction kinetics follow with

$$[\dot{B}'] = a'[C']$$

$$[\dot{C}'] = b'[B'C']$$

$$[B'\dot{C}'] = c'[B'][C']$$

with

$$[\dot{B}'] = \alpha'[B']^{x'}$$

$$[\dot{C}'] = \beta'[C']^{u'}$$

where all of the parameters are fixed constants. Dimension analysis of these relationships gives

$$x' = \frac{3}{2} \quad \text{and} \quad u' = \frac{4}{3}$$

As a result, the order of autocatalysis of protein synthesis in the hypercycle is found to be

$$\nu_{hyp} = \frac{3}{2}$$

The information content of each amino acid residue with respect to those polypeptides found in the hypercycle envisioned as an evolutionary successor of the protohypercycle leads to

$$H(\nu = \nu_{hyp}) < \nu_{hyp} H(\nu = \nu_{proto})/\nu_{proto}$$
$$< \nu_{hyp} H(\nu = 0) \simeq 5.21 \text{ b per amino acid residue}$$

The succession of inequalities implies that the source matrix of hypercycles is already latent in the source matrix of protohypercycles, the latter of which is again latent within the source matrix of thermal proteins.

This result guarantees, at least on an information-theoretical ground, the possibility that the source matrix of information with regard to contemporary proteins may be within the source matrix of hypercycles having autocatalysis of the order of $^3/_2$ for protein synthesis. For the source matrix of hypercycles can cover the matrix of contemporary proteins if $H(\nu = \nu_{hyp})$ is really greater than 4.153 b per amino acid residue.

The source matrix of thermal proteins is potential in incorporating a wide variety of protein synthesis mechanisms to be actualized. If one represents the information character of the source matrix only in terms of the information content of each amino acid residue, the potential capacity of realizing *de novo* mechanisms of protein synthesis would fall away from the characterization. In order to save the materialistic capacity of varying the way proteins are synthesized, it is necessary to recognize that nonanthropomorphic information processing is inherently involved in varying the way the production process proceeds. It is incorrect and not legitimate to say that thermal proteins having the information content of 3.47 b per amino acid residue would be informationally less inclusive compared with contemporary ones having 4.153 b per amino acid residue.[12] Quite to the contrary, thermal proteins can be informationally more inclusive than contemporary ones in having a latent wider spectrum of possible protein synthesis mechanisms, and the spectrum is constantly constrained as it is transformed into the actual.[13] The source matrix of thermal proteins is more inclusive than that of contemporary ones, since the later source matrix subject to the more specific production process is simply a subset of the earlier source matrix accompanied by the more inclusive potential production process.

We thus find that the sequence from thermal proteins through the protohypercycle up to the hypercycle can give a coherent evolutionary sequence of the mechanisms of protein synthesis.[14] The endogenous increase in the order of autocatalysis of protein synthesis is a result of the working of material flow equilibration that serves as the motor of nonanthropomorphic information processing. This materialistic process also provides a moving reference against which anthropomorphic information is specified.

IV. TRANSCRIPTION

It is sometimes useful to examine the source matrix of information for proteins as subject to an invariant production process, though this scheme is simply a theoretical artifact because it is isolated from the inevitable interference from evolutionary and developmental processes. The usefulness of such an invariant source matrix of information is in its feasibility for identifying what sort of polymers is the primary information carrier capable of instructing the production of all of the other polymers involved. One can associate the polymeric carrier having the largest information content with the primary information carrier and others having the smaller information content with its derivatives. Insofar as the production process remains invariant, the measurable anthropomorphic information in terms of the information content can distinguish which polymers are most fundamental in the hierarchy of production and which are their derivatives.

The invariant production process leaves the source matrix of information or the ensemble of products invariant. There is no further generation of information. The only information process that is present is transcription. The production of derivatives from the primary information carrier that is most inclusive informationally is the process of transcription.

Thermal polymerization of amino acids suggests that the mixture of free amino acids itself serves as the information carrier to instruct the sequence of polymerization. The autocatalytic polymerization of the first order in polypeptide B as presented in Equation 1 in the previous section obtains the information of polymerization from its own products.

The protohypercycle in Equation 2, on the other hand, has a more complex structure by incorporating polynucleotides in its reaction cycle. In the small concentration limit, one can obtain

$$[\dot{B}] = \alpha[B]^{4/3}$$

$$[\dot{C}] = \beta\left(\frac{4\alpha^2}{3bc}\right)^{3/2}[B]$$

Polypeptide B requires one and a fraction of one third of the same polypeptide for directing its own polymerization. In contrast, polynucleotide C requires only one of polypeptide B for the instruction of its own polymerization. The results show that polypeptide B is much more inclusive informationally than polynucleotide C in their cyclic operation of the protohypercycle. One thus finds that, although polypeptide B and polynucleotide C each requires both for its own synthesis, the primary carrier of information which is required for the working of the cyclic polymerization is polypeptide B, instead of polynucleotide C. This is due to the fact that polypeptide B is more inclusive informationally. The polymeric information translation proceeds from polypeptide to polynucleotide, and not vice versa.

A similar analysis can be applied to the hypercycle in Equation 3. In a small concentration limit, one also obtains

$$[\dot{B}'] = \alpha'[B']^{3/2}$$

$$[\dot{C}'] = \beta'\left(\frac{3b'c'}{4\beta'^2}\right)^2[B']^2$$

It turns out that polynucleotide C′ is more inclusive informationally than polypeptide B′, since polynuceotide C′ requires two of polypeptide B′ for instructing its own polymerization whereas polypeptide B′ requires one and a fraction of one half of polypeptide B′ for the instruction. This fact tells us that the polymeric information translation proceeds from polynucleotide C′ to polypeptide B′ in the cyclic polymerization process because of the higher inclusiveness of information on the part of polynucleotide C′.

The direction of polymeric information translation between polynucleotide and polypeptide is reversed during the transition from the protohypercycle to the hypercycle.[11] The information translation from polynucleotide to polypeptide observed in the hypercycle is in accord with the one associated with contemporary coded protein synthesis. However, what we have observed is that the contemporary information translation can be an evolutionary successor of the one proceeding from polypeptide to polynucleotide. The information translation from polypeptide to polynucleotide can evolve from an autocatalytic polymerization of polypeptide in the absence of polynucleotide. Furthermore, the autocatalytic polypeptide can have its evolutionary origin in thermal polymerization of amino acids.

The information translation requires an identification of the extent to which each molecule serves as an information carrier. It is thus legitimate to say, at the least, that the whole ensemble of the preexisting products constitutes the information carrier for endogenously instructing how the subsequent production proceeds. In particular, the information carrier that instructs thermal polymerization of amino acids is the mixture of free amino acids itself, and the information used in this context can be interpreted as being anthropomorphic if the thermal polymerization remains invariant. Each amino acid in the mixture can carry a definite amount of information content during thermal polymerization. As a matter of fact, when the production process remains invariant, one can utilize anthropomorphic information by associating the relative frequency of finding each product appearing in the invariant ensemble of products with its information content. The necessary premise is the invariance of the relative frequencies of product in the ensemble, otherwise the reference against which invariant anthropomorphic information is measured would not be available.

A simple autocatalytic cycle of making polypeptides of the first order autocatalysis renders the polypeptide to be the information carrier to instruct the polymerization insofar as the autocatalytic cycle remains invariant. Those polypeptides in the simple autocatalytic cycle constitute an internally generated boundary condition for the subsequent production. The invariant production process lets the polymeric information carrier serve as a guiding reference for the invariant recapitulation process. The primary polymeric information carrier that is most inclusive under an invariant recapitulation process specifies how all of the other derivatives are generated therefrom. Once the most inclusive polymeric carrier is provided along with its production machinery, the production of the other less inclusive derivatives is already latent within the machinery.

Information content in the anthropomorphic sense can be identified by referring to an invariant source matrix of information. However, information transcription is a materialistic production process that is necessarily nonanthropomorphic. Nonanthropomorphic information processing involves the generation of information when viewed from the anthropomorphic perspective, though it is no more than the enhancement of being constrained materialistically. If the transcription is accompanied by the generation of information, it would not be feasible to claim the presence of information to be transcribed. Only when the information transcription belongs to an invariant recapitulation process can one say that information to be transcribed is present, not to mention the inevitable interference from nonanthropomorphic information processing. A polynucleotide can be said to carry information to be transcribed during the process of coded protein synthesis only by referring to the underlying recapitulation process. Unless such an invariant recapitulation process is available, the literal information transcription from polynucleotide would not remain legitimate and the information other than that involved in the mere transcription would also interfere in the process.

Nonanthropomorphic information processing is by itself nonrecapitulating because of its intrinsic irreversibility. The recapitulation process would remain as simply a theoretical artifact. Even if it has been the practice to associate protobiological and biological development with the recapitulation process purely on a theoretical basis by isolating it from evolutionary process, real development is more than recapitulation and serves as a temporal carrier of evolution. Information transcription that assumes the presence of information to be transcribed reduces to hypothesizing protobiological and biological development as a recapitulation process.

V. INDEFINITENESS

Nonanthropomorphic information processing derives from the law of motion of the one-to-many mapping type. The inseparability of the law of motion and its boundary conditions means that both of them are indefinite in their implications for each other. Unique to nonanthropomorphic information processing is the successive application of constraints on the indefiniteness carried by both the law of motion and the associated boundary condition. The successive constraint originates in the process materializing conservedness that constantly eliminates those interaction configurations that would violate conservation laws *a posteriori*. Accordingly, it is imperative to admit that every molecule in an interacting many-molecule system moves under the boundary condition which is constituted by all of the other interacting molecules. The indefiniteness associated with each molecule thus originates in the boundary condition that delimits the molecule from its neighborhood.

It is of course possible to theoretically conceive a boundary condition upon a set of many interacting molecules, but such an exogenous imposition is necessarily anthropomorphic. The present anthropomorphic intervention does not solve the problem of how the boundary condition upon each interacting molecule is generated and evolves. If the boundary condition happens to be stationary, it might be possible to associate it with the one conceived theoretically. But, the real problem is to examine the nature of endogenously generated boundary conditions before they become the object of theory.

The indefiniteness every molecule carries is due to its delimitation from the surroundings. And the very origin of the indefiniteness is the finiteness of the propagation velocity of interaction changes in the medium. If the propagation velocity were infinite, the boundary condition upon all of the participating molecules would be definite and determinate at every moment without leaving any internal tendency to induce changes endogenously. Theoretical imposition of boundary conditions of any sort just coincides with the case that calls for an instantaneous propagation of interaction changes in the medium. In reality, however, the propagation velocity of interaction change of any kind does not exceed light velocity. Successive repercussion of interaction changes is propagated in the medium indefinitely, and thus induces internal changes of boundary condition upon every interacting material element because the boundary condition is a form of interaction with its exterior.

In addition, every interacting atom and molecule is plastic enough to accommodate itself to an indefinitely wide variety of boundary conditions. Quantum mechanics, in fact, observes that the quantum mechanical wave function of any material aggregate is responsive to any change in its boundary condition. The endogenous capability of generating an indefinitely wide variety of boundary conditions and the intrinsic plasticity of those boundary conditions make every interacting atom and molecule a carrier of the process materializing conservedness, that is to say, a carrier of nonanthropomorphic information processing.

The indefiniteness of boundary condition is actualized within its plasticity on the real time axis. The plasticity yields variability in the manner the production process proceeds.[15] The ensemble of real products serving as the boundary condition for subsequent production cannot be equated with the source matrix of anthropomorphic information. Mutally supportive are the finiteness of the propagation velocity of interaction changes in the medium, the inseparability of the law of motion and its boundary conditions, and the successive sequence from product to production. The association of a real product with its information content in the anthropomorphic sense necessarily breaks the consistent tripartite circle. Anthropomorphic information processing presumes a complete separation between the law of motion and its boundary condition. Accordingly, the law of motion is taken to be of the one-to-one mapping type. The law of motion of the one-to-one mapping type would faithfully transform what the boundary condition provides. The boundary condition is constituted by the ensemble of products. Thus, the distinction between nonanthropomorphic and anthropomorphic information is rooted in the law of motion, whether of the one-to-many or of the one-to-one mapping type.

The indefiniteness of the boundary condition remains legitimate only from the perspective of the law of motion of the one-to-many mapping type. If one understands the boundary conditions constituted by the preexisting products within the scheme of the law of motion of the one-to-one mapping type, nothing indefinite could intervene. We thus recognize that nonanthropomorphic and anthropomorphic information processing are incommensurable as much as the law of motion of the one-to-many mapping type and that of the one-to-one type are. A decisive factor is whether one is entitled to take the boundary condition as being separable from the underlying law of motion. This problem is very theoretical in its origin. It is one thing to make a theoretical assumption about the fixedness of boundary conditions, but quite another to examine whether the boundary condition would be separately fixable in reality. In fact, what we have found as being real is the inseparability of the law of motion and its boundary conditions. This inseparability is the source of indefiniteness, the latter of which again serves as the source of the possible in the nonanthropomorphic information processing. The transformation from the possible to the actual is tenable also because of the inseparability of the law of motion and its boundary condition.

An amino acid molecule is indefinite in its implication because its is by itself incapable of determining the boundary condition it is subject to. If one takes the view that what an amino acid molecule is is definitely determined without referring to its boundary condition as its term literally suggests, the information processing would turn out to be necessarily anthropomorphic. Any change in the production process involving amino acid molecules would have to ask external agents to intervene. The inseparability of the law of motion and its boundary conditions, on the other hand, requires that every constituent element be indefinite and allows all of the other interacting elements to determine its boundary conditions. Appreciation of indefiniteness of any material element implies that even if it is called by a given and fixed name, every element keeps changing its contribution to the boundary conditions on all of the others interacting with it. Every material element varies its implication depending upon the context in which it is placed. This aspect exhibits a decisive contrast compared with anthropomorphic information processing, in which the context for defining information remains invariant and it is not necessary to constantly refer to the context.

The definiteness of anthropomorphic information originates in a theoretical artifact allowing the context under which information operates to remain invariant. It is thus ironic that the basis of the definiteness in theory is the indefiniteness in reality. Nonanthropomorphic information processing, on the other hand, keeps changing the context in which each product appears. As a matter of fact, the indefiniteness each material element carries is due to the variable context under which it is placed.

VI. SUMMARY

Information is classified into two subgroups, anthropomorphic and nonanthropomorphic. Anthropomorphic information presumes an invariant source matrix of information, about which an intelligent receiver or observer is knowledgeable. The presence of such an intelligent receiver is imperative to anthropomorphic information. On the other hand, the law of motion of the one-to-many mapping type underlying the successive sequence from product to production allows each material element to be a materialistic information receiver that is nonanthropomorphic. The materialistic information receiver maintains the capacity of being constrained, as much as the intelligent information receiver keeps the internal capacity to be informed. Nonanthropomorphic information processing, though purely materialistic in its operation, provides a transient reference against which anthropomorphic information is definable. That is to say, nonanthropomorphic information processing keeps varying the source matrix of information in the anthropomorphic sense.

Polypeptide synthesis from amino acids can be associated with anthropomorphic information processing if the production machinery remains invariant. The invariant machinery provides a reference against which the information content of each product is measured. In contrast, if the production machinery can be variable, the corresponding polypeptide synthesis is of nonanthropomorphic type. Transition from thermal polymerization of amino acids to simple autocatalytic polypeptide synthesis, if feasible, and then up to the heterocatalytic polypeptide synthesis involving polynucleotides, is an evolutionary indication of nonanthropomorphic information processing.

The carrier of nonanthropomorphic information is the whole ensemble of preexisting products that necessarily constitutes the endogenous boundary condition for the subsequent production, whereas one can view each product as a carrier of anthropomorphic information if the source matrix of information is supposed to remain invariant. A polymer can be said to carry information to be transcribed only if there is such an anthropomorphic observer who perceives the presence of a polymeric information carrier as such.

Nonanthropomorphic information based on the capacity of being constrained on the part of material aggregates requires the presence of materialistic indefiniteness. The very existence of such an indefiniteness originates in the inseparability between the law of motion and its boundary condition. Every atom and molecule is indefinite about its boundary condition, and this indefiniteness serves as a source of nonanthropomorphic information processing.

REFERENCES

1. **Gatlin, L. L.,** *Information Theory and the Living System*, Columbia University Press, New York, 1972.
2. **Shannon, C. E. and Weaver, W.,** *The Mathematical Theory of Communication*, University of Illinois Press, Urbana, 1949.
3. **Matsuno, K.,** Is matter inanimate?: protobiological information from within, *Origins Life*, 14, 489, 1984.
4. **Rheinberger, H. J. and McLaughlin, P.,** Darwin's experimental natural history, *J. Hist. Biol.*, 17, 345, 1984.
5. **Harada, K. and Fox, S. W.,** Characterization of functional groups of acidic thermal polymers of α-amino acids, *BioSystems*, 7, 213, 1975.
6. **Matsuno, K.,** Evolutionary changes in the information content of polypeptides, *J. Theor. Biol.*, 105, 185, 1983.
7. **Matsuno, K.,** Operational description of microsystems formation in prebiological molecular evolution, *Origins Life*, 10, 39, 1980.
8. **Nakashima, T. and Fox, S. W.,** Synthesis of peptides from amino acids and ATP with lysine-rich proteinoid, *J. Mol. Evol.*, 15, 161, 1980.

9. **Eigen, M.**, Self-organization of matter and the evolution of biological macromolecules, *Naturwissenschaften*, 58, 465, 1971.
10. **Eigen, M. and Schuster, P.**, *The Hypercycle: A Principle of Natural Self-Organization*, Springer-Verlag, New York, 1979.
11. **Matsuno, K.**, Natural self-organization of polynucleotides and polypeptides in protobiogenesis: appearance of a protohypercycle, *BioSystems*, 15, 1, 1982.
12. **Yockey, H. P.**, Self organization origin of life scenario and information theory, *J. Theor. Biol.*, 91, 13, 1981.
13. **Fox, S. W. and Matsuno, K.**, Self-organization of the protocell was a forward process, *J. Theor. Biol.*, 101, 321, 1983.
14. **Dose, K.**, Prebiotic polypeptides and the origin of biological information, *Naturwissenschaften*, 70, 378, 1983.
15. **Henshaw, P. S.**, A positive determiner in evolution, *J. Soc. Biol. Struct.*, 7, 235, 1984.

Chapter 7

PROTOREPRODUCTION

I. INTRODUCTION

Reproductive material units, such as molecules and cells, emerge as a consequence of the successive feedback from product to production. Molecular and cellular reproduction is a nonanthropomorphic information processing. The sequence of letting the preceding products be the endogenous boundary conditions for the subsequent production is likely to yield products of like kind. The evolutionary transition from the emergence of protoreproductive units to contemporary reproductive ones points to the temporal enhancement of reproductive fidelity. One dimensional replication due to complementary base pairing requires a whole set of three-dimensional machinery that zippers together complementary base pairs and then opens up the zipper for subsequent replication. Two-dimensional replication due to intercalating synthesis of silicate layers requires a three-dimensional mechanism of detaching the daughter layers from the matrix layer.

Reproductive molecules necessitate, in their own making, a particular spatial configuration of the preceding products. The emergence and evolution of protoreproductive molecules assume protocells or precells that house mechanisms of molecular replication. Protocells as open material aggregates constantly apply self-constraints to themselves due to the underlying material flow equilibration, in which they are also reproductive as growing and multiplying. Material flow equilibration enhances the fidelity of both protocellular and of molecular replication with time. Protocellular and molecular replication proceeding in protocells can serve as a means of constructing an energy transduction mechanism. Formation of phospholipid-like molecules, the resulting vesicles, and of the mechanism of proton translocation in a closed-membrane system can be a consequence of the successive feedback from product to production. The origin of the genetic code should result from the interplay between protocellular and molecular replication.

II. ORIGINS OF PROTOREPRODUCTION

Nonanthropomorphic information processing that permits the preceding products to be the boundary conditions for the subsequent production gives rise to what looks like a reproductive or protoreproductive process. This is because when the preceding products serve as a boundary condition for making products of a similar kind, those products would increase exponentially with time and because those with the largest production rate per unit time would dominate the process. Consider, for instance, the sequence in which polymer A serves as a boundary condition for the production of polymer B, the latter of which newly serves as a boundary condition for making polymer C, and so on. When the production sequence is closed, replicating polymers would arise. Furthermore, among the possible closed production cycles, the one with the largest production rate would dominate.

The emergence of protoreproductive units that are localized spatially points to a particular phase of the successive feedback from product to production. In the natural production process, the whole set of the preexisting products can join in constituting the boundary condition for the subsequent production. The boundary condition develops irreversibly with time. The rationale of the underlying irreversibility is the law of motion of the one-to-many mapping type supplemented by intrinsically partially fixed boundary conditions.

One candidate for the protoreproductive units on the early Earth has been hypothesized to be layer-silicates because of their ubiquity.[1] Montmorillonite and its homologues that are

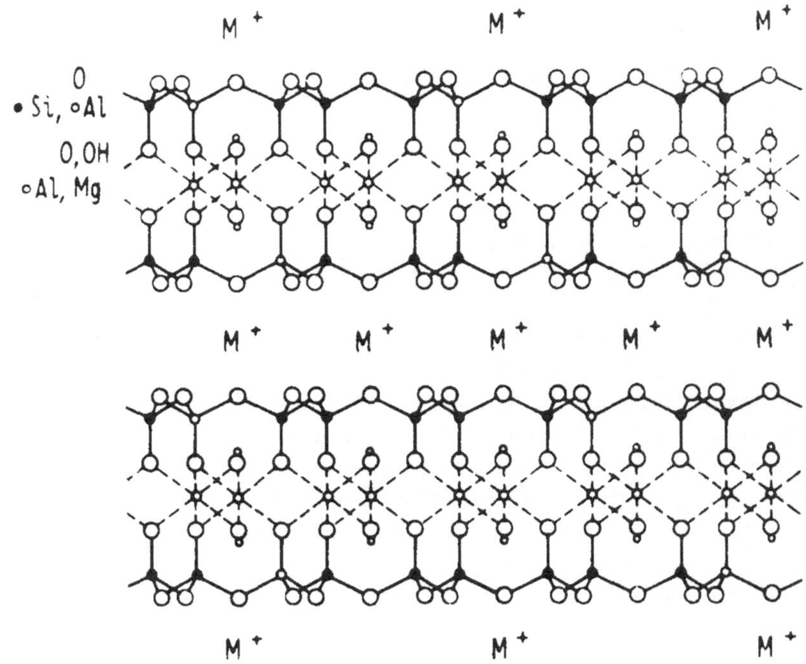

FIGURE 1. A single layer of montmorillonite. M^+ represents a univalent cation.

silicates with layered structures increase the distance between the parallel layers when water molecules penetrate into the interlayer region. As the spacings increase, the crystals then disintegrate into the individual layers. Individual layers of a disintegrated crystal may act as nuclei for new crystals when they are introduced into an appropriate culture medium containing the necessary ingredients. On decreasing the electrolyte concentration, those newly synthesized layers may also undergo disintegration, thus yielding a self-duplicating system. A feasible vehicle for this purpose is the intercalating synthesis of new montmorillonite layers. The present intercalating synthesis is more than just a theoretical surmise. It has been experimentally demonstrated by carefully controlling the sodium to potassium ratio and by devising a cyclic control of the electrolyte concentration so as to enable replica layers to detach from the matrix layer.

The intercalating synthesis of montmorillonite layers, as depicted in Figure 1, is certainly an ingenious model that is experimentally demonstrable on the onset of protoreproductive units. However, when one asks how the boundary condition that makes the intercalating synthesis of montmorillonite layers possible could establish itself in the natural setting, an intractable problem would arise. Whereas the matrix layer for making replica layers is certainly a product of the preceding production and serves as an endogenous boundary condition, the external condition necessary for actualizing cycles of replication such as repetitive cycles of thawing, or rains and droughts are definitely needed for proper functioning of replication due to the intercalating synthesis (cf. Figure 2) of montmorillonite layers.

Exogenous boundary conditions inevitably intervene in replication processes. This type of exogenous influence over replication is ubiquitous both spatially and temporally on an evolutionary time scale. Light energy from the sun constitutes one type of the most significant boundary conditions of exogenous origin on terrestrial biology. Nevertheless, there is a significant difference between organic replication of cells and molecules in the realm of biology and an inorganic replication of montmorillonite layers. The difference is in the fact that the inorganic replication does not equip by itself a mechanism for enhancing the re-

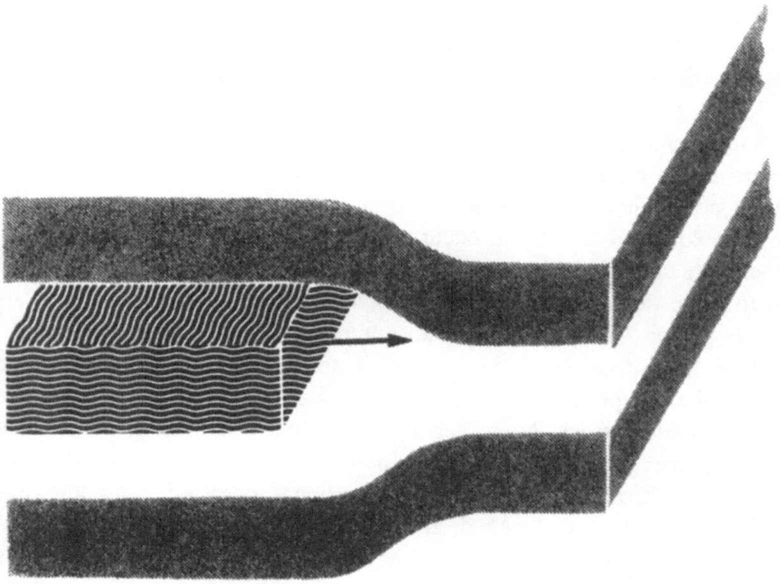

FIGURE 2. Intercalation and intercalating synthesis of a silicate layer. (From Weiss, A., *Angew. Chem. Int. Ed. Engl.*, 20, 850, 1981. With permission.)

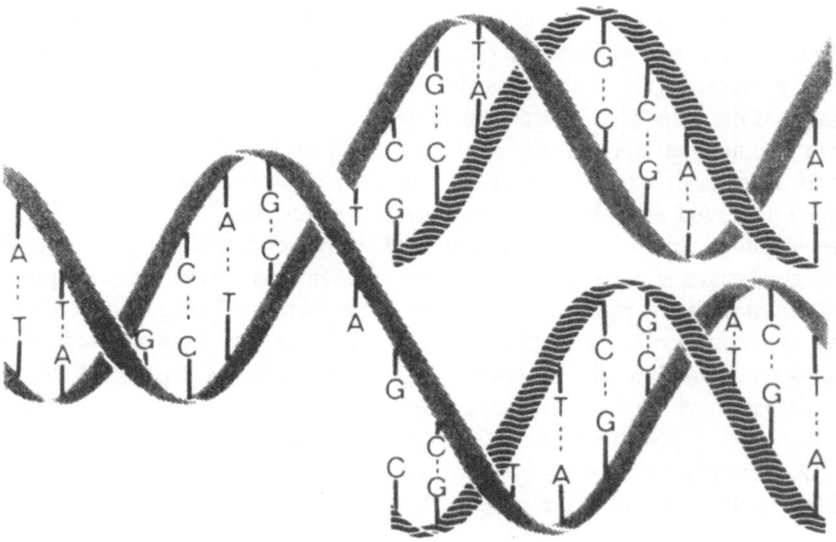

FIGURE 3. Residue-by-residue replication of DNA (or RNA) molecule. Dark strand represents the matrix, and the shaded one is a replica.

productive fidelity unless aided by an event of exogenous origins. The intercalating synthesis of montmorillonite layers just witnesses the lessening of the reproductive fidelity over the first 22 ~ 23 generations of replication.

One more case of protoreproduction is an inorganic replication of polynucleotides on templates due to complementary base pairings (cf. Figure 3).[2] In fact, if metal ions such as zinc and lead are available along with templates, even though short, polymerization of up to 50 nucleotides has been experimentally demonstrated without being accompanied by proteinous enzymes. The preceding products can serve as a template for the subsequent polymerization. However, the inorganic replication of polynucleotides does not yet incorporate into itself a mechanism for enhancing the reproductive fidelity while it keeps elongating the polymer length.

It is of course theoretically conceivable that replicating systems could be devised without being accompanied by the mechanism of enhancing the reproductive fidelity. In case it is theoretically supposed that variations are random in their genesis in the sense that every single variant has equal probability of occurrence, the likelihood of having a variation that raises a reproductive unit and enhances the reproductive fidelity would be extremely unlikely as compared to those that denigrate the reproductive fidelity and then destroy those reproductive units altogether. The problem is with randomness and random variants.

The concept of randomness enters into the origin and evolution of protoreproductive units as two fundamental assumptions about the origin of variation. One is that a frame of reference is available which makes it possible to generate variation. The other is that each variant is equally likely to appear. In the genesis of random variation, a complete separation is assumed between the generation of variations and the variations generated, with the generator of random variations remaining unchanged. If it could change in a random way, this would come to presuppose a higher frame of reference for the random change. On the other hand, if the generator can change in a predetermined way, this would deny random variation itself. The assumption that every variation is equally probable *a priori* is nothing more than a mere reflection of the ignorance of the observer who has not yet examined all of the possibilities in an exhaustive manner, while claiming that all of the possible variants are already present. Random variation does not obey either the law of motion of the one-to-many mapping type nor that of the one-to-one type. The law of motion of the one-to-many mapping type supplemented by intrinsically partially fixed boundary conditions does not allow the exhaustive enumeration of all of the possible variants in advance because of the intrinsic indefiniteness of the associated boundary conditions. In addition, the law of motion of the one-to-one mapping type supplemented by completely fixed boundary conditions by its very definition excludes such a possibility of the ignorance on the part of observer because of the presumed unlimited controllability over boundary conditions.

Random variation would be acceptable only if the generator of variation remains unchanged. A good example of this kind is statistical mechanics at or near thermal equilibrium. What equilibrium statistical mechanics teaches us is that the most likely generator of variation is the one that can maximize the number of equiprobable variants. This is a static argument and cannot account for how the generator of variation could develop with time. Hence, statistical mechanics is applicable only to those generators which can remain stationary. Although it might be possible to extend statistical mechanics to the thermodynamically nonequilibrium case including open systems, the basic paradigmatic statement remains unaltered. Even in the nonequilibrium scheme, how the generator itself develops is left untouched.[3]

One could imagine that both the neo-Darwinian random variation whose genesis does not presume the effect of *a posteriori* natural selection and the neutral random variation which remains indifferent to natural selection have their own supportive facts so long as the generator of variation is supposed to remain unchanged. However, a fixed generator of variation is nothing more than a theoretical artifact. In particular, whether the generator remains unchanged is not a matter of hypothesis, but a real question that must be answered. What we have recognized is that so long as any living system and its precursors down to protoreproductive units are identified as self-generated open material aggregates, the generator of variation keeps restricting and constraining its products. What is more, a self-generated open material aggregate is the generator of variation itself.

A self-generated open material aggregate acting as a generator of variation is always transforming itself into a more constrained form. Once a self-generated open material aggregate appears, an enhancement of constraining in the generation of variation makes evident the irreversibility of developmental and evolutionary time. Random and nonrandom variations are sharply distinguishable if one refers to time. Random variation means simply that every possible variant is equally likely to occur, independent of the time when it really occurs. In contrast, nonrandom variation due to material flow equilibration keeps constraining the members constituting the ensemble of possible variants, whose exhaustive enumeration could not be attainable. The likelihood of each variant depends on when it occurs. Nonrandomness always refers to time. For instance, suppose hypothetically an identical variant happens to occur at two different time points. What this nonrandomness suggests is that the one that occurred at the later time point has the greater likelihood. This observation indicates that the carrier of nonrandomness is not each variant, but the self-generated open material aggregate itself as the generator of variation. It is time which can bring about the uneven likelihood of occurrence among possible variants. Such a nonrandomness inevitably implies something predetermined since evolution does in retrospect appear to be unidirectional. What is predetermined is the process of self-constraining in the generation of variation.

If nonrandomness were static in the sense that the distribution of uneven likelihood among possible variants is fixed and conceivable in advance, there would be no evolution since the most likely variant can appear at any arbitrary time with the same likelihood. This denies the historical character of evolution. Evolution is always under the influence of what it has already accomplished. For this reason, we must refer to the generator of variation, not to variations generated, in order to explicate nonrandom variations in evolution. What is evolutionary is the generator of variation. The static generator of variation, conceived within fixed boundary conditions, is not evolutionary in this regard. The dynamic nonrandomness of variation points to the generator of variation as an active agent. The origin and evolution of such an active agent are in accord with a physically permissible process that reduces to material flow equilibration in self-generated open material aggregates.

Variation always materializes within a certain constraint in self-generated open material aggregates which keeps restricting its own extent with time. It should be stressed that even if we have not yet had much experience in dealing with self-limiting constraints or endogenously varying boundary conditions in physics, it would be premature to say that physics precludes the possibility of self-limiting constraints. What we have observed in evolution in fact is just the opposite.

Random variation is conceivable under fixed boundary conditions. However, fixed boundary conditions are not compatible with the law of motion of the one-to-many mapping type supplemented by indefinite boundary conditions. Random variation could survive in theory only in an approximate sense that the indefiniteness observed with boundary conditions could remain unchanged over a limited time interval arbitrarily chosen. A proper appreciation of the law of motion of the one-to-many mapping type supplemented by indefinite boundary conditions does not fail to provide a firm theoretical basis for the occurrence of nonrandom variation. Giving rise to protoreproductive units and enhancing their reproductive fidelity could be an instance of nonrandom variations that are intrinsically irreversible and intensify their own specificity with time. Perfect reproductive fidelity, even if conceivable in theory, would be no more than an asymptotic limit of enhancing the extent of self-constraint.

The onset of protoreproduction refers to a particular phase of material evolution such that reproductive fidelity can be defined thereafter and the fidelity begins to enhance its specificity from then on. The materialistic underpinning is within the law of motion of the one-to-many mapping type. Protoreproductive units thus appear in the successive sequence from product to production in an endogenous manner. Enhancement of reproductive fidelity is expected only in the same matrix that gives rise to the protoreproductive units in the first place, because the matrix functions as a material carrier of the endogenous sequence from product to production in a successive manner. Consequently, if a reproductive unit of exogenous origin happens to be embedded in a material setting that differs from its former matrix, its successive reproduction cannot necessarily be guaranteed and the reproductive fidelity may deteriorate.

Of course, if the surrogate matrix *in vitro* happens to simulate the native one *in vivo* reasonably well, as in the case of viral RNA Qβ-replicase complex,[4] replication of RNA can confine its reproductive fidelity to a sharply delimited range.[5] The factors that control the replication of RNA include temperature, ionic strength, pH, supply of precursors, and the presence of replication inhibitors or of template degraders such as nucleases. The real problem, however, is the fact that though the entities constituting the boundary condition of the native matrix for the subsequent production are produced totally *in vivo*, the boundary condition constituting the surrogate one *in vitro* is forcibly isolated from the endogenous succession from product to production and is simply imposed as such. Externally controlled replication must seek the impetus for enhancing the reproductive fidelity also in external agents. On the other hand, reproductive fidelity and its temporal enhancement in the material system refer exclusively to the endogenous scheme of forming the boundary condition for the subsequent process.

Experiment on replication is thus a double-edged sword. If an experimental boundary condition is so finely contrived as to guarantee the process of replication, this would deprive the experimental setup of the endogenous capacity for letting the reproductive fidelity enhance itself internally. Conversely, if the experimental boundary condition is so loosely realized as to leave a lot of factors unspecified, it would be most unlikely to expect controlled experimentation. The unique characteristic of experiment on replication is in tracing variations in the boundary conditions. The necessary condition for this type of experimentation is, in the first place, that boundary conditions have to be variable, since, at the least, the natural process on Earth has witnessed a variable sequence of boundary conditions in the evolutionary process that has taken place there.

The decisive question of an experiment on replication is how to identify the mechanism that induces variations in the reproductive fidelity. In this regard, random variation cannot be invoked. If random variation were a major impetus for introducing changes in the evolution of reproductive units, the replication rate as well as the reproductive fidelity would be subject to random variation. Although an overall effect of natural selection can be summarized as the preservation of those reproductive units with the greater replication rate, it does not necessarily tend to preserve those with the finer reproductive fidelity because of the very nature of random variation that lacks any interdependence between replication rate and reproductive fidelity. Consequently, nonrandom variation has to be properly invoked.

Material flow equilibration as the law of motion of the one-to-many mapping type acting at an open material aggregate that is viewed as the generator of nonrandom variation, on the other hand, lets both replication rate and reproductive fidelity be interdependent. The origin and evolution of protoreproductive units are accordingly part of the process of generating nonrandom variation that is inherent in open material aggregates. The problem of replication rate and reproductive fidelity is reduced to how surviving open material aggregates can be made possible, as actually occurred on Earth.

III. MULTIPLICATION OF PROTOCELLS

Molecular replication and cellular multiplication are inseparable from each other when they function in contemporary biological systems. Cellular aggregates provide the boundary condition upon the molecular replication taking place there, while those aggregates are the construct of the preceding molecular production. Accordingly, the condition prerequisite to the evolutionary emergence of molecular replication is the formation of the required boundary conditions that are spatially localized. Those spatially localized boundary conditions can be viewed as material aggregates open to material flow, since it is the boundary conditions which internally regulate the amount of material or resource flows both entering and leaving. Protocells or precells are thus required as the boundary condition that could give rise to and maintain replicating molecules therein.

The problem of which appeared first in the course of evolution, molecular replication or cellular multiplication, is rather ill-stated. Molecular replication requires cellular structures as the boundary condition for its own action. What must be examined is how protocellular aggregates could evolve so as to make out of themselves the boundary condition which could come to yield molecular replication.

Any material aggregate, whether it may be atom, molecule, or a complex of molecules, is open to material flow when it is in the process of increasing or decreasing the amount of material accumulation. The boundary separating the material aggregate from its exterior is self-generative in the sense that the boundary is formed through a self-consistent interaction between its interior and exterior. What is more, if a material aggregate thus formed can have any significance for the intrinsic evolutionary capability of matter at all, it cannot be a mere passive entity shaping its own boundary with the exterior only by responding to an externally controlled boundary condition.

For instance, the honeycomb-like hexagonal cells due to the heat flow-convection flow instability of Benard shape their structures by responding simply to the contour of the container in which the fluid is placed. These Benard cells, although certainly open to material flow, are exogenous in their generation and do not maintain the internal capability of changing their own structures by themselves with time.[6] A material aggregate which has an intrinsic evolutionary capability is one that can successively change its interaction with the exterior not only by responding to the impetus from outside but also by influencing the exterior as a reaction to the former process in a recurrent manner. If the material aggregate fails to induce a change in the exterior that could again influence the aggregate successively, it would remain simply an entity passive to the exterior. The evolutionary cause would have to be sought in the external boundary condition.

A principal reason why open material aggregates maintain their intrinsic evolutionary capability is that these aggregates not only react to the impetus from their exterior but also induce some change in their surroundings in response to the preceding impetus. During the formation of a polymer from monomers, for instance, it can change the way of interaction with its exterior as the elongation progresses because the aggregated monomers altogether specify how and in which way the next monomer attaches. Consequently, what is crucial to the intrinsic evolutionary capability of matter is identified as its possession of internal freedom that remains intact without being exhaustively controlled by external means or agents. The internal freedom is simply the capability of matter to change its internal state while exerting an influence upon its surroundings. The endogenous capability of an open material aggregate is inherent to the process materializing flow continuity or in material flow equilibration, which is an actualization of the law of motion of the one-to-many mapping type supplemented by indefinite boundary conditions. The internal freedom that can never be exhaustively specified has its source in those indefinite boundary conditions.

At this point, it is important to recognize that the influence of external sources acting upon matter and the subsequent reaction of matter upon its surroundings never occur in a simultaneous manner. Because of the sequential nature of the reaction, a material aggregate changes its internal character by itself although this is initiated by an impetus from the surroundings. A consequence is a subsequent variation in the surroundings caused by the change in interaction with the aggregate. The successive generation of variation and reaction between matter and its surroundings is thus responsible for producing and constantly varying open material aggregates.

In terms of evolution, the idea of a primordial soup of atoms and abiotic molecules provides the most significant model of an open material aggregate.[7] The combined process of material association and dissociation can give rise to material units which are open to material flow, provided the material aggregates which are formed can participate in further association with other elements during their inevitable partial degradation.

Any open material aggregate specifies both its own association energies with other molecular units and its dissociation energies into smaller units. But, both the association and dissociation energies vary as the underlying material aggregate changes itself. These changes in the association and dissociation energies are due to the reaction of the material aggregate to its surroundings as responding to the preceding molecular association and dissociation. The molecular association energy changes as the association or dissociation proceeds. Similarly, the molecular dissociation energy changes as the association or dissociation proceeds. Open material aggregates thus have the capability of constantly changing their own states through the change in the molecular association and dissociation energies which respond to the preceding association and dissociation. Open material aggregates formed in the primordial soup are intrinsically active toward their surroundings in the sense that they successively change their interaction with the surroundings in response to the preceding molecular association or dissociation or both. Open material aggregates endogenously generate and vary their boundaries with the surroundings through successive changes in the energies of material association and dissociation.

One form of generating open material aggregates in the primordial soup is thermal polymerization of amino acids. During the process of polymerization, the growing thermal polyamino acid is in a state of open material aggregate. These thermal polyamino acids or proteinoids can in turn serve as the constituent elements comprising protocellular microparticles or proteinoid microspheres in aqueous milieu as another type of open material aggregates. The material association process differs between the thermal polymerization of amino acids and the formation of microspheres from thermal polyamino acids in aqueous milieu. The peptide bond synthesis during thermal polymerization is due to covalent interaction, whereas the material accretion underlying the production of microspheres includes the noncovalent-type interaction that is anisotropic. The change in the character of interaction between the growing open material aggregate and its surroundings during material association is thus endogenous.[8] In particular, thermal polyamino acids having lipid-like properties, including anisotropy,[9] are responsible for forming membranes and spheres.

Thermal polyamino acids whose formation would be quite conceivable in the primordial soup have lipid-like properties.[10] Likewise, phospholipid-like material would have arisen with ease under primitive Earth conditions.[11] This would suggest an early origin of vesicles made of phospholipid-like material as the simulated formation of microspheres from lipid-like thermal polyamino acids indicates. Phospholipid vesicles are only very slightly permeable to charged ions, contrary to the modern cell membranes that are permeable to particular kinds of molecules such as proteinaceous or organic molecules of complex structure. The closed nature of the vesicles and the relative impermeability to protons would have satisfied the requirement for the development of a chemiosmotic energy-coupling system.[12]

An electrochemical gradient for some ionic species would have been maintained across the membrane of bilayer vesicles. Electrical excitability observed with proteinoid microspheres made of basic and acidic proteinoids attest to the maintenance of electrical-charge disequilibrium across the bilayer membrane.[13,14] This demonstration of charge separation with bilayer vesicles is indicative of an evolutionary precursor of modern chemiosmosis as the major energy-generating process in contemporary biological systems. Although the contemporary proton motive gradient attributed to electrogenic extrusion of protons through the membrane lets proton flow drive various active transport processes including formation of ATP from ADP and inorganic phosphate, the problem from the evolutionary perspective is

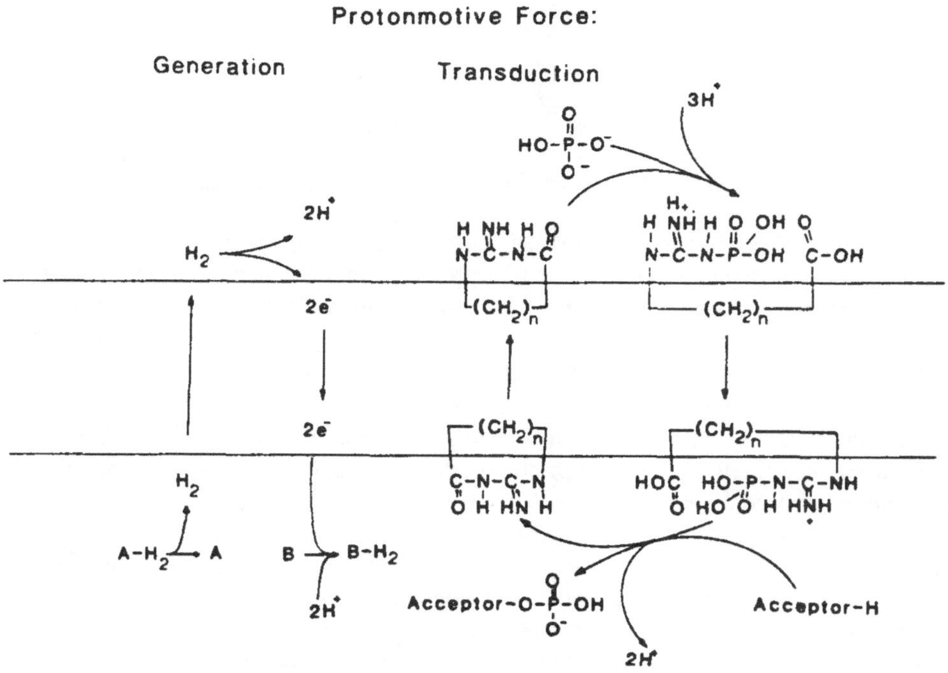

FIGURE 4. A hypothetical primitive mechanism for protonmotive force. High-energy phosphate bonds are generated through the bilayer membrane. (From Koch, A. L., *J. Mol. Evol.*, 21, 270, 1985. With permission.)

with how the electron flow resulting in a proton motive force developed. Charge separation across the closed bilayer membrane of proteinoid microspheres is in fact a manifestation of a type of motive force for charged particles. Basic and acidic proteinoids can supply charged particles that come to be separated spatially in aqueous milieu. As a matter of fact, at an early evolutionary stage such a charge separation supposedly would have been a form of the asymmetric distribution of an oxidant and reductant across the cellular membrane.[15] A candidate for such a model is shown in Figure 4.

Charge separation is one derivative characteristic of material accumulation in which material elements are electrically charged. In particular, material accumulation is a consequence of material flow equilibration proceeding in open material aggregates. This is because material flow equilibration is self-constraining in its operation and because material accumulation is a form of self-constraint through the loss of the kinetic degrees of freedom each material element individually had before joining in the correlated macroscopic motion in material aggregates. Material accumulation is thus intrinsic to the law of motion of the one-to-many mapping type supplemented by indefinite boundary conditions that have the capacity of accommodating a wide variety of possible material aggregates. Charge separation associated with material accumulation at open aggregates now functions as a motive force for charged particles, thus actuating various active transport processes.

As charge separation at the vesicle became large enough, active material accumulation would have been enhanced and the osmotic pressure inside would have increased above the pressure outside. An increase in osmotic pressure would have made it easier for charged molecules with hydrophobic properties to leave a hydrophilic environment and to form additional phospholipid-like materials.[16] This would have caused formation of a new inner bilayer inside the bilayer vesicle through invagination as shown in Figure 5. Spontaneous invaginations of bilayers inside the vesicle and their growth could have resulted in division of the parental vesicle.[18] Protocellular multiplication could have followed.

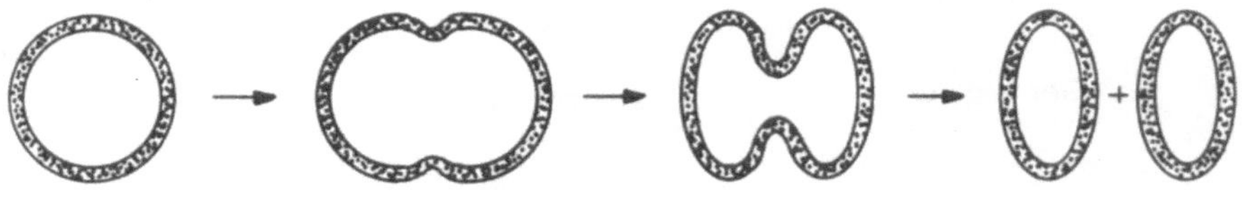

FIGURE 5. A schematic sequence of invagination at a protocell.

Experimental observations on simulated protobiogenesis suggest that protocellular multiplication was abundant. Budding of smaller protocells and the generation of endoparticles inside a parental protocell are common in experiments with proteinoid microspheres.[19] Protocellular multiplication is a primitive form of cellular reproduction. However, it is important to notice that such protoreproduction, which is not yet accompanied by molecular replication, differs from hypothetically perfect reproduction. The protoreproduction does not assume perfect fidelity between the successive generations, whereas perfect cellular reproduction does. The fidelity of cellular protoreproduction is determined through material flow equilibration proceeding in protocells. The equilibration is self-constraining and induces symmetry-breaking to dissolve the degeneracy latent in those protocells.

Later protoreproductive units are less degenerate internally as compared with the earlier ones due to the successive application of self-constraint originating in material flow equilibration. This implies that the earlier protoreproductive units are more inferior in their cellular reproductive fidelity in the sense that they are more degenerate and more inclusive internally compared with the later ones. Cellular reproductive fidelity increases as the self-constraint due to material flow equilibration accumulates with time, since the fidelity indicates the extent to which reproductive units are constrained generation after generation.

Protoreproduction proceeds through the form of open material aggregates. This mode of protoreproduction, due to its openness, sheds some light on how the replication of macromolecules sets in and proceeds inside open material aggregates as a particular instance of the successive feedback from product to production. The replication of polymers always assumes, within of course a limited degree of molecular reproductive fidelity, a cyclic reaction process in which the product of one stage acts as a template or an enzyme for the polymerization process of the next stage while the product at a certain later stage happens to coincide with that of an earlier stage. A new problem of the interplay between cellular and molecular reproductive fidelities thus arises.

Let us examine such a case: a reaction cycle having N steps of polymerization where the product of the N-th step is identical, within a limited fidelity, to the template or the enzyme for the polymerization of the first step.[20] If the inflow of energized monomers is not conditioned or influenced by the N-step reaction cycle in order to survive, then a probability p_i that the externally controlled inflow of monomers as a resource of the i-th step polymerization provides the right amount required for the polymerization over a unit time interval is always less than unity, $p_i < 1$. If the kinetic and internal energies of the incoming monomers are too large or too small for the required polymerization, they cannot participate in the reaction cycle. The interaction among energized monomers always induces in themselves a widespread energy distribution. Since our working hypothesis has been that the reaction cycle itself can neither condition nor influence the amount of energy which the inflow of monomers carries with itself at each reaction step, the probability that the inflow carries the right amount of energy for a given polymerization process is always less than unity. Furthermore, if the amount of incoming monomers of each kind exceeds or falls below the amount required, the proper functioning of the reaction cycle will be disturbed.

The probability that the reaction cycle maintains itself over a unit time interval is $p = p_1p_2 \cdots p_N$ (<1). Consequently, the probability that the reaction cycle survives over the time interval T, p^T, tends to vanish as T increases. Unless the inflow of monomers is properly regulated by the reaction cycle so that it survives, the cycle itself cannot survive for a long time. The reaction cycle which fails to influence the entering material resources cannot maintain the replication process after it exhausts the supplied resources, and it fails to survive even for a while if the necessary resources are not supplied. What is significant to an evolutionary precursor of the biological machinery of polynucleotide and polypeptide replication is its own capability of influencing the interaction with the surroundings. Successive operation of reaction cycles is crucial to the protobiological and biological replication of polynucleotide and polypeptide.

The successive functioning of the reaction cycle does require a process that adjusts the interaction with the surroundings and admits only those monomers that fulfill the condition for maintaining the reaction cycle. One such agent that can internally regulate the mode of interaction with the surroundings is an endogenously generated material aggregate open to material flow. This aggregate generates the boundary which separates it from the surroundings via the interaction with its surroundings. The necessary condition for the successive replication of macromolecules is the capacity of the underlying reaction cycle to influence the entering monomers. Open material aggregates have this capacity of varying their boundary conditions. Protocellular units satisfy the necessary condition for serving as a carrier of the surviving reaction cycle thanks to their endogenous capacity for variable boundary conditions, since the reaction cycle requires a temporally variable boundary condition depending upon its growth. Furthermore, open material aggregates can be protoreproductive by themselves, not to mention the fact that they could have potential capacity of housing reaction cycles in themselves.

Protoreproduction of open material aggregates is in fact a result of both the ease with which open material aggregates are formed under a suitable protobiological condition and the specificity due to the ongoing symmetry-breaking that keeps constraining those aggregates by dissolving the previous internal degeneracy. The metabolism and the maintenance of a specific function that processes matter is a necessary condition for protocellular reproduction.[21] Material flow equilibration of open material aggregates is certainly a physical embodiment of both the metabolism and the specificity associated with the process of self-constraining. Protoreproductive cellular units that can serve as a necessary condition for giving rise to the reaction cycle of replicating molecules could have preceded molecular replication.

Protocellular multiplication of bilayer vesicles that could be made of phospholipid-like[22,23] or lipid-like material generated protobiologically proceeds through charge separation and active transport processes.[24] Forces that underlie the active transport processes are responsible for structuring and multiplying protocells, because of the ubiquity of the mechanical force involved in forming any spatial structures. Unique to the force acting in the bilayer vesicles is the fact that it is involved in both charge separation and active transport processes. Charge separation, as with bilayer vesicles, powers both active transport processes and the subsequent charge separation, for the separation continues to survive. The force due to charge separation thus acts not only as an impetus for driving the derivative transport processes but also as the impetus for materializing that very force in a sequential manner. There is nothing that violates the second law of mechanics relating force to the resulting mechanical displacement. However, the electrochemical process yielding charge separation is special in that the propagation of charge separation in space does require a force to generate it and that the force originates in the preceding charge separation. Force acts as an impetus to change force.

We note that force is classified into two subclasses. One is for mechanical displacement to which the second law of mechanics addresses itself. The other is for the induction of force underlying the actualization of the third law of mechanics. Force for the induction of force acts as a boundary condition for materializing the force for mechanical displacement. Consequently, the force for the propagation of charge separation serves to structure the boundary condition for the force imputed to charge separation. Force for the induction of force is propagated in the medium as a major factor determining the endogenous boundary condition on charge separation taking place locally. The propagation velocity remains of course finite, being less than light velocity, because of the involvement of electrochemical reactions that proceed and propagate at a finite velocity. The finiteness of the propagation velocity of force for the induction of force leads to force equilibration that refers to the process materializing null-conservedness of forces proceeding at a finite velocity, because the third law of mechanics stating the null-conservation of forces of action and reaction must be fulfilled irrespective of the nature of forces. Both material flow equilibration and force equilibration occur in protocells.

Force for the induction of force, when operated through its temporal sequence, could increase its strength exponentially with time if the increment of induced force is proportional to the force for the induction in the process of force equilibration. Force for the induction of force due to charge separation, though it draws on electrochemical reactions as its driving energy source, participates in forming the boundary condition. It is this boundary condition which is responsible for structuring forces for the mechanical displacement of the medium and for chemical bindings such as covalent and noncovalent bindings. Invagination of bilayers inside the lipid-like bilayer vesicles and their multiplication attest to the action of force for the induction of force or, equivalently, force equilibration as an agent for generating and varying boundary conditions.

As not only a consequence of, but also a prerequisite to, charge separation, the accompanying force for the induction of force develops an internal tendency to accommodate material resources to enhance charge separation. This is because the induction of charge-separation force is proportional to the force that originates in the preceding charge separation. Both production and evolution of phospholipid-like material are thus aided by the very process of charge separation. A necessary outcome is that the materials with the lesser charge separation may be alternated with those with the greater charge separation. Protocellular multiplication can serve as a material carrier to develop lipid-like or phospholipid-like materials from the available resources.

IV. ENERGY TRANSDUCTION: ATP PROCESS

Molecular replication which could have originated in protocells is required to endogenously control not only the intake of monomeric resources to be polymerized but also the amount of energy intake for the polymerization. Lipid-like or phospholipid-like bilayer vesicles have their own energy-generating mechanism.[15]

Charge separation across the bilayer membrane generates a protonmotive force that can drive the active transport of phosphate and the formation of high energy phosphate bonds if some electrogenic carriers are available so that the functional end may be sufficiently lipid soluble to penetrate the bilayer.[25] These electrogenic carriers could have been quite conceivable protobiologically or even might have been created abiogenically, since examples of them include an analogue of demethylated creatimine with a sufficiently long alkyl chain terminated by the dehydration product of a carboxyl and a guanidine group. When inorganic phosphate and protons extruded electrogenically are available on the outer surface of the bilayer vesicle, the functional end could change its conformation after binding phosphate and excess protons so as to enable the hydrophilic and charged phosphate-containing portion of the molecule to penetrate in the hydrophobic region of the membrane. The presence of a protonmotive force could have let the molecular complex move from the outer to the inner surface, with the consequence that the reversed conformational change liberated the excess protons. The transport of phosphates would have been coupled with that of protons and could have resulted in the formation of high-energy phosphate bonds within the vesicle or the protocell. Since the high-energy bond, when liberated, can supply the energy required for polymerization of monomers, the charge-separated bilayer vesicles could have produced an endogenous environment in which molecular multiplication could proceed.

Charge separated bilayer vesicles made of lipid-like or phospholipid-like materials equipped with the abiotic electrogenic carriers are just one of the likely means for how the energy required for molecular replication came to be generated inside cells evolutionarily. Although one may think of many other likely scenarios for the origin and evolution of biological energy transduction, one lesson we can learn from the likely case of charge separated bilayer vesicles is that the overall energy transduction is influenced by the energy consumed in the cell.

The conservation of energy asserts that the energy inflow entering an arbitrary interface has to be equilibrated with the energy outflow leaving there. Energy equilibration is in action during the process of energy transduction. If the energy consumed inside the cell for molecular multiplication taking place there increases spontaneously, the formation of high energy phosphate bonds has to increase accordingly in order to compensate for the deficiency. Energy equilibration comes to enhance the turnover rate of the proton-copuled phosphate transport by the electrogenic carriers between the outer and the inner membrane surface so as to meet the increased energy requirement inside the cell. However, the variation in the energy to be transducted cannot be propagated in the medium instantaneously. The conformational change of the electrogenic carrier is involved in the energy transduction, and it takes time of an order that would be typical for electrochemical reactions. Energy equilibration is thus propagated at a finite velocity.

One consequence of energy equilibration associated with the conformational change in the electrogenic carrier is successively induced variations in the amount of charge separation per unit time. The force responsible for charge separation comes to be under the influence of its derivative force. The derivative can influence the generative force, but force equilibration must be observed since the third law of mechanics stating the null conservation of forces of action and reaction has to be satisfied in any case. If force equilibration proceeded at an infinite velocity, the distinction between the generative and the derivative forces would disappear because both could be defined in a simultaneous manner.

On the other hand, however, if it is rightly admitted that the propagation velocity of force equilibration is finite as demonstrated in the force involved in the conformational change of the electrogenic carrier of energy transduction at the bilayer vesicle, the generative force for charge separation cannot remain intact as an invincible boundary condition on the subsequent process. The derivative force controlling phosphorylation, for example, can influence the generative force for charge separation so as to fulfill the third law of mechanics *a posteriori* unless a prohibitive condition intervenes. The generative force for charge separation remains intact right at the moment that the derivative force controlling phosphorylation happens to vary. But the aftereffect of the variation in the derivative force will eventually reach the generator of force for charge separation. The combined effect is fulfillment of the third law of mechanics. The induced change in the generative force for charge separation can again induce a variation in the derivative force for phosphorylation. This sequence of repercussions can continue to survive indefinitely.

Both energy and force equilibrations at those charge-separated bilayer vesicles share the same conformational change of the electrogenic carrier as a vehicle of transferring the interaction change in the medium. If the force acting upon the medium were totally counterbalanced by the force of reaction from the displacement movement of the medium, the third law of mechanics would be fulfilled anywhere in the medium and force equilibration would proceed at an infinite velocity. Conversely, if the force acting upon the medium is counterbalanced not solely by the force of reaction from the induced movement of the medium but also by the variation in the force of action, force equilibration that proceeds in the medium at a finite velocity points just to the propagation of variations in the force of action induced by that very force a moment ago.

The energy to drive the force of action can also impart its portion to the propagation of force equilibration, in addition to imparting it to the displacement movement of the medium. Force can be transferred without inducing displacement of the medium during force equilibration. Force equilibration involves the generation of force in addition to the forced displacement of the medium in the light of the third law of mechanics that has to be fulfilled *a posteriori* in any case. The propagation of force in the form of force equilibration is a kind of energy flow in which both the energy flows entering and leaving an arbitrary interface are equilibrated. Energy equilibration is thus associated with force equilibration.

Force in the process of force equilibration can participate in the deformation of the medium as well as the transport of molecules of various types. The invagination of bilayer membranes leading to multiplication of the bilayer vesicle can be an instance of the deformation motion induced at the vesicle, since the charge separation occurring there can activate the process of force equilibration that could generate a force large enough to induce the deformation motion. Furthermore, the transport of molecules, though imputed to the motive force to carry molecules, is necessarily accompanied by molecular flow. Material flow equilibration that underlies the actualization of material flow continuity is involved in the molecular transport.

Molecular multiplication that could proceed inside the vesicle requires the right amounts of both molecular and energy intakes. If there is any imbalance between the two types of intake, the surviving molecular multiplication could be disturbed. The adjustment of molecular and energy intakes can be endogenously controlled by the corresponding forces that are under the influence of force equilibration. In fact, although force equilibration by itself is uncertain as to how the force acts on the molecular and the energy intakes, it functions, at the least, so as to compensate for the imbalance between the two types of intake, if any. This is because each of the molecular and the energy flows entering and leaving an arbitrary interface have to be equilibrated independently. Force equilibration is involved in both energy and material flow equilibrations in such a manner that both the force to transfer the high-energy phosphate bonds through the bilayer membrane and the force to take in monomers to be polymerized with the aid of the energy thus taken in are internally determined so as to eliminate any imbalance between the two coupled types of intake flow.

Molecular multiplication in retrospect turns out to be a consequence of the supply of the right amounts of energy and molecules. However, the lasting molecular multiplication necessitates the mechanism that guarantees the supply of the right amounts of energy and molecular resources. This mechanism must be one such that if the supply of any ingredient is in short or excess supply in relation to others, the endogenous compensation process to let all of the supplies be used is necessary. Otherwise an indefinite continuation of molecular multiplication could be hampered. Charge-separated bilayer vesicles made of lipid-like or phospholipid-like materials do in fact equip themselves with the capability of adjusting the supplies of necessary resources through varying the forces causing their transport by means of force equilibration. Molecular multiplication requires two conditions to be met. One is the resources, both energetic and materialistic. The other is the control mechanism that can really keep the molecular production process going. Charge-separated bilayer vesicles satisfy just these two requirements.

In particular, the triad of force, energy, and material flow equilibration operates at the vesicle in an interdependent manner. This implies that whatever the resources are, both of the material and the energy intakes are so regulated as to eliminate any imbalance between the two. Molecular multiplication can be a consequence of the force equilibration that eliminates imbalance between the two types of intake, since molecular multiplication is a form of materializing the triad of force, energy, and material flow equilibration by consuming all the intakes of both molecules and energy for the production process to follow without leaving any leftover. The coordinated process of the threesome of force, energy, and material flow equilibrations that is inherent in charge-separated bilayer vesicles has an endogenous capacity of maintaining continual molecular production in the vesicles. Molecular multiplication is a natural consequence of equilibrations in force, energy, and material flow.

Phosphorylation of electrogenic carriers in the charge-separated bilayer vesicles can be accompanied by a wide variety of high-energy-phosphate compounds. Among these, the adenosine triphosphate/adenosine diphosphate (ATP/ADP) coupling could have become dominant evolutionarily, in which the phosphoryl donor is ATP, not ADP. The present evolutionary dominance of ATP/ADP coupling raises the question of why is it that ATP would have established itself as the primary phosphorylating agent instead of, for instance, ADP. It can also be asked that why is the adenosine moiety, instead of guanosine or cytosine, involved.

It seems difficult to seek a definitive answer to the second question. However, it has been demonstrated experimentally that among the five pyrimidine and purine bases that are found in contemporary organisms, adenine is most easily synthesized in the laboratory that simulates prebiotic conditions.[26] Of course, the primordial adenosine moiety-binding site could have not been highly specific enough for adenosine.

The origin and evolution of the adenosine moiety could be traced back to a reasonable abundance of adenine nucleotides in the primordial settings. Evolutionary enhancement of the specificity between the adenosine moiety binding site and adenosine could have been materialized through the self-constraining process of equilibration in force, in energy, and in material flow. In fact, the evolutionary significance of adenine nucleotides is shown by other adenine derivatives such as Coenzyme A, flavin adenine dinucleotides, and puridine nucleotides.

The evolutionary dominance of ATP/ADP coupling requires an answer to the question of why ADP cannot serve as a phosphorylating agent on par with ATP, since both ATP and ADP can work as a phosphoryl donor equally effectively at least as far as thermodynamics is concerned. ADP/AMP(adenosine diphosphate/adenosine monophosphate) coupling that makes ADP a phosphoryl donor would have been equally likely as ATP/ADP coupling within the scheme of thermodynamics.

A clue to the distinction between ATP and ADP as a candidate for phosphoryl donor is the fact that in contemporary organisms, all enzyme-catalyzed reactions involving ATP are accompanied by a divalent cation such as Mg^{2+}. The true substrate is not ATP, but ATP-Mg complex.[27] The difference between ATP and ADP is in the chelating capability with Mg^{2+}. ATP is more effective than ADP as a chelating agent. The mechanistic advantage of chelation ATP-Mg^{2+} over uncomplexed ATP can overcompenstate its thermodynamic disadvantage so that the hydrolysis of ATP-Mg^{2+} is less favorable than that of uncomplexed ATP thermodynamically in terms of the free energy involved. The ADP/AMP coupling does not exhibit sufficient affinity for the divalent cation, thus the mechanistic structure of chelation that can be expected from ATP-Mg is not the case for ADP/AMP.

Both the primordial abundance of adenine nucleotides and the mechanistic structure of ATP in its chelation of divalent cation Mg^{2+} can provide a rationale for ATP becoming the primary phosphoryl donor for the transmembrane electrogenic carrier during evolution. Once phosphoryl donors for electrogenic carriers transferring high-energy phosphate bonds across the membrane and once charge-separated bilayer vesicles are available, the energy transduction and generation necessary for molecular multiplication can appear in the protocellular material aggregate.

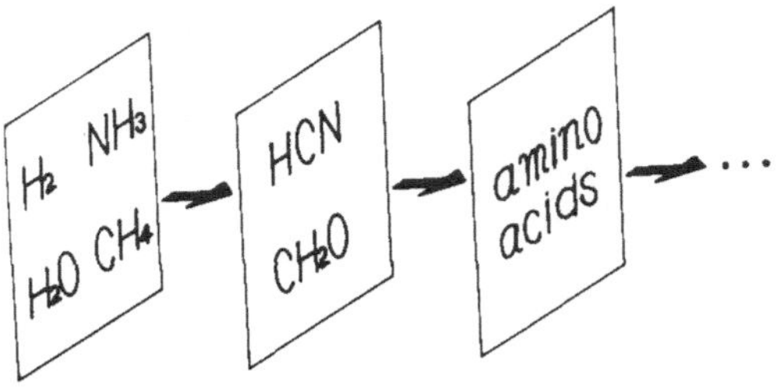

FIGURE 6. A schematic sequence of molecular synthesis. Prior products constitute the boundary condition for the production process to come.

V. MOLECULAR REPLICATION

Molecular multiplication of biological significance, when examined from an evolutionary perspective, first necessitates the synthesis of amino acids and nucleotides.[26] If the initial atmosphere of primitive Earth was strongly reducing as the Miller-Urey experiment suggests, the reducing gas mixture of CH_4, NH_3, and H_2O, further accompanied by H_2 or not, can result in a large yield of amino acids when the energy is supplied in the form of an electric discharge. One characteristic aspect of the products was that they were not a random mixture of organic compounds, but rather limited to a relatively small number of compounds compared with all of the conceivable hypothetical alternatives. The internal constraint limiting the actual products is due partly to the successive sequence of feedback from product to production giving rise to endogenous variations in the boundary condition. The amino acids were formed not directly from those component elements CH_4, NH_3, and H_2O, but from the solution reactions of intermediates such as hydrogen cyanide and aldehydes as schematically demonstrated in Figure 6.

When the boundary condition is transformed from the reducing gas mixture to the solution of hydrogen cyanide and aldehydes,[28] one condition is persistently fulfilled, namely, material flow equilibration. Only those material aggregates or reactants that have been through material flow equilibration can survive. The hydrogen cyanide and aldehydes are just examples of surviving material aggregates for a relatively long period, as exhibiting themselves as a consequence of material flow equilibration acting there.

The primordial syntheses of amino acids and their precursors would be possible even if the initial gas mixture[29] was mildly reducing like the CO, CO_2, N_2, and H_2 mixture, though the yields are considerably limited compared with the case of strongly reducing gas mixture. The amount of hydrogen required for this type of synthesis in the mildly reducing atmosphere is, however, considerably higher and it is not certain whether primitive Earth provided an atmosphere having such a high concentration of hydrogen. If the early atmosphere was not rich in hydrogen and not reducing, the reduction of carbon dioxide and nitrogen, which could be expected from volcanic outgassing, would have to be completed by reducing agents other than molecular hydrogen.[30] The reduction of carbon dioxide and nitrogen as a necessary step for providing raw materials for synthesizing molecules of biological significance would be possible if the right electron donors were present on primitive Earth. One possibility would be iron. In fact, the ferrous ion in the crust on the primitive Earth could have served for this purpose. Another possibility would be the photochemical reducing of N_2.

Experimental synthesis of amino acids from the simulated primitive atmosphere has not yet been conclusive. Nevertheless, all of the available circumstantial evidences suggest the likelihood of synthesizing amino acids and their precursors on primitive Earth. This consequence, however, is not limited to amino acids. Nitrogenous molecules such as purines and pyrimidines that constitute an indispensable part of nucleic acids along with both sugars and phosphorous, are produced from hydrogen cyanide that also serves as a precursor of amino acids. Adenine as a purine is readily formed from hydrogen cyanide oligomers. Likewise, cytosine as a pyrimidine is formed from cyanoacetylene and cyanates. Although these form only a partial list of simulated synthesis of molecules of biological significance, we can say at least that there has been enough experimental and geohistorical evidence suggesting that amino acids and nucleic acids, or their precursors could have been available on primitive Earth.

The next decisive step had to do with how to condense organic molecules produced under the primordial settings into large polymers. Functional integration of those organic molecules has to occur. Here, inorganic chemistry comes on the scene, with the observation that metal ions can participate as templates and catalysts for various reactions including the prototypes of photosynthesis and respiration. The metal ions are indispensable elements constituting the prototypes of metallic enzymes.

In this regard, the interplay between organic molecules and clays would have become significant on primitive Earth, as demonstrated in the clay-mediated oligomerization of biomonomers,[6] e.g., in the formation of polypeptides and oligonucleotides. The adsorption of organic molecules onto clays is a key factor in condensing these molecules for polymerization. It is thus conceivable that clays could have played a significant role as a condensing agent in evolution, but the time interval over which those clays properly functioned as a condensing agent would have been quite limited because of their lack of endogeneity in varying boundary conditions. Clays do not change their crystalline structure as a consequence of the polymerization of condensing organic molecules. In contrast, material evolution is a process of constantly varying the boundary condition for subsequent production to follow. The clay mechanism, if it actually operated, would have to have been substituted by others in the course of development.

Continual oligomerization and polymerization of amino acids and nucleotides necessarily imply the transition from simple organic molecules to an organized production system insofar as the production process survives. What is more, the organized production system turns out to yield its replica in time, of course, within a certain limited reproductive fidelity. A question thus arises with regard to which type of replication, either polypeptides or polynucleotides, could have emerged first.

If the first replicating polymer was protein, though within a limited reproductive fidelity, one would face the problem of how to construct the protein replication without it being accompanied by nucleic acids. In particular there is a demonstratable example exhibiting that the synthesis of certain peptide antibiotics is accomplished by a group of specific proteins without the help of nucleic acids. This is a form of catalytic cycle only of proteins, which would certainly serve as a good model suggesting the protein first hypothesis. But what is at stake about this catalytic cycle of proteins is that they are simple and not specific enough to catalyze the synthesis of more complex proteins. The problem is how the peptide catalytic cycle could evolve with time up to the point at which it can incorporate nucleic acids.

On the other hand, the first replicative polymers would have been polynucleotides if complementary replication of polynucleotides proceeded without being accompanied by proteinous enzymes. But, one problem with the nucleic-acid-first hypothesis is that the polymerization of nucleotides does necessitate, effectively, the removal of one water molecule as one nucleotide is attached to the growing polynucleotide, while the detachment of complementary strands for the synthesis of the next generation requires the addition of water. Unless the dry and wet conditions are superbly controlled in an alternating manner, the successive replication of polynucleotides would be greatly hampered.

One remedy to overcome the lack of controllability over the boundary condition, that is inherent to the nucleic-acid-first hypothesis, is to ask both proteins and nucleic acids to appear simultaneously as the members that construct their replication cycle. It has already been confirmed that there exists a preferential affinity of polylysine for adenine-thymine rich DNA and polyarginine for guanine-cytosine rich DNA.[31] This observation has been further confirmed by the presence of correlations between hydrophilic and hydrophobic properties of amino acids and nucleotides. And, again, the problem with the protein-nucleic-acid-simultaneity hypothesis is a lack of enough specificity between amino acids and nucleotides on the monomeric or, at most, the oligomeric level, inasmuch as the peptide catalytic cycle cannot be specific enough as to incorporate nucleic acids. The process of enhancing the specificity between proteins and nucleic acids would have to be invoked.

The three possibilities of protein-first, nucleic-acid-first, and protein-nucleic-acid simultaneity hypotheses, as they stand, are inadequate and cannot provide a definitive answer to the question of which molecules were the first replicative ones. In fact, there is one drawback that is common to all three of the hypotheses. The drawback is within the boundary condition on each type of replication process. What constitutes the boundary conditions for the peptide catalytic cycle not accompanied by nucleic acids, and what for successive replication of polynucleotides not accompanied by proteins, and what for the replication of protein-nucleic acid complexes? If the boundary conditions were totally exogenous in origin, each of the three hypotheses about the onset of replication systems would face intractable difficulties. But, if one pays due attention to the fact that boundary conditions are endogenous[32] and variable in their own right, as we have already observed, a totally different perspective arises.

The protein-first, nucleic acid-first, and protein-nucleic acid simultaneity hypotheses all have to be examined in the light of endogenous boundary conditions. As a matter of fact, a principal factor governing endogenous boundary conditions at the protobiological level is the emergence of charge-separated bilayer vesicles, that are confirmed both experimentally and theoretically. These vesicles certainly constitute spatial boundary conditions upon various chemical reactions proceeding there, while their forms vary with these reactions.

Endogenous boundary conditions as in the case of charge-separated bilayer vesicles admittedly influence what is being formed. However, the endogeneity in shaping what is forming is very specific, compared with the case of exogenous boundary conditions. The endogeneity of boundary conditions can let what looks like boundary conditions vary as responding to the reaction from what has been formed. When one examines the variable interaction between what is forming and what is being formed in terms of intervening material flow, the process of material flow equilibration comes on the scene.

Once the endogeneity of boundary conditions is guaranteed, the temporal enhancement of specificity turns out to be a natural consequence of equilibration, for material flow equilibration operates intrinsically in a self-constraining manner by eliminating in time those impossible interaction configurations that were latent within the preceding configurations, as referring to the condition of material flow continuity. The symmetry-breaking character inherent in the process of equilibration induces an enhancement of material accumulation in the protocellular material aggregates by constraining or freezing the kinetic degrees of freedom that each molecule has had. Unless restricted by prohibitive conditions, the endogeneity of boundary conditions can enhance with time the extent of specificity of material accumulation. The rationale is such that insofar as one understands the contemporary intricate coupling between proteins and nucleic acids is an indication of a very specific mode of material accumulation, the endogeneity of boundary conditions can take care of its materialization no matter how long it may have taken.

Insofar as it is admitted that the contemporary protein-nucleic acid complex is no more than a mode of material accumulation of extraordinary specificity, the necessary premise for gaining such an evolutionary specificity is the preservation of the condition that guarantees the endogeneity of boundary conditions. The problem of evolutionary enhancement of specificity now reduces to the problem of how to guarantee and why it is possible at all to preserve the endogeneity of boundary conditions.

Material flow equilibration as an agent for forming open material aggregates or what looks like spatial boundary conditions is an interaction process that fulfills the condition for material flow continuity. This process of fulfilling the condition for flow continuity is expressible equivalently as the process of seeking the final condition internally, although the term seeking may sound a bit anthropomorphic. It is, however, important to recognize that the internal seeking of the condition for material flow continuity is undoubtedly materialistic and non-anthropomorphic, because it is simply a physical characteristic of material flow continuity in the realm of the law of motion of the one-to-many mapping type.

Unless all of the conceivable endogenous changes in the interaction configuration are destined to fail in fulfilling the condition for material flow continuity or unless the law of motion is limited to or approximated by the type of one-to-one temporal mapping, the internal process of seeking and satisfying the condition for material flow continuity can survive and the endogeneity of boundary conditions can be guaranteed on this basis. In fact, contemporary organisms associated with complex protein-nucleic acid couplings attest to the endogeneity of boundary conditions. Terrestrial biological evolution provides at least one historical example telling us that the endogeneity of boundary conditions has been guaranteed over the period of roughly 4.5 billion years.[33]

The endogeneity of boundary conditions is, however, by no means evident by itself as demonstrated in the observed facts that an analogue of terrestrial organisms or their precursors has not been identified, for instance, on the lunar and the Martian surfaces. If the endogeneity of boundary conditions fails or if the law of motion acting there is well approximated by that of the one-to-one mapping type, the boundary conditions will have to be exogenous. A crucial question to be asked now is whether and how a given environment can guarantee the endogeneity of boundary conditions or, in short, why there are organisms on the planet Earth and not on Mars. This is the problem on the fitness of environment.

We do not yet know where to draw the demarcation line that distinguishes between the possible and the impossible in the environmental capacity of varying boundary conditions endogenously when a certain environment is given. The fact is that the fitness of the terrestrial environment is such that the endogeneity of boundary conditions has been guaranteed at least over the last 4.5 billion years. The endogeneity of boundary conditions survives to the extent that the environment is plastic enough to change some of its conditions in response to the reaction from what has been subordinate to the previous environment, e.g., an oxidization of the primitive reducing atmosphere of the Earth due to the onset of photosynthetic reaction. The law of motion of the one-to-one mapping type cannot influence the environment under which it has been set. One thus recognizes that the endogeneity of boundary conditions is attributed to the law of motion of the one-to-many mapping type because the latter refers to the dynamics of what looks like boundary conditions. The fitness or, equivalently, the plasticity of the environment is due to the environmental capacity of not exhausting its variability in response to the reactions from its interior. A main difference between the planet Earth on the one hand and the planets Mars and Venus on the other with respect to the fitness of environment is that both Mars and Venus would have exhausted their endogeneity of environmental boundary conditions, even if they once existed.

The endogeneity of boundary conditions shows that not all of the attributes for specifying those boundary conditions are reaction free and remain indifferent to what is being formed inside. Of course, the orbit of Earth around the Sun constitutes reaction-free and exogenous boundary conditions on the Earth, since the orbit is controlled with sufficient accuracy by the classical law of motion of the one-to-one mapping type. But, what does matter is that the terrestrial environment has maintained some of its attributes that develop as following the law of motion of the one-to-many mapping type while rejecting its approximation to that of one-to-one mapping. The emergence and formation of charge-separated bilayer vesicles are evidence of an attribute pointing to the endogeneity of boundary conditions, because the environmental condition differs before and after the appearance of these vesicles.

One of the firm bases upon which the process of molecular multiplication and replication can be examined thus turns out to be the endogeneity of boundary conditions. We know at least one historical example in which the endogeneity of boundary conditions has been fulfilled on the planet Earth. Further elucidation of the endogeneity condition will require an examination of why the endogeneity condition would have not been kept on the planets Mars and Venus and of how large or small the window of the environmental condition would be that can give rise to the endogeneity of boundary conditions. One would be asked to evaluate the extent to which the Earth's possible alternate orbits around the sun could still maintain the endogeneity of boundary conditions as the actual orbit does. In short, the question is whether the Earth could maintain the endogeneity of boundary conditions even if it took the orbit much nearer to Mars or to that of Venus when it was born about 4.5 billion years ago. A definitive answer remains to be seen. Nevertheless, the endogeneity of boundary conditions can serve as a promising working hypothesis for further examination of molecular multiplication and replication.

The endogeneity of boundary conditions will serve as a useful theoretical device, but it will also raise a new problem if it is examined experimentally. It is of course possible to design an experimental setup or boundary condition that can simulate each decisive step of evolution in nature. However, what is at stake here is how it would be possible to design an experiment that would allow the endogeneity of boundary conditions. Physical experimentation usually takes it for granted that uncontrollable factors are kept as minimal as possible or, equivalently, that the endogeneity of boundary conditions is lessened as much as possible. Experiments that can lead to molecular multiplication and replication, on the other hand, have to be such that the endogeneity of boundary conditions is kept as large as possible.

The endogeneity of boundary conditions can yield protocellular material aggregates in the primordial setting, either natural or simulated. If the component polymers are thermal proteins or proteinoids, the resulting charge-separated vesicles that are formed in the aqueous milieu will provide sites for further chemical reactions and syntheses. Proteinoid microspheres as a model of protocells make their own peptides if energized free amino acids are available in the presence of basic thermal polyamino acids or proteinoids. Catalytic polymerization of amino acids can proceed on or inside proteinoid microspheres. Internal multiplicative constraints due to material flow equilibration functioning in proteinoid microspheres induce both an enhancement of material accumulation and an increase in the specificity of molecular configuration there. A possible way of achieving this at an early stage of evolution is the transition from open-looped to closed-looped catalytic polymerization of amino acids. If the catalytic polymerization is open looped, no polyamino acid serving as a catalyst for the formation of a different polyamino acid can be made either by its own products or by their derivatives. However, if a certain catalyst is made either by its own products or by their derivatives, the polymerization constitutes a closed-looped cycle, thus making the overall polymerization autocatalytic.

Once an autocatalytic process sets in, material accumulation inside the protocellular material aggregates increases exponentially with time, at least initially. Autocatalytic polymerization provides a more constrained protocellular process because of the decrease in the number of degrees of kinetic freedom due to the increase in specific material accumulation. Material flow equilibration underlies the transition from acyclic to cyclic polymerizations. Moreover, protocells are intrinsically capable of producing molecular species that could further constrain the internal molecular configuration. After the transition to cyclic polymerization, both the growth rate and the structuring due to the decrease in the number of degrees of kinetic freedom increase. Those cyclic polymerizations with the smaller growth rates are phased out, thus increasing the specificity of molecular configuration in the remaining cyclic polymerizations with the greater growth rate.

Cyclic polymerization as a reaction cycle is a primitive form of molecular multiplication which is in turn dependent on the protoreproductive process at the cellular level that provides the endogeneity of boundary conditions. If molecular multiplication proceeded without the accompanying cellular process that guarantees the endogeneity condition, there would be no enhancement of structuring.

The onset of a reaction cycle inside a protocellular material aggregate is in accord with the symmetry-breaking due to the decrease in the number of degrees of kinetic freedom originating in material flow equilibration. The protoreproductive material aggregate enhances the fidelity of its cellular reproduction each generation due to the endogenous symmetry-breaking and the associated dissolution of previous degeneracy. In fact, the incorporation of a reaction cycle allowing replicating polymers is a consequence of enhancing the cellular reproductive fidelity associated with the protoreproductive open material aggregate. The molecular reproductive fidelity of replicating polymers is also enhanced through the endogenous operation of symmetry-breaking because of the successive dissolution of previous degeneracy and because of the successive enhancement of constraint.

The protoreproduction of open material aggregates is a result of both the ease with which open material aggregates are formed under a suitable protobiological condition and the specificity due to the ongoing symmetry-breaking accompanying the dissolution of internal degeneracy. The metabolism and the maintenance of a specific function that processes matter is a necessary condition for protocellular reproduction. Material flow equilibration of open material aggregates is certainly a physical embodiment of both the metabolism and the specificity associated with material flow. Protoreproductive cellular units as carriers that provide the endogeneity of boundary conditions precede the reaction cycle of replicating polymers.

One of the intriguing questions about the reaction cycle is how it evolves with time. Although it is possible to describe a form of arbitrary reaction cycle in terms of so-called reaction-kinetic dynamics, the crucial part of this question is how the description of the reaction cycle itself changes with time. In this regard, open material aggregates carrying the reaction cycle provide some answer. The boundary condition of the reaction cycle is specified by the underlying open material aggregate. What is more, the boundary condition itself changes through material flow equilibration. This implies that open material aggregates are prerequisites for both the emergence and the evolution of the reaction cycle. A more satisfactory description of the reaction cycle must be accompanied by the dynamics of an open material aggregate which serves as the carrier of the cycle. Open material aggregates constitute the environment under which the reaction cycle materializes. The change in the reaction cycle that has been initiated by the change in interaction between the supporting material aggregate and its exterior can again influence the supporting carrier. This is because the change in interaction is propagated in the medium at a finite velocity. The emergence and the evolution of the reaction cycle is thus inseparable from the open material aggregates that support it. The connecting thread is material flow equilibration.

Symmetry-breaking accompanied by the dissolution of previous degeneracy is also pivotal in organizing the reaction cycles appearing in an open material aggregate. The intricacy of interlocking reaction cycles is possible only when these are incorporated in an open material aggregate. The symmetry-breaking due to material flow equilibration can allow independent reaction cycles, if any, to transform into an interlocked larger cycle. The interlocking is a physical realization of the symmetry-breaking that deprives those independent cycles of the degeneracy in their mutual disposition inside the open material aggregate. The reaction cycles, once interlocked, cannot operate independently of each other and thus lose the previous degeneracy observed within their arbitrary mutual disposition.[34]

A typical reaction cycle is that of polypeptides and polynucleotides. Open material aggregates can initiate the process of polypeptide replication. The plausibility of an abiotic peptide synthesis cycle has already been suggested. Likewise, residue-by-residue replication of polynucleotide is another possibility for the reaction cycle.[35,36] An open material aggregate can serve as a carrier to give rise to the interlocked replication of polypeptides and polynucleotides with each other. An interlocking of small reaction cycles can give rise to a larger one. The emergence of such a hierarchical reaction cycle is a consequence of the functioning of the symmetry-breaking intrinsic to open material aggregates. An example of the hierarchical reaction cycle is Eigen's hypercycle between polypeptides and polynucleotides, modeled after contemporary coded protein synthesis.[37] An evolutionary precursor to this hypercycle is also proposed as a protohypercycle.[38] What is significant at this point is the potential capability of open material aggregates to interrelate both polypeptide and polynucleotide replications in a correlated cyclic manner.

The hierarchical reaction cycle also induces an enhancement of the molecular reproductive fidelity of polypeptides and polynucleotides when compared with those found in each of the uncorrelated reaction cycles. The enhancement comes from the dissolution of intrinsic degeneracy due to the symmetry-breaking as a result of forming a hierarchy of reaction cycles. The molecular reproductive fidelity of polymers thus observed differs from the cellular reproductive fidelity of open material aggregates. However, both originate in the same process of material flow equilibration inherent to open material aggregates. Protocellular processes are prerequisite to the onset and evolution of macromolecular multiplication. Enhancement of molecular reproductive fidelity presumes an increase in the cellular reproductive fidelity of open material aggregates.

VI. THE ORIGIN OF THE GENETIC CODE

Protoreproductive cellular units can eventually cause macromolecular multiplication in their interior if resources are available through the endogeneity of boundary conditions. The underlying process is symmetry-breaking originating in material flow equilibration of open material aggregates. An onset of macromolecular multiplication takes advantage of protoreproductive units, whenever and wherever available. Under these circumstances there are two possibilities of protoreproductive units, one for polypeptides and the other for polynucleotides. What laboratory experiments have suggested is that polypeptides synthesized thermally can form protoreproductive cellular units in an aqueous milieu. On the other hand, however, it still remains to be seen whether polynucleotides synthesized with or without templates could form protoreproductive cellular units in the absence of polypeptides. This does not, however, exclude the likelihood that polynucleotides could enter the protoreproductive cellular units made of polypeptides or phospholipids at a later evolutionary stage.

Thermal proteins originate in the self-sequencing of amino acids.[10] If the polymerization of amino acids proceeds only in the form of their self-sequencing, the protocellular reproduction of open material aggregates made of thermal proteins would be at most heterotrophic because of the fact that the formation of polypeptides takes place independent of cellular protoreproduction. The heterotrophic protoreproductive units enhance the material accumulation by means of the accretion of polypeptides made elsewhere. However, the heterotrophic protoreproduction is not the only process. The side-chain specificity of thermal proteins provides the possibility that the thermal proteins can work as catalysts for further polypeptide synthesis.[39] A reaction cycle of polypeptide synthesis can be expected to follow accordingly.

The protoreproductive cellular units incorporating a cyclic polypeptide synthesis in themselves is now autotrophic because they can synthesize their own polypeptides. Since the autotrophic reproductive units are autocatalytic in their polypeptide synthesis, the material accumulation is greatly enhanced compared with the case of nonautocatalytic synthesis as a natural consequence of material flow equilibration. The heterotrophic reproductive units

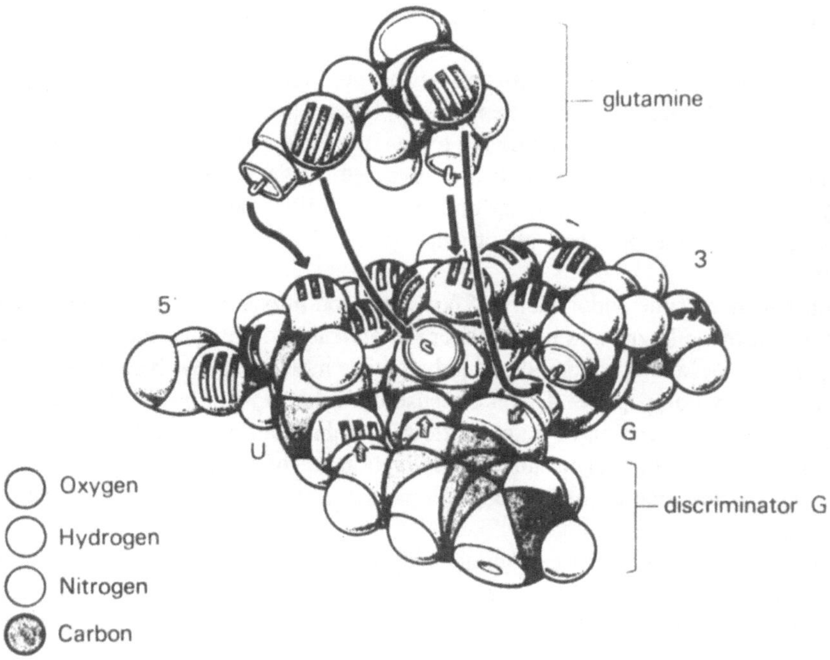

FIGURE 7. A key-lock relationship between the complex of four nucleotides, three anticodonic bases and one discriminator base, and the corresponding amino acid. The case of glutamine is presented. (From Shimizu, M., *J. Mol. Evol.*, 18, 297, 1982. With permission.)

can be alternated with the autotrophic ones because of the greater rate of material accumulation of the latter. The symmetry-breaking due to material flow equilibration acts so as to maintain consecutive accumulation of matter in open material aggregates. Consequently, only those aggregates with the greater accumulation rate survive. The alternation of the heterotrophic reproductive units with the autotrophic ones is just an example of the symmetry-breaking associated with open material aggregates.

Autotrophic protoreproduction certainly shows that the boundary conditions for resource intake of both molecules and energy become more endogenously internalized, compared with the heterotrophic ones. An autotrophic reaction cycle of polypeptides is, however, not specific enough to catalyze more complex polypeptides and polynucleotides, as we have already seen in the modern relics of such a simple reaction cycle as certain antibiotics called prions.[40,41] A necessary condition for evolutionary incorporation of polynucleotides into the simple reaction cycle of polypeptides is the presence of an affinity between amino acids or polyamino acids and polynucleotides.

An indication of affinity between amino acids and nucleotides is the attractive interaction between some amino acids and their anticodonic homopolynucleotides.[42] More specifically, a direct positive correlation of hydrophobicity between amino acids and their anticodonic nucleotides has been confirmed, except for two amino acids, tryptophan and tyrosine, that are the most hydrophobic of all while their anticodons are quite hydrophilic. However, as far as the remaining 18 different kinds of amino acids of biological significance are concerned, the presence of chemical affinity between these amino acids and their anticodonic nucleotides would suggest a likely incorporation of anticodonic nucleotides into the preexisting simple reaction cycle of only polypeptides.

One more evidence of the mutual affinity between amino acids and their anticodonic nucleotides comes from their specific molecular configurations in space that just fit in with each other like a key-lock relation (as demonstrated in Figure 7). This structural fitness in space of amino acids and their anticodonic nucleotides applies also to tryptophan and tyrosine.[43]

The conclusion we can derive from this is the presence of dynamic and static affinity between amino acids and their anticodonic nucleotides. Nevertheless, the mere presence of such a chemical affinity is not good enough to guarantee the origin and evolution of complex genetic apparatus. Another decisive factor in evolution is the endogeneity of boundary conditions. A crucial problem is how the endogeneity of boundary conditions, if guaranteed, could utilize in time the presence of intrinsic chemical affinity between amino acids and nucleotides.

One consequence of the endogeneity of boundary conditions is that polypeptide synthesis proceeding in protocellular units has an internal tendency to enhance the order of its autocatalysis, that is the average number of polypeptide molecules involved as a catalyst for making an arbitrary polypeptide. This is because the enhancement of material accumulation is in accord with the operation of material flow equilibration that constrains the number of degrees of kinetic freedom, and because the production rate of polypeptides to be accumulated is proportional to the amount of polypeptides present, raised to the power of the order of autocatalysis.

Intrinsic chemical affinity between amino acids and nucleotides suggests that if nucleotides or polynucleotides are available, the likelihood will be greatly enhanced that the polypeptide synthesis mechanism can incorporate polynucleotides into itself. The synthesis of polynucleotides from nucleotides can proceed with the aid of polypeptides working as catalysts. At the same time, the synthesis of polypeptides from amino acids is also under the influence of those polynucleotides synthesized by polypeptides because of their mutual affinity. One can thus visualize the emergence of a reaction cycle of polynucleotides synthesized by polypeptides if a polypeptide synthesized under the influence of a certain polynucleotide can catalyze the synthesis of the polynucleotide of the same type directly or indirectly with the aid of some of the derivative polypeptides. This will give an interdigitation of both polypeptide and polynucleotide synthesis cycles.

A simple polypeptide synthesis cycle without an interdigitation of polynucleotides indicates that one type of polypeptide can work as a catalyst for the synthesis of another type of polypeptide and that all the related polypeptides constitute a catalytic reaction cycle. In this case, the average number of polypeptides working as a catalyst for making an arbitrary polypeptide is unity. On the other hand, however, if the two reaction cycles of polypeptides and polynucleotides are interdigitated with each other, the average number of polypeptides working as a catalyst for making one polypeptide will become greater than unity. The order of autocatalysis will become greater than unity. This is because the synthesis of one polypeptide requires both a polypeptide working as a catalyst and the intervening polynucleotides, the latter of which in their own production also require polypeptides as their catalysts. The endogeneity of boundary conditions has been identified as enhancing the order of autocatalysis. Consequently, the interdigitated reaction cycle with the greater order of autocatalysis in polypeptide production dominates the other cycles with the smaller order of autocatalysis.

The interdigitated reaction cycle of polypeptides and polynucleotides requires more than one polypeptide molecule to produce or to instruct the making of one polypeptide. This exhibits a marked constrast compared with the simple cycle only of polypeptides alone, in which one polypeptide molecule instructs the making of one polypeptide. More specifically, the interdigitated reaction cycle requires more than one amino acid residue to direct one residue in the production process of polypeptides, whereas the simple cycle of polypeptides hypothetically requires only one amino acid residue to instruct one residue. One thus observes that, as far as information expressed in the anthropomorphic sense is concerned, the interdigitated reaction cycle can have greater information content per each amino acid residue compared with the similar information content associated with the simple cycle consisting only of polypeptides. The origin of the superficial increase in the information content is internal. The utilization of more than one amino acid residue in instructing one amino acid residue in polypeptide synthesis is a consequence of the endogenous symmetry-breaking originating in material flow equilibration or, equivalently, a consequence of the endogeneity of boundary conditions, which intrinsically favors open material aggregates with the greater accumulation rate.

The interdigitated reaction cycle of polypeptides and polynucleotides suggests that both polypeptides and polynucleotides can work as information carriers to instruct further polymeric synthesis. Then, the problem arises regarding which is informationally more inclusive, polypeptides or polynucleotides. One can estimate the average number of polypeptide molecules required for instructing the making of one polypeptide, since the contribution from polynucleotides can be equivalently translated in terms of polypeptides because of the cyclic nature of the polymeric synthesis. The same estimation can also apply to polynucleotides. If the average number of polypeptide molecules required for the instruction of making one polypeptide is greater than the average number of polypeptides required for instructing the making of one polynucleotide molecule, one can say that polypeptide is informationally more inclusive than polynucleotide because of the participation of a greater number of polypeptide molecules in the synthesis of one polymer. In this case, the polypeptide will turn out to be a more inclusive information carrier than the polynucleotide.

The original polymeric information translation starting from a simple reaction cycle of only polypeptides proceeded from polypeptide to polynucleotide, not vice versa. On the other hand, if the average number of polypeptides required for instructing the making of one polynucleotide molecule is greater than the similar number for making one polypeptide, the polynucleotide will now become informationally more inclusive. The polymeric information translation will proceed from polynucleotide to polypeptide.

A theoretically proposed protohypercycle which is autocatalytic to the order of $^4/_3$ in the polypeptide is an example of the coupled or interdigitated reaction cycle of polypeptide-polynucleotide complex with the translation from polypeptide to polynucleotide, in which the average number of polypeptides required for making one polypeptide is $^4/_3$ while the number for making one polynucleotide is one. Another theoretical model of the interdigitated reaction cycle with the opposite translation from polynucleotide to polypeptide is Eigen's hypercycle with its order of autocatalysis of $^3/_2$ in polypeptide, in which the average number of polypeptides required for making one polynucleotide is two while the number for making one polypeptide is $^3/_2$. As far as only the protohypercycle and the hypercycle are concerned, the latter hypercycle would be an evolutionary latecomer because of the greater order of autocatalysis in polypeptide and because of the reversed translation from polynucleotide to polypeptide. The nonanthropomorphic informational capacity of the instruction for making the hypercycle is already latent in the protohypercycles. Or equivalently, the evolutionary potential of the protohypercycles is within both the endogeneity of boundary conditions and the intrinsic chemical affinity between amino acids and nucleotides.

The evolutionary significance of the mechanisms involved within the genetic code requires scrutiny of the origins of polynucleotides[44] such as DNAs, messenger RNAs, ribosomal RNAs, and transfer RNAs and polypeptides[45] such as ribosomal proteins, tRNA-synthetases, and RNA-polymerases. One of the integrated views contributing to the solution of this problem is to draw on the endogeneity of boundary conditions and the intrinsic chemical affinity between amino acids and nucleotides.[46] The endogeneity of boundary conditions gives rise to material flow equilibration at open material aggregates that induces symmetry-breaking internally. The symmetry-breaking can enhance the interdigitation of various reaction cycles that could otherwise exist independent of each other.

Accordingly, if nucleotides or polynucleotides are available in addition to amino acids or polypeptides, the interdigitation of reaction cycles of both polypeptide and polynucleotide synthesis could be realized because of the very nature of the symmetry-breaking. One can thus visualize a materialization of the interdigitated reaction cycle of polypeptides and polynucleotides as a candidate for an evolutionary precursor of the genetic code apparatus. The transition from the direct interaction between some amino acids and their anticodonic homopolynucleotides that has been confirmed experimentally to the indirect interaction between amino acids and their anticodons is in fact an example of enhancing the interdigitation of further reaction cycles of incorporating the intervening adaptors.

VII. SUMMARY

Molecular replication that can enhance its reproductive fidelity with time necessitates the endogeneity of boundary conditions in the sense that the intake of resources is internally regulated just so as to respond to what is actually required in the molecular synthesis. Protocellular material aggregates that are open to material flow are a particular form of materializing the endogeneity of boundary conditions, in which what is forming is constantly varied as a response to the reaction from what is being formed. Material flow equilibration associated with the endogeneity of boundary conditions serves as an agent to enhance the fidelity of molecular replication taking place there due to its self-constraining capability. Although molecular replication not accompanied by the endogeneity of boundary conditions, such as layer-replication of montmorillonite and residue-by-residue replication of polynucleotide in the absence of enzymic proteins, is conceivable, its reproductive fidelity would necessarily be under interference of exogenous origin and thus lacks the internal capability of enhancing its own specificity with time.

Protocellular material aggregates such as charge-separated bilayer vesicles can also incorporate an energy transduction mechanism. High-energy phosphate bonds can participate in constructing and maintaining these protocellular aggregates by supplying the necessary energy for molecular synthesis taking place there. As molecular synthesis proceeds inside a protecell, it can multiply. Molecular replication can emerge and evolve in these multiplying protocells. Chemical reaction cycles as carriers of molecular replication can emerge in protocells. Interdigitation of various reaction cycles is a form of symmetry-breaking that points to the evolution of molecular replication. The genetic code apparatus is a very specific form of interdigitation of various reaction cycles of both polypeptides and polynucleotides. The intrinsic chemical affinity between amino acids and nucleotides underlies the presence of interdigitated reaction cycles of polypeptides and polynucleotides. The actual process that leads to and increases the complexity of the interdigitated reaction cycles composed of amino acids and nucleotides is based on the endogeneity of boundary conditions.

REFERENCES

1. **Weiss, A.,** Replication and evolution in inorganic systems, *Angew. Chem. Int. Ed. Engl.*, 20, 850, 1981.
2. **Lohrmann, R., Bridson, P. K., and Orgel, L. E.,** Efficient metal-ion catalyzed template-direct oligonucleotide synthesis, *Science*, 208, 1464, 1980.
3. **Lochak, G.,** Irreversibility in physics: reflections on the evolution of ideas in mechanics and the actual crisis in physics, *Found. Phys.*, 11, 593, 1981.
4. **Mills, D. R., Peterson, R. L., and Spiegelman, S.,** An extra-cellular Darwinian experiment with a self-duplicating nucleic acid molecule, *Proc. Natl. Acad. Sci. U.S.A.*, 58, 217, 1967.
5. **Biebricher, C. K.,** Darwinian selection of self-replicating RNA molecules, in *Evolutionary Biology*, Vol. 16, Hechet, M. K., Wallace, B., and Prance, C. T., Eds., Plenum Press, New York, 1983, 1.
6. **Paecht-Horowitz, M.,** The influence of some physico-chemical factors on self-organization in prebiotic systems, in *Molecular Evolution and Protobiology*, Matsuno, K., Dose, K., Harada, K., and Rohlfing, D. L., Eds., Plenum Press, New York, 1984, 189.

7. **Florkin, M.,** Ideas and experiments in the field of prebiological chemical, in *Comprehensive Biochemistry,* Vol. 29B, Florkin, M. and Stotz, E. H., Eds., Elsevier, Amsterdam, 1975, 231.
8. **Matsuno, K.,** Compartmentalization of self-reproducing machineries: multiplication of microsystems with self-instructing polymerization of amino acids, *Origins Life,* 10, 361, 1980.
9. **Wolken, J. J.,** Self-organizing molecular systems, in *Molecular Evolution and Protobiology,* Matsuno, K., Dose, K., Harada, K., and Rohlfing, D. L., Eds., Plenum Press, New York, 1984, 137.
10. **Fox, S. W.,** Protenoid experiments and evolutionary theory, in *Beyond Neo-Darwinism,* Ho, M. W. and Saunders, P. T., Eds., Academic Press, London, 1984, 15.
11. **Broda, E.,** *The Evolution of Bioenergetic Process,* Pergamon Press, Oxford, 1975.
12. **Mitchell, P.,** *Chemiosmotic Coupling and Energy Transduction,* Glynn Research Ltd., Cornwall, England, 1968.
13. **Przybylski, A. T., Stratten, W. P., Syren, R. M., and Fox, S. W.,** Membrane action, and oscillatory potentials in simulated protocells, *Naturwissenschaften,* 69, 561, 1982.
14. **Matsuno, K.,** Electrical excitability of proteinoid microspheres composed of basic and acidic proteinoids, *BioSystems,* 17, 11, 1984.
15. **Koch, A. L.,** Primeval cells: possible energy-generating and cell-division mechanisms, *J. Mol. Evol.,* 21, 270, 1985.
16. **Nickerson, K. W.,** A hypothesis on the role of pressure in the origin of life, *J. Theor. Biol.,* 110, 487, 1984.
17. **Douglas, J.,** Hypothetical entropy-driven mechanism for self-regulation of the size and division of primitive cells, suggesting the origin and nature of mesosomes, *J. Theor. Biol.,* 109, 475, 1984.
18. **Whatley, F. R.,** Problems in the rate of evolution in biological systems, *Phil. Trans. R. Soc. London Ser. A,* 303, 611, 1981.
19. **Fox, S. W., McCauley, R. J., and Wood, N.,** A model of primitive heterotrophic proliferation, *Comp. Biochem. Physiol.,* 20, 773, 1967.
20. **Matsuno, K.,** Protobiology: a theoretical synthesis, in *Molecular Evolution and Protobiology,* Matsuno, K., Dose, K., Harada, K., and Rohlfing, D. L., Eds., Plenum Press, New York, 1984, 433.
21. **Rosen, R.,** Relational biology and the origin of life, in *Molecular Evolution and Protobiology,* Matsuno, K., Dose, K., Harada, K., and Rohlfing, D. L., Eds., Plenum Press, New York, 1984, 421.
22. **Hargreaves, W. R. and Deamer, D. W.,** Origin and early evolution of bilayer membranes, in *Light Transducing Membranes,* Deamer, D. W., Ed., Academic Press, New York, 1978, 23.
23. **Rao, M., Eichberg, J., and Oro, J.,** Synthesis of phosphatidylcholine under possible primitive earth conditions, *J. Mol. Evol.,* 18, 196, 1982.
24. **Deamer, D. W. and Oro, J.,** Role of lipids in prebiotic structures, *BioSystems,* 12, 167, 1980.
25. **Odum, J. M. and Peck, H. D.,** Hydrogen cycling as a general mechanism for energy coupling in sulfate reducing bacteria, Desulfovibrio sp., *FEMS Micro. Lett.,* 12, 47, 1981.
26. **Miller, S. and Orgel, L. E.,** *The Origin of Life on the Earth,* Prentice-Hall, Englewood Cliffs, N.J., 1974.
27. **Wilson, J. E.,** Some thoughts on the evolutionary basis for the prominent role of ATP and ADP in cellular energy metabolism, *J. Theor. Biol.,* 111, 615, 1984.
28. **Ferris, J. P. and Hagan, W. J.,** HCN and chemical evolution: the possible role of cyano compounds in prebiotic synthesis, *Tetrahedron,* 40, 1093, 1984.
29. **Greenberg, J. M.,** Chemical evolution in space, *Origins Life,* 14, 25, 1984.
30. **Towe, K. M.,** Environmental condition surrounding the origin and early archean evolution of life: a hypothesis, *Precamb. Res.,* 16, 1, 1981.
31. **Hartman, H., Lawless, J. G., and Morrison, P., Eds.,** *Search for the Universal Ancestors,* National Aeronautics and Space Administration, Washington, D.C., 1985, 96.
32. **Whyte, L. L.,** *Internal Factors in Evolution,* George Braziller, New York, 1965.
33. **Henderson, L. J.,** *The Fitness of the Environment,* Macmillan, New York, 1913.
34. **Fox, R. F.,** The uroboros, in *Molecular Evolution and Protobiology,* Matsuno, K., Dose, K., Harada, K., and Rohlfing, D. L., Eds., Plenum Press, New York, 1984, 413.
35. **Orgel, L. E.,** Selection in vitro, *Proc. R. Soc. London Ser. B,* 205, 435, 1979.
36. **Kuhn, H. and Waser, J.,** Molecular self-organization and the origin of life, *Angew. Chem. Int. Ed. Engl.,* 20, 500, 1981.
37. **Eigen, M.,** Self-organization of matter and the evolution of biological macromolecules, *Naturwissenschaften,* 58, 465, 1971.
38. **Matsuno, K.,** Natural self-organization of polynucleotides and polypeptides in protobiogenesis: appearance of a protohypercycle, *BioSystems,* 15, 1, 1982.
39. **Root-Bernstein, R. S.,** Amino acid pairing, *J. Theor. Biol.,* 94, 885, 1982.
40. **Kimberlin, R. H.,** Scrapie agent: prions or virinos?, *Nature (London),* 297, 107, 1982.
41. **Prusiner, S. B.,** Novel proteinaceous infectious particles cause scrapie, *Science,* 216, 136, 1982.

42. **Lacy, J. C. and Mullins, D. W.,** The genetic anticode: the role of thermal proteinoids in development of hypothesis, in *Molecular Evolution and Protobiology,* Matsuno, K., Dose, K., Harada, K., and Rohlfing, D. L., Eds., Plenum Press, New York, 1984, 267.
43. **Shimizu, M.,** Molecular basis for the genetic code, *J. Mol. Evol.,* 18, 297, 1982.
44. **Follmann, H.,** Deoxyribo nucleotide synthesis and the emergence of DNA in molecular evolution, *Naturwissenschaften,* 69, 75, 1982.
45. **Kacser, H. and Beeby, R.,** Evolution of catalytic proteins, or on the origin of enzyme species by means of natural selection, *J. Mol. Evol.,* 20, 38, 1984.
46. **Root-Bernstein, R. S.,** On the origin of the genetic code, *J. Theor. Biol.,* 94, 895, 1982.

Chapter 8

EQUILIBRATION IN DEVELOPMENT

I. INTRODUCTION

The onset of protoreproduction naturally induces the process of development that controls how a reproductive unit grows in each generation. Development may seem teleological in its outlook, but is undoubtedly materialistic. The superficial teleology associated with the developmental process is reconciled with a law of motion accompanied by its own boundary condition, though with the reservation that the law of motion is intrinsically inseparable from the boundary condition. The law of motion underlying developmental process is of one-to-many temporal mapping. Actualization of the law of motion of the one-to-many mapping type is a process of equilibration such that development acts so as to constantly remove disequilibrium which, if it existed, would violate physical conservation laws. The process of removing disequilibrium never fails to subsequently give rise to the similar process of removing disequilibrium due to its inevitable spillover. The cleavage process is an indication of force equilibration in which the process materializing null-conservedness of forces both to act upon and to be acted upon proceeds, whereas proliferating cell divisions are due to material flow equilibration in the latter of which the process materializing flow continuity is in action.

Equilibration either in force or in material flow, or in energy renders each cell a decision-making unit during developmental process. The law of motion of the one-to-many mapping type witnesses decision making in materialistic process. Development is pluripotent in anticipation, but has constrained its extent of being potent in retrospect. Regional specification during development manifests the law of motion of the one-to-many mapping type, implying that an earlier stage of development is already latent with what will have come to actualize at a later stage. Only the lapse of time is required for eliminating other alternatives or possibilities from their realization. A young organism being latent with the later possibilities make itself function intrinsically as a generator of variations, though constantly constrained by the process of equilibration.

II. INTERNALLY REGULATIVE PRINCIPLE

The reproductive process takes place in time, and the specificity of the molecular configuration in a reproductive unit varies or is enhanced with time. Development as a process referring to the way an organism grows shows that regional specification[1] in the molecular configuration advances.[2] Regional specification may seem to be an objective finality in anticipation, but in retrospect is a consequence of regulation.[3,4] Objective finality and regulation go hand in hand in development. Nevertheless, development as a materialistic process has to dispense with those terms with an anthropomorphic connotation. Superficial objective finality and regulation have to be reexamined so as to eliminate the apparent anthropomorphic implication.

Development is materialistic in processing energy, force, and molecular flow in organisms that exist as molecular aggregates. A necessary logical consequence of this is that the conservation of energy, the null conservation of forces of action and reaction, along with the conservation of molecules, are fulfilled at any stage of the developmental process. These physical conservation laws assume that the processes materializing conservedness proceed at a finite velocity, for any interaction change for the conservation laws is propagated in the medium at a velocity not exceeding light velocity. The process materializing conservedness lets the corresponding conservation law be its objective finality, but the present objective finality is unfailingly materialistic and nonanthropomorphic for the same reason that physical conservation laws of energy, force, and material flow are purely materialistic.

The process materializing conservedness is internally teleological and the corresponding conservation law serves as an internal objective finality. One thus finds that the process materializing conservedness which is internally teleological in anticipation turns out to be internally regulative in retrospect by eliminating the impossible that would otherwise have failed to fulfill the underlying conservation law *a posteriori*. The process materilizing conservedness is based on the process of equilibration in such a way that any disequilibrium, which would otherwise violate the corresponding conservation law, is eliminated by interaction adjustments generated internally in the medium. These interaction adjustments are then followed by subsequent adjustments of a similar nature through the propagation of indefinite repercussions in the medium. Equilibrations in energy, in force, and in material flow are internally regulative during development. The process of equilibration is in fact a coordinative condition that underlies the transformation of the possible into the actual by constantly limiting the extent of the formerly possible. The process reduces equivalently to eliminating the impossible that had been latent previously.

Internal regulation presumes internal objective finality, and the mode of such internal functioning is admittedly materialistic and nonanthropomorphic. Furthermore, since both the materialistic regulation and objective finality are an attribute of the law of motion, they have to be examined in the light of the law of motion and its boundary conditions. What is unique to the process of equilibration is the inseparability of the law of motion and its boundary conditions. For conservation laws not only constitute the boundary conditions, but also intervene in the law of motion through the processes materializing conservedness.

The process of equilibration points to the intrinsic variability of organisms because the process materializing conservedness survives indefinitely through successive repercussions of similar equilibrations being propagated in the medium. The intrinsic variability implies that the material carrier of equilibration is infinitely degenerate with what will come to materialize, since the capacity of generating variations is not specific enough to identify in advance what will come. The infinite degeneracy tells us that the organism as a material carrier of equilibration serves as a source of future possibilities. What is actualized is the source of future possibilities envisioned therefrom, and the source is constantly constrained with time due to the continual elimination of the impossible latent with the very source. Equilibration as an internally regulative principle at work in development is coordinative as characterized by both its intrinsic variability and the ongoing elimination of the impossible in reference to physical conservation laws. The law of motion transforms what looks like boundary conditions into a more constrained form through its one-to-many temporal mapping.

Development operates in a continual fashion through successive reproduction, but development in protoreproduction definitely differs from that in biologically full-blown reproduction. The difference lies in the origin of and evolution of the attributes that characterize the contemporary developmental process.

The fate map as a diagrammatic director for identifying how each region of the contemporary embryo moves, changes, and transforms with time has been a useful theoretical tool in analyzing the developmental process. The fate map changes as development progresses.[5,6] The utility of the fate map is due to the absence of completely random mixing of cells in their division.[7] The mixing implies that a cell in a certain region in an embryo may become something that is quite different from the one expected from the cell in the same region of another embryo of the same type. If the mixing of cells were completely absent, the fate map would tell us how cell divisions proceed with unlimited precision. But, this is not the case in reality. The absence of completely random mixing of cells allows the fate map to have some predictive capability, of course within a limited precision.

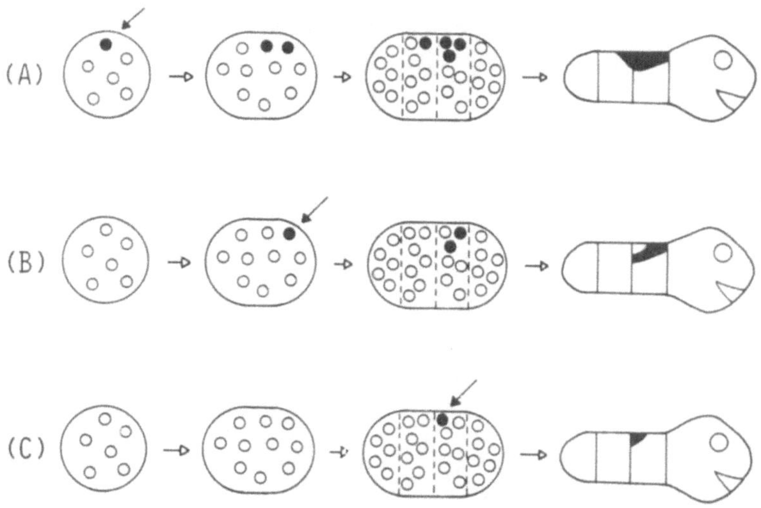

FIGURE 1. Fate map in embryogenesis and the dissolution of degeneracy during development. (A) A cell is labeled in the early embryo and its progeny span the boundary between two segments in the adult. (B) A cell is labeled before determination, but because of its position, its progeny do not span the segment boundary. (C) A cell is labeled after determination and its progeny are also confined to one segment. The present clonal analysis indicates that embryogenesis is a process of dissolving the internal degeneracy latent in the embryo: (From Slack, J. M. W., *From Egg to Embryo: Determinative Events in Early Development,* Cambridge University Press, Cambridge, 1983. With permission.)

One thus recognizes that the fate map is an expression of the law of motion of the one-to-many mapping type, because the one-to-many temporal mapping remains indefinite in anticipation, though definite in retrospect. The fate map in development should be referred to within the context of the law of motion of the one-to-many mapping type supplemented by intrinsically partially fixed boundary conditions.

The fate map in the presence of cell mixing to a certain extent in turn lets each region of tissue have the potency or potential capacity of making many different things of itself if put in suitable environments, as conceptually demonstrated in Figure 1. In this sense, the developmental process can be seen as an instance of restricting intrinsic potency. Imposition of such a restriction may seem to presume the presence of a materialistic determinant that identifies what particular terminal type the cell which inherits it will turn out to be. Cytoplasmic determinants have been proposed as candidates to fulfill this role.[8,9] A conceptual demonstration of cytoplasmic determinants is depicted in Figure 2. However, the materialistic underpinning of cytoplasmic determinants has not yet been identified. Although it is often assumed that those determinants would consist of messenger RNA, the association of mRNA that codes particular proteins specific to particular cells with cytoplasmic determinants seems unlikely, because the developmental process from the fertilized egg to terminally differentiated cells is loose in the sense of admitting many other alternative paths, in contrast to the fact that the association of mRNA with the corresponding proteins is extremely tight. Cytoplasmic determinants that can be deduced from experiments in developmental biology thus witness an attribute of the law of motion of the one-to-many mapping type acting there.

Regional specificiation of differentiating cells implies more than what materialistic determinants in isolation can do. Internal control of regional specification is a more of interactions that can induce a particular type of regionalization in the embryo, whereas those determinants in isolation lacks the capacity of controlling how the signal to induce such a regionalization is actualized and transmitted. The inductive signal assumes the competence of the embryo which increases its complexity as responding to it.[10,11] This type of inductive

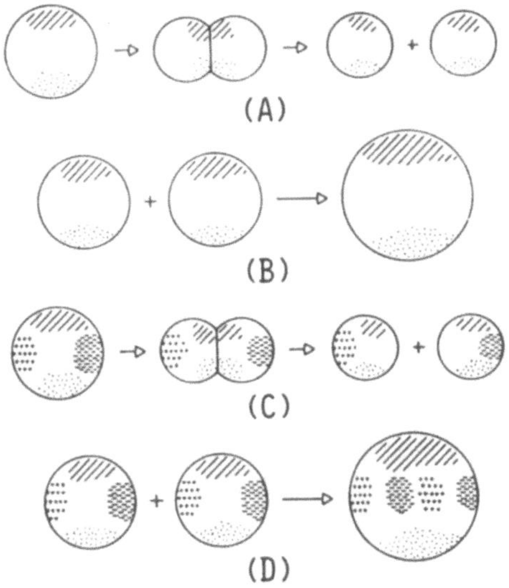

FIGURE 2. Hypothetical experiments indicating the functions of both embryonic regulation and cytoplasmic determinants. A and C show blastomere isolation experiments and B and D show the fusion of two eggs to form a giant embryo. In either case individuals with a normally proportioned anatomy can only arise if the determinants are symmetrically disposed along the plane of separation or joining. In A and B this is the case and in C and D it is not. Embryogenesis is an endogenous symmetry-breaking process. Symmetry-breakings of exogenous origin as indicated in C and D interrupt the normal process of endogenous symmetry-breaking. (From Slack, J. M. W., *From Egg to Embryo: Determinative Events in Early Development,* Cambridge University Press, Cambridge, 1983. With permission.)

interaction means that the competent tissue is faced with making choices. The range of choices available to the competent tissue is an expression of potency intrinsic to the embryo. The competence is successfully lost as the tissues continue to make choices.

The inductive signal and the competence of the embryo are adequate to depict how regional specification proceeds. At this point, the reality of inductive signals may be questioned on the ground that although the successive differentiation of cells requires some form of signals as an impetus to activate, the signals themselves cannot vary the developmental pathway to be actualized out of many possible alternatives. However, such a built-in, tight developmental pathway has repeatedly been refuted. The inductive signals and the developmental competence of the embryo cannot be completely separated from each other.

The real question is about what mechanisms are responsible for transmitting inductive signals and how the responses to them are actualized. One possibility for transmitting an inductive signal is the diffusive propagation of certain chemical substances from the inductive source to the responding area. Still, the problem with the diffusive propagation is that it is quite unlikely to allow chemical substances to cross the plasma membranes as rapidly as they actually do. One way, to solve this problem is to assume the presence of gap junctions, channels connecting one cell with another that let small molecules go through by passive diffusion.[12] If gap junctions are real, one might imagine that the embryo which is internally connected through channels at its early stage would be divided up in time so as to prevent each region from interfering with others by transmitting and responding to inductive signals. But, experiments suggest that such a division occurs much later than would be expected from the hypothesis of gap junctions. The timing of inductive interactions cannot be fully deduced on the basis of either the diffusive propagation of chemical substances carrying inductive signals or the gap junctions for the communication of those signals.[13]

The fate map, cytoplasmic determinants, and induction, however, represent some basic attributes of developmental process, though each of them has not yet acquired its sound materialistic underpinning. Nevertheless, a common denominator is the operation of the law of motion of the one-to-many mapping type. The transmission of inductive signals and the associated restriction under the competence of a region of tissue that responds to the former signals in fact show that there are many possibilities in anticipation, but only one in retrospect. The internally regulative capacity of organisms is within the law of motion of the one-to-many mapping type which is founded upon the process materializing conservedness. Equilibrations in energy, in force, and in material flow are just a materialistic expression of the law of motion of the one-to-many mapping type and accordingly serve as an internally regulative principle acting in the developmental process.

An early embryo is a materialistic carrier of an internally regulative principle that also works in its subsequent later development. This relationship between the early embryo and the internally regulative principle in development tells us the fact that a clear-cut separation between the regulative principle and its boundary condition is not feasible. The differentiating embryo does not allow a fixed boundary condition in its time-extensive morphological space. What looks like a varying spatial boundary condition during the developmental process just shows how the internally regulative principle works. It is thus illegitimate to claim the presence of something both definite and invariant that can control the developmental process.

The fate map of development, insofar as it remains legitimate in its own right, has to be variable and indefinite, since it changes with time in a manner such that the prediction of how cells will be further differentiated with unlimited precision is impossible as a matter of principle. If the prediction of development with unlimited precision were possible, a literal preformationism would be called for. But this is simply not the case in reality. The fate map itself is indefinite about what it would imply, apart from the common characteristic that the map would become more detailed though more rigid and less flexible with time. The fate map is not a static entity: its appearance shows the contrary. It represents a process that constantly constrains the extent of indefiniteness it carries. The fate map by no means constitutes a static boundary condition imposed on the developmental process. Conversely, the developmental process yields what the fate map tries to represent.

Cytoplasmic determinants, although their existence has been identified to a limited extent in experiments, also do not constitute a static boundary condition upon the developmental process. There is no such thing in development as something that would guarantee complete determination of how cells will be differentiated without leaving the slightest degree of indefiniteness. This implies that those cytoplasmic determinants in reality refer to the process that lessens its extent of being indefinite with time. Similarly, the transmission of inductive signals and the competence of a region of tissue that responds to them do not refer to a static boundary condition. Neither those inductive signals nor the responding competence can be identified in an exhaustive manner.

The developmental process shows that cell differentiation starting from the fertilization of an egg is subject to a time-extensive spatial boundary condition that is variable in a manner not controllable externally. It is thus noted that development is a dynamic process influencing the boundary condition as much as being subject to it. What remains indefinite with regard to development is the extent to which it influences the boundary condition that has conditioned the very developmental process.

The indefiniteness common to the fate map, cytoplasmic determinants, and induction in development is found in the intrinsic uncertainty in partitioning the dynamic process into the two parts of influencing and of being influenced by the very boundary condition, though at the least physical conservation laws have to be obeyed there. Unless some prohibitive means to prevent it from influencing its boundary conditions is available, the dynamic process can vary the boundary condition as a particular mode of its operation. The boundary condition as a unique form of interaction between what is forming and what is being formed is, in principle, internally variable through the process of interaction. Fixed boundary conditions that remain indifferent to what is going on are very anthropomorphic in their origin because the presence of some external agents to fix them is implicit. Of course, fixed boundary conditions are acceptable within the schemes of the law of motion of the one-to-one mapping type, but at the expense of the process of equilibration that actually accompanies the process materializing conservedness. The latter process is intrinsically of a one-to-many temporal mapping type.

The fate map, cytoplasmic determinants, and induction that are invoked in describing the developmental process are after all specific attributes referring to how the dynamic process influences and is influenced by its boundary condition. Unless fixed boundary conditions are available, any attribute referring to their dynamics must remain indefinite in its implication because of the lack of unmoved definite reference to rely upon. The source of the indefiniteness associated with the fate map, cytoplasmic determinants, and induction is sought in the arbitrariness in influencing and in being influenced by their boundary conditions, while fulfilling physical conservation laws.

The developmental process cannot be deciphered in terms available only to the law of motion of the one-to-one mapping type supplemented by completely fixed boundary conditions. Terms having an indefinite implication such as "potency" and "making choices" are inevitable and cannot be reduced to those applicable only to the law of motion of the one-to-one mapping type. The indefiniteness in separating influencing and being influenced by boundary conditions, which can be viewed as a principal attribute of the law of motion of the one-to-many mapping type, is fundamental to the developmental process. Development is a process of influencing the boundary condition upon further development. The present convoluted nature of development shows how internal regulation proceeds. Equilibrations in energy, in forces, and in material flow are just a minimum necessary condition on how development successively influences the boundary condition on further development. Conservation laws in energy, in forces, and in material flow must be observed in any case. The internally regulative principle as a form of equilibration addresses itself to how the boundary condition upon a dynamics varies with the dynamics.

Fate map, cytoplasmic determinants, and induction may sound teleological because each of them is an antecedent representation of what will materialize. Objective finality underlying any teleological development assumes a law of motion of its own to specify how the antecedent representation of it is transformed in time. If the law of motion is of a one-to-one mapping type, objective finality will completely be equated with the initial boundary condition applied to the law of motion. There is no indefiniteness about its objective finality and every detail of the future development is unambiguously predetermined with unlimited precision, because the one-to-one correspondence between the objective finality and the initial boundary condition is strictly guaranteed.

On the other hand, however, if the law of motion is of a one-to-many mapping type, it can always be asked why this particular one out of many possible alternatives is actually chosen, in spite of the fact that there does not exist any prohibitive condition to preclude all of the others. This situation might seem to suggest that the future could influence the present. But, the extent to which the future influences the present in the materialistic operation of the law of motion of the one-to-many mapping type is quite restricted compared to other general cases in which teleology may survive. The law of motion of the one-to-many mapping type lets the future influence the present only in the sense that those alternatives that will have turned out to violate physical conservation laws are eliminated. Accordingly, the purposiveness and objective finality that are inevitably associated with the fate map, cytoplasmic determinants, and induction in development are strictly materialistic in their origin and by no means anthropomorphic. The developmental process in biology is an instance of nonanthropomorphic teleology.

III. PERPETUAL DISEQUILIBRIUM AND EQUILIBRATION

The developmental process in three-dimensional morphological space exhibits its dynamic character represented by the law of motion of the one-to-many mapping type in the form of a field property, because these basic attributes of development such as the fate map, cytoplasmic determinants, and induction are actualized in a distributed manner in the morphological space. The field property also includes, among others, polarity,[14,15] indicating the regional difference in making choices during cell differentiation and an embryonic field that defines and specifies the spatial distribution of biochemical substances controlling regional specification of the differentiating embryo.[16,17] If one raises the question of how the field property of the developmental process comes to be observable, the interplay between the procedure of measurement and the law of motion of the one-to-many mapping type will have to be scrutinized.

The law of motion of the one-to-many mapping type necessitates the process of internal measurement that guarantees the fulfillment of physical conservation laws by internally identifying and realizing interaction adjustments needed for it. Internal identification and realization thus accompany internal measurement that exhibits itself as a materialistic and nonanthropomorphic consequence of fulfilling physical conservation laws under the legitimate condition that any interaction changes are propagated in the medium at a finite velocity. Internal realization always follows internal identification and both do not occur simultaneously. An aftereffect of internal realization of fulfilling physical conservation laws in a local region is to cause successive internal identification of the need for interaction adjustments for conservation in its neighborhood. This indefinite sequence of identification and realization is foreign to external measurement, which is necessarily anthropomorphic.

External measurement that presumes a specially contrived measurement apparatus prepared by external agents, on the other hand, takes it for granted that physical conservation laws in energy, in forces, and in material flow are infallible in what is being measured. For the external apparatus is by definition not supposed to disturb what is being measured irrespective of what is going on between the two. As far as external measurement is concerned, physical conservation laws are infallible. In particular, if the law of motion is of a one-to-one mapping type instead of a one-to-many, there is no room for internal measurement and only external measurement survives. For physical conservation laws are evident in the boundary condition and the function of the law of motion is simply to transform what has been conserved through one-to-one temporal mapping. The process preserving conservedness follows.

External measurement is, however, limited compared to internal measurement in that the influence from what is measuring to what is being measured is neglected. Nevertheless, measurement relevant to the law of motion of the one-to-one mapping type is intrinsically external, and the only problem is how to devise a measurement apparatus that would minimize the influence upon what to be measured.

Internal measurement is more inclusive than external measurement. This will become evident if the law of motion of the one-to-many mapping type is the case. External measurement is incompetent to identify the cause of the process materializing conservedness, because the cause itself would manifest an internal difference between *a priori* conservedness and *a posteriori* conservation laws. Internal measurement identifies the cause for and then implements the process materilizing conservedness in such a manner that the cause is not measurable externally.

If one calls the difference between internal identification of and internal realization of interaction an internal disequilibrium, internal measurement proceeds so as to eliminate such internal disequilibrium.[18] External measurement is not capable of identifying internal disequilibrium. If a quantity that is not measurable externally is introduced, one might think that it would undermine physical theories of materialistic processes. But this charge is superficial.

The reason is that external measurement is very limited in the sense that it is impossible to evaluate the extent to which the measurement apparatus disturbs what is being measured, especially in the process materializing conservedness between what is measuring and what is being measured. One might ask that another measurement apparatus measures the mutual interference, but infinite regression of apparatus would be inevitable for its completion.

Internal disequilibrium as a state attribute of internal measurement is not measurable externally. If one pays attention only to externally measurable quantities, the impetus causing the process materializing conservedness cannot be invoked and the process could be seen as the one transforming what has already been conserved. Externally observable is the process preserving conservedness. Hence, if conservation laws manifest themselves only in the form of transforming what has been conserved, the initial supposition of dealing with only externally measurable quantities would be verifiable. The real problem, however, is that there is no strict condition prohibiting quantities other than those measurable only externally, because external measurement constitutes only a limited class of measurement in general.

Internal identification preceding internal realization concerns itself with internal disequilibrium, though the state of internal disequilibrium is not measurable externally. The internal state induces the process of equilibration to eliminate such a disequilibrium, otherwise one could measure a violation of conservation laws externally. We are thus lead to observe that there are internal agents that can recognize internal disequilibrium and act so as to remove it accordingly. The presence of such internal agents is a direct consequence of physical conservation laws that can be measured as such externally, especially within the scheme of the law of motion of the one-to-many mapping type.

Internal agents for identifying internal disequilibrium definitely differ from anthropomorphic agents for doing measurement only externally. The difference lies in the fact that those internal agents are destined to influence what has previously influenced them in a successive manner indefinitely so as to fulfill physical conservation laws, whereas the anthropomorphic agent maintains intact the conservedness of physical quantities on the part of what is being measured. Such an anthropomorphic measuring agent is exceptional in the realm of exclusively materialistic processes. If the anthropomorphic agent requires the state of preserving physical conservation laws at any moment, this would dispense with internal disequilibrium that internal agents can identify as a cause for the process materializing conservedness. The origin of discrepancy is within the assumption that the state of preserving physical conservation laws be identifiable at any moment externally.

The law of motion of the one-to-many mapping type lets physical conservation laws externally measure only as a form of an end result of the process materializing conservedness. In contrast, internal agents can cope with both the state of internal disequilibrium and the process materializing conservedness identified as the process of equilibration. Internal disequilibrium at one place serves as an impetus for equilibration there, which subsequently causes internal disequilibria in the neighborhood.

Describing the field of internal disequilibrium in the morphological space may not seem self-consistent because internal disequilibrium is relevant to internal agents capable of internal identification and realization, and not to the observer who tries to describe it externally. But, insofar as one understands that physical conservation laws concern themselves only with the process materializing conservedness instead of the state of being conserved at every moment, it will become possible to assimilate the external agent with the internal one that is capable of identifying internal disequilibrium. The field description of internal disequilibrium will become possible only in the assimilation of the external measuring agent with the internal ones, with the reservation that the preservation of physical conservation laws is not invoked as the state attribute definable at every moment. Of course, such an assimilation and the emphasis on process at the expense of state are relevant only to the law of motion of the one-to-many mapping type and not to that of the one-to-one mapping type in the latter of which conservation laws apply to both process and state. The difference between the laws of motion of the one-to-many mapping type and of the one-to-one is qualitative, especially with regard to the implementation of physical conservation laws. The fact that the process instead of the state carries physical conservation laws in the scheme of the law of motion of the one-to-many mapping type in turn allows the field of internal disequilibrium as an expression of internal state to be internally capable of varying itself so as to fulfill those conservation laws.

The developmental process in biology is an instance of the law of motion of the one-to-many mapping type as confirmed by various experiments. What is more, the law of motion has to be of the one-to-many mapping type in recognition of the fact that any interaction change is propagated in the medium at a finite velocity not exceeding light velocity. A practical problem is whether the interaction change for fulfilling physical conservation laws is propagated in the medium at a velocity much greater than other velocities of interest. Developmental process shows that the propagation velocity of equilibration cannot be set at a value much greater than other velocities of interest such as those of typical molecular moment. This is because the fate map, for instance, changes with development like the field of internal disequilibrium does.

The field of internal disequilibrium is common to the operation of the law of motion of the one-to-many mapping type. In view of the fact that the field of internal disequilibrium underlies what one calls the fate map, cytoplasmic determinants, induction, polarity, and the embryonic field,[19] one can recognize that each of these developmental attributes is an expression of both the state of internal disequilibrium and the associated process of equilibration. Physical conservation laws are recognized only in the form of the process materializing conservedness. The state of being conserved is incompatible with what is referred to as developmental variability, since the process for being conserved generates such variability.

The field of internal disequilibrium is also viewed as the field of internal agents for equilibration. The root of these internal agents of identification and realization is within the law of motion of the one-to-many mapping type. There is nothing similar to Maxwell's demon, whose introduction is extremely anthropomorphic. The materialistic underpinning of internal agents is immanent. The only problem is whether the presence of such internal agents is duly appreciated as it should be. If the law of motion of the one-to-mapping type is well approximated by that of the one-to-one mapping type, the elimination of those internal agents would forcibly be called for. In fact, the law of motion of the one-to-one mapping type supplemented by completely fixed boundary conditions serves as a good approximation in many branches of physics, though complete control of boundary conditions is abandoned in many cases because of technical difficulties. The developmental process, on the other hand, persistently defies the approximation of the law of motion by that of the one-to-one mapping type. An appreciation of internal agents for identification and realization is simply an outcome of abandoning the law of motion of the one-to-one mapping type. Physical conservation laws in the making, not in the product, are responsible for giving rise to such internal agents.

A principal character of internal agents is to make choices so as to fulfill physical conservation laws *a posteriori*. Although making choices may sound anthropomorphic, this capacity inherent to internal agents is strictly materialistic in the sense that the law of motion of the one-to-many mapping type takes the capacity of making choices for granted from the very beginning. The omnipresence of internal agents is applicable to both protobiologically primitive and biologically full-blown developmental processes. The only difference is on the materialistic carrier of internal agents that maintains internal coherence as an organized unit while choices are being made. An internal agent identifies the need for and then implements the process of equilibration. Although internal identification assumes that influence is being propagated from elsewhere, internal realization at the internal agent as embodied in the process of equilibration turns out to influence other agents nearby. This mutual influencing among internal agents can bring about a coherent association of these internal agents. Internal agents become more interdependent in the course of constraining their former potency. The field of internal disequilibrium varies with the succession of equilibration at internal agents, but never disappears altogether.

The fate map of an embryo at its early developmental stage is intrinsically more indefinite compared with the similar map envisioned at a later stage. This does not imply that the fate map is no more than an explanatory device lacking a materialistic basis. Quite to the contrary, the fate map is an expression of the field of internal disequilibrium that is intrinsically degenerate with internal disequilibria yet to come. The process of equilibration constantly dissolves the degeneracy. The fate map is always in the process of lessening its extent of indefiniteness. The fate map thus must be understood within the framework of the law of motion of the one-to-many mapping type. If one examines the fate map that becomes more detailed and less plastic with time within the context of the law of motion of the one-to-one mapping type supplemented by completely fixed boundary conditions, the map might seem to point to an intervention from some vitalistic agents. The inappropriateness of this interpretation results from taking the law of motion of the one-to-one mapping type for granted. What seems vitalistic in developmental processes, even including such as entelechy of Driesch, can gain a materialistic underpinning within the scheme of the law of motion of the one-to-many mapping type supplemented by intrinsically, partially fixed boundary conditions.

An intrinsic limitation of the law of motion of the one-to-many mapping type is the impossibility of identifying both the state and the process of an entity in motion in a simultaneous manner. Physical conservation laws are legitimate only in process. This process dominance at the expense of state is common to any attribute of the developmental process. Cytoplasmic determinants, if legitimate as they should be, have to refer to the process in time instead of the state definable at every moment. The state description is necessarily incomplete, and the process description has to be invoked. The field of internal disequilibrium as a description of internal state cannot remain legitimate unless the process of equilibration accompanies it. Cytoplasmic determinants are not entities identifiable in isolation, but an attribute of the field of internal disequilibrium accompanied by the process of equilibration.

The field of internal disequilibrium is more than an expression of the distribution of biochemical substances in a spatially limited region. It also incorporates the potency of varying the field itself. The field of internal disequilibrium remains ambiguous and indefinite because of the degeneracy with its future developments. Nevertheless, this ambiguity is inevitable in view of the fact that the law of motion of the one-to-many mapping type cannot determine its future development unambiguously in advance. For instance, one can think of a particular embryonic field such as the limb field that will form a limb from the mesodermal mantle roughly at the time of gastrulation. This might suggest that it would not be appropriate to refer to a limb field before the limb really emerges. But, if one understands the embryonic field in the sense of a field of internal disequilibrium, the field latent with indefinite future possibilities will turn out to be inevitable. Although it is and has to be indefinite and ambiguous as to its future development, the embryonic field is unfailingly materialistic in its origin. The field of internal disequilibrium perpetuates disequilibrium in spite of the involvement of successive equilibration. This perpetual disequilibrium serves as an impetus to vary the field of internal disequilibrium. The embryonic field in the state of perpetual disequilibrium, in principle, defies the description of the state with unlimited precision. Indefiniteness is inevitable when one tries to describe the process subject to the law of motion of the one-to-many mapping type.

IV. GENOTYPE-PHENOTYPE INTERDEPENDENCE

Forces to structure forms in the morphological space are an attribute of the field of internal disequilibrium in forces, and the process materializing null-conservedness of forces of action and reaction is involved there in the form of force equilibration. Similarly, any organism as an aggregate open to material flow is subject to material flow equilibration as a dynamic attribute of the field of internal disequilibrium in material flow, which is internally regulative in the form of the process materializing flow continuity. The process of equilibration inherent to the field of internal disequilibrium, however, does not stand alone. It is admittedly under the influence of genes and genomes. DNA constitutes a boundary condition upon how developmental process proceeds.[20] Still, the developmental process is not only influenced by but also influences its boundary condition[21] because of the underlying law of motion of the one-to-many mapping type.

The genes reshaped in an organism later in its development can differ from the earlier ones in that the boundary condition upon reshaping genes is variable with development. The later genes are only a part of the possible ones latent within the same organism at an earlier stage. This is nothing more than an expression of the law of motion of the one-to-many mapping type. When we say that genes are latent with the possible genes that are to appear at a later stage of development, it simply means that genes are variable *a priori* due to the ongoing process of equilibration operating in the carrying organism whereas the genes reshaped as being subject to equilibration are only those satisfying physical conservation laws *a posteriori*. *A priori* variability of genes in organisms manifests itself, for instance, in intracellular exchanges of gene segments identified as transposable genic units or jumping genes.[22] Being different from the classical mechanism of mutation that includes chromosomic disjunctions and crossovers, the intracellular variability shows that genes reshape themselves even during the very process of somatic cell divisions. Genes in the possible originate in an *a priori* variability due to the process of equilibration at the level of organism, and genes in the actual refer to an *a posteriori* constraint imposed due to the equilibration. Both dynamics to be influenced by and to influence boundary conditions are involved.

Phenotype is an expression of genotype in the morphological space. But, phenotype is not strictly subordinate to genotype. Genes can be variable during development in the sense that the phenotype in the products constitutes the boundary condition upon genes in the making under the common constraint of equilibration. Genotype and phenotype are interdependent in their mutual causation through equilibration. Both evolution and development in the form of morphogenesis share the common law of motion of the one-to-many mapping type supplemented by intrinsically partially fixed boundary conditions.

The law of motion of the one-to-many mapping type is convoluted in its implication. It maintains intrinsic variability, defying its exhaustive identification in advance, while being subject to variations thus generated. This characteristic applies to both development and evolution. The origin of variation cannot be separated from the process of transforming evolving organisms, which is contrary to the neo-Darwinian dichotomy between variation and transformation. If variation can be separated from the transformation process, it might be conceivable that the transformation can be approximated by that of one-to-one temporal mapping by letting any deviations from it be variations. In fact, such an approximation is, in a practical sense, unassailable because variations are defined as a form of deviation from the transformation of one-to-one temporal mapping. The validity of such a transformation cannot be questioned as a matter of principle because other types of transformation are barred from consideration. However, the situation will be drastically changed if it is recognized that the law of motion is exclusively of the one-to-many mapping type.

An appreciation of non-one-to-one temporal mapping in development has been fostered by paying attention to such terms as the fate map, cytoplasmic determinants, induction, and polarity which points to the capacity for variability while being inevitably indefinite in their implication. The same type of appreciation is also expected for evolution. This, however, does not imply that the neo-Darwinian dichotomy between variation and transformation is inappropriate where the generation of variation is supposed to be independent of what is being transformed in time. The neo-Darwinian scheme does function properly as an approximation of reality to any arbitrary accuracy one can think of,[23] since it would be feasible to engineer a sophisticated transformation of one-to-one temporal mapping that almost simulates the real transformation.

Quantitatively speaking, there is no argument about the neo-Darwinian dichotomy. The real problem is, however, whether the qualitative nature of the temporal transformation operating in reality in evolution is of the one-to-one mapping type or of the one-to-many. Once this question is stated, the way out of it is uncompromising in admitting that variation and transformation are inseparable in evolution, contrary to the neo-Darwinian claim. This does not, of course, exclude the occurrence of variation whose origin is separated from the transformation under consideration. What is at stake, on the other hand, is the recognition of the fact that transformation cannot proceed without generating variations. Variability in evolution shares the same functional basis with variability in development.

V. DEVELOPMENTAL SYMMETRY-BREAKING

The field of internal disequilibrium in development that induces the process of equilibration as an elementary process of evolution is inherent to the law of motion of the one-to-many mapping type. One characteristic mode of operation of equilibration in the developmental process is symmetry-breaking due to sequential multiplication of constraints originating in the process materilizing conservedness applied to differentiating cells. A later embryo is more particular and less symmetrical compared with the earlier one because of the dissolution of the degeneracy preserving the higher symmetry. An example is the onset of regional differentiation in a uniform initial state, like the formation of the dorsoventral axis in the amphibian embryo.[24]

One can interpret the occurrence of symmetry-breaking within the framework of the law of motion of the one-to-one mapping type by seeking an external interference of symmetry-breaking perturbations upon its boundary condition. Nevertheless, this does not imply that symmetry-breaking in development is a consequence of applying symmetry-breaking boundary conditions to the symmetry-preserving law of motion. Development in reality does follow the symmetry-breaking law of motion, in spite of the fact that the symmetry-preserving law of motion can simulate a real developmental process to an arbitrary degree of accuracy when supplemented by suitable symmetry-breaking boundary conditions.[25,26] Even if a simulation of products is accomplished with arbitrary accuracy, this does not allow us to conclude that the simulation of the process could also have been done with the same accuracy. The validity and the appropriateness of process simulation must be examined in its own right. At stake is whether the law of motion acting in a developmental process is of the one-to-one mapping type or of the one-to-many. We have already noted that the developmental process is a subject matter obeying the law of motion of the one-to-many mapping type that persistently defies process-simulation following the one-to-one temporal mapping.

Expecting a set of well-defined equations of motion, such as reaction-diffusion equations, to describe a developmental process reduces to asking for a law of motion that is of the one-to-one mapping type, for the equations of motion unambiguously specify the forces of action that have necessarily presumed the accompanied forces of reaction, also unambiguously even if implicit in their expression. If the force of action is determined, this automatically implies the determination of the force of reaction because of the third law of mechanics stating the null conservation of the acting and reacting forces. Determination of the force of action at every moment which takes similar determination of the force of reaction for granted makes the process materializing null-conservedness of forces superfluous or allows it to be propagated at an infinite velocity. If the process materializing conservedness is set aside, one would face a formidable problem of how to deduce the origin of symmetry-breaking perturbations placed on the boundary condition that would induce the observed symmetry-breaking in development. The law of motion of the one-to-one mapping type cannot cope with the process of symmetry-breaking by its very definition. If the process of symmetry-breaking is examined seriously, the law of motion of the one-to-many mapping type has to be invoked.

VI. CLEAVAGE PROCESS

A significant process during early embryonic development is cell divisions, either accompanied by cell growth or not. If eggs are laid and develop in isolation, early development proceeds in such a manner that cell divisions take place without increasing dry weights: daughter cells are produced by dividing the mother cell in half. In most cases, those eggs carrying both embryonic and extra-embryonic regions let their embryos grow by dividing their cells while at the expense of the yolk mass in the latter extra-embryonic region.[1]

Cell divisions that are ubiquitous in early embryonic development are under the influence of the relevant fate map. The fate map finds its materialistic underpinning in the field of internal disequilibrium that induces the process materializing conservedness. The field of internal disequilibrium pointing to the difference between internal identification and realization underlying the actualization of physical conservation laws is conceivable at least in terms of energy, forces, and material flow. The associated internal processes of equilibrium in energy, in forces, and in material flow are interconnected in principle, since no physical conservation laws can be violated *a posteriori*. Nevertheless, one can ask which field of internal disequilibrium or which process of equilibration would be primarily significant in each phase of development. In particular, cell divisions in early embryonic development, which are our current concern, refer to those forces that structure forms in the morphological space.[27] The field of internal disequilibrium in forces is thus involved, in which the field internally induces the process materializing null-conservedness of forces of action and reaction. This observation is, however, by no means intended to undervalue the significance of equilibrations in energy and in material flow. The point of emphasis is on the apparent involvement of the field of internal disequilibrium in forces and of force equilibration in cell divisions.

A typical example of cell divisions is holoblastic cleavage formation of a single fertilized egg. Internal tensile force acts upon the cell.[28] One can define a set of closed contour lines of internal tensile force upon the cell surface such that the tangential line at each point along a closed contour line is in parallel with the internal tensile force there. Holoblastic cleavage formation accomplished by cell divisions of a single fertilized egg into two, four, eight, sixteen . . . proceeds along with variations in closed contour lines of internal tensile force both in shape and in strength as demonstrated in Figure 3. Among those closed contour lines comprising the cell surface, the maximum contour line of internal tensile force can be uniquely defined such that the internal tensile force along the line is locally maximum in strength in its neighborhood.

In order to see how force equilibration is involved in the holoblastic cleavage formation, let us examine a small line segment of the maximum contour line, pulled at both ends by the corresponding internal tensile forces. When the resultant force acting upon the line segment does not vanish, the segment will deform, i.e., either expand or contract, due to the second law of mechanics. However, the deformation motion depends upon whether the underlying law of motion is of the one-to-one mapping type or of the one-to-many.

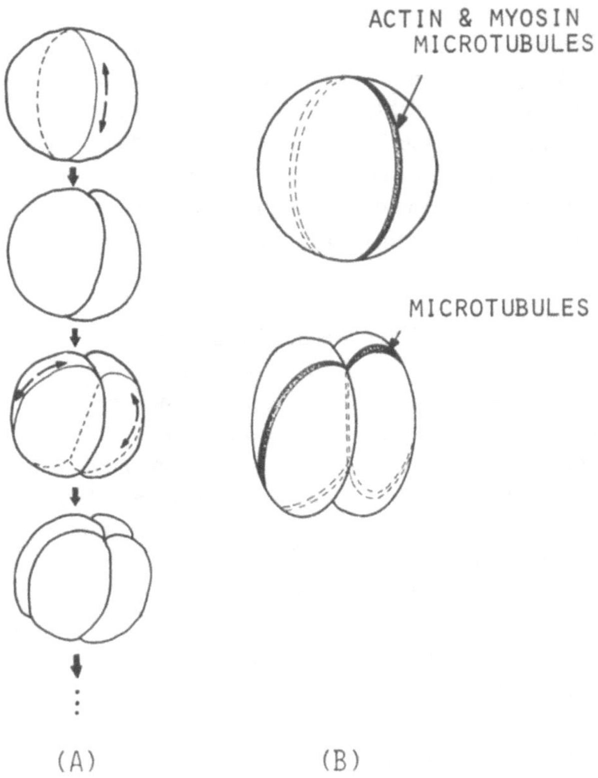

FIGURE 3. Holoblastic cleavage process. Maximum contour line of internal tensile force as depicted in A can increase its intensity with time through force equilibration, and accordingly a cleavage is formed. This process is successively alternated. The materialistic underpinning of force equilibration that can generate contractile force is microtubules of actin and myosin that come to to be concentrated along the equator, as depicted in B.

If the law of motion is of the one-to-one mapping type, the resultant force upon the segment acting as an impetus for inducing displacement, once fixed, will uniquely determine the displacement. The nonvanishing resultant force acting on the segment is counterbalanced solely by the force of reaction from the induced displacement. This is no more than a statement of the third law of mechanics. Force equilibration as the process materializing null-conservedness of forces of action and reaction is to proceed at an infinite velocity, for the unique relationship between the induced displacement and the force to act upon simultaneously determines the counterbalancing force to be acted upon.

On the other hand, if the law of motion is of a one-to-many mapping type as the process materializing conservedness in reality implies, the induced displacement of the line segment upon the cell surface would not be uniquely determined even if the nonvanishing resultant force serves as an impetus to induce such a displacement. The resultant internal tensile force to act upon the segment will be counterbalanced not only by the force to be acted upon from the induced displacement but also by inducing further changes in the internal tensile forces to act upon. Characteristic of the law of motion of the one-to-many mapping type is that forces that serve as an impetus to change forces themselves in a sequential and propagative manner so as to actualize the third law of mechanics. What is being propagated in the process of force equilibration at a finite velocity along the closed contour line is variations in the internal tensile force acting there. The field of internal disequilibrium in tensile forces on the cell surface is accompanied by the process of force equilibration, which recurrently serves as an impetus to induce the field of renewed internal disequilibrium. This sequence lasts indefinitely.

Force equilibration proceeding at a finite velocity lets the internal tensile force acting on the cell surface change its strength in a propagative manner for the sake of null-conservation of forces. If the increment of internal tensile force due to propagating force equilibration round the maximum contour line is positive, its strength would increase indefinitely by repeating force equilibration round the closed contour line. Its initial growth is exponential with time. The maximum contour line would be the one that maintains a positive increment of internal tensile force due to the force equilibration round the closed contour line, otherwise the one with a positive increment could alternately succeed the maximum contour line of internal tensile force.

The maximum contour line of internal tensile force, however, cannot increase its strength indefinitely. An increase in the internal tensile force can result in cleavage formation along the contour line. After cleavage formation, one of the remaining contour lines will become the maximum contour line, force equilibration along which can cause a further cleavage formation. By following the similar force equilibration, more complicated cleavages could be formed.

The materialistic underpinning of the maximum contour line of internal tensile force is assumed by microfilaments and microtubules that participate in form-shaping movements of cells and cytoplasm.[29] The microfilaments are made of actin and myosin, and accordingly have contractile properties in the presence of ATP. In fact, myosin is polymerized into filaments in cells and is concentrated in the cleavage furrow of dividing cells. The filaments are oriented parallel to the band of actin filaments that encircles the equator of the cell.[30]

Force equilibration along the microfilaments is such that when calcium is released, resulting in its concentration above micromolar levels, the solation of the cytoplasm may take place by reducing the number of crossbridges in the scaffolding of cytoskeleton and activating actin contraction. The present contraction propagates along the microfilaments, and accordingly causes further calcium release from bound sites and solation. The propagation of contraction wave underlies the force equilibration. Furthermore, the contraction is irreversible and morphological in the sense that after contraction, the cell remains contracted because the calcium regulation mechanism restores calcium rapidly so that the gel state is recovered by crosslinking and an altered form results.[31] Alternation of the maximum contour line of internal tensile force along the equator of dividing cells is thus a particular case of actomyosin complex under a proper mechanism of calcium regulation.

The manner in which the maximum contour line is successively alternated in cleavage process is under genetic influence. The cleavage process due to force equilibration is an indication of symmetry-breaking in the morphological space and in fact force equilibration induces such a symmetry-breaking because of its self-constraining capability of dissolving the preceding degeneracy. The genetic influence is to specify in a more particular manner how force equilibration breaks the symmetry property intrinsic to the preceding degeneracy. Both holoblastic and meroblastic cleavage processes, for instance, exhibit symmetry-breaking, but the meroblastic one is less symmetrical. The difference originates in genes.

Force equilibration yielding a cleavage process breaks the symmetry of developing organisms in two ways. One is temporally intrinsic in raising symmetry-breaking due to the very operation of the law of motion of the one-to-many mapping type, and the other is temporally cumulative in that the whole memory, either fine or coarse, of force equilibrations in the past influences the current force equilibration. Force equilibration functions irrespective of the memory of the past equilibrations. Still, the particular manner of equilibration is further influenced by its own past memory. The static aspect of genes refers to the fact that they are a materialistic carrier memorizing how force equilibration has proceeded in the progenitor organisms. On the other hand, the dynamic aspect is due to the fact that the stored past memory is capable of influencing the current force equilibration, though in a manner such that the past memory alone cannot completely predetermine the present.

The fact that the past memory alone cannot exclusively determine how the current force equilibration proceeds is the intrinsic variability of the equilibration. Genes in isolation cannot be dynamic, since their variability is forcedly separated. Only those genes in developing organisms are variable unless external agents intervene. Consequently, force equilibration as an internal and endogenous agent giving rise to forms and structures in the morphological space is more than what genes can prescribe definitely based upon the memory stored in them. It can also generate variations beyond what the identifiable memory stored in genes prescribes. Nevertheless, all of the variations that force equilibration generates do not induce the corresponding variations in genes. It should be noted that variations in genes that occur in developing organisms are only a part of variations due to force equilibration. This is just a dynamic expression of the lack of unique correspondence between soma and germ cell,[32] especially with regard to the origin and transmission of variations, e.g., an acquired phenotypic variation cannot be inherited.

The developmental process starting from cleavage formation in a fertilized egg is beyond what genes as a materialistic controller drawing on past memory can specify, in particular, in giving rise to variations not yet identified in the memory. Even the cleavage process in early embryonic development is no exception. The intrinsic variability of genes is not within genes themselves, but within the developmental process. The developmental process, in principle, can exhibit characters not to be found in genes in isolation.

VII. GROWTH

The cleavage process in early embryonic development is in most cases accompanied by growth. Indicative of cellular growth are stem cells that serve as a progenitor of all of the derivative cell devisions therefrom.[33] Material flow is necessarily involved in the cellular growth process because of material intake that is inevitable if growth is to occur. One thus recognizes that the process materializing flow-continuity, or material flow equilibration, is in operation. The field of internal disequilibrium in material flow induces material flow equilibration. Disequilibrium between entering and leaving material flow at an arbitrary interface within developing embryos is internally identifiable. The succeeding internal realization of material flow continuity is accompanied by endogenous interaction adjustments. However, internal identification of disequilibrium in material flow and the succeeding realization of flow continuity does not occur simultaneously. There is a time lag between the two, no matter how small it may be. This finite time lag is externally observable as variations in the material accumulation there. The field of internal disequilibrium in material flow is thus related to external measurement done *a posteriori*. In this sense, internal disequilibrium in material flow is physically real even if the disequilibrium itself is not externally measurable as such. A positive flow disequilibrium results in an *a posteriori* increase in the material accumulation there, whereas a negative one does just the opposite.

Material flow equilibration intrinsic to the field of internal disequilibrium in material flow is further characterized by both the magnitude of flow disequilibrium, either positive or negative, and the direction of equilibration to be propagated. Consequently, if there is such a local region covered by a set of closed loops along which material flow equilibration is propagated while keeping flow disequilibrium positive, this region can exhibit its growth because of the increase in material accumulation. Cellular growth in a developing embryo can be derived from material flow equilibration keeping flow disequilibrium positive.

Material flow equilibration itself is internally driven by the preceding flow disequilibrium, and leaves the renewed flow disequilibrium accordingly. This sequence continues to survive in developing embryos. Nevertheless, material flow equilibration that keeps flow disequilibrium positive is no more than a very special case of equilibrations in general. The manner in which material flow equilibration acting in developing embryos leaves the subsequent flow disequilibrium is under genetic influence. If material flow equilibration keeping flow disequilibrium positive ceases to function beyond a certain time interval, the growth process will be interrupted from that time.

Whether material flow equilibration driven by positive flow disequilibrium leaves in sequence the similar positive disequilibrium or the negative one is controlled mostly by genes and genomes, but not completely. What genes leave as not being strictly controllable is the intrinsic variability in material flow equilibration, like force equilibration in cleavage process. Cellular growth subject to material flow equilibration that keeps flow disequilibrium positive thus can exhibit a character that has not been exhaustively preprogrammed.

VIII. SYNCHRONIC AND DIACHRONIC PROCESSES

Equilibration in development is diachronic in its operation due to the fact that it serves as a temporal nexus, but its consequence is synchronic in fulfilling physical conservation laws because of their ubiquity in time. Development is thus an interplay of the two different processes, synchronic and diachronic. A synchronic process is intrinsically deprived of its capacity of generating *de novo* variations, otherwise the synchronic ubiquity would be destroyed. Everything involved in the synchronic ubiquity is unambiguously identifiable as such. Nothing indefinite is allowed to intervene in the synchronic ubiquity. In fact, the law of motion of the one-to-one mapping type supplemented by completely fixed boundary conditions is an instance of synchronic process because everything involved in motion is completely identifiable in the boundary conditions. The superficial diachronic character of the law of motion of the one-to-one mapping type admittedly reduces to the synchronic one.

On the other hand, however, the diachronic character observed in the law of motion of the one-to-many mapping type supplemented by intrinsically partially fixed boundary condition does remain indefinite in its implication because of the indefinite degeneracy within what is transformed in time. The diachronic process of dissolving the indefinite degeneracy in sequence can be seen as a process of generating variation. If only the recorded history of development is at issue, it might be possible to single out those characters of synchronic ubiquity because of the fact that everything in the record has already used up its variability. But, what is missing here is the very competence of intrinsic variability. The diachronic dissolution of indefinite degeneracy in time is contrasted with the synchronic ubiquity of definiteness, and they are incommensurable.

Noting that the law of motion of the one-to-many mapping type is in operation in reality, especially in the biological realm, one has to pay due attention to the primary significance of the diachronic dissolution of indefinite degeneracy instead of the synchronic ubiquity of definiteness. More than that, the diachronic dissolution does not exhaust the indefiniteness within what is transformed in time. The synchronic ubiquity of definiteness is not appropriate to developmental processes in biology. This implies that even if deterministic events and courses are identified in the developmental process, they are inevitably accompanied by the indefiniteness or intrinsic variability of their own. If a physical theory is an attempt to aim at identifying the synchronic ubiquity of definiteness by setting aside the mechanism of generating *de novo* variations, developmental processes in biology will be beyond such a theory. Only when one admits that there is an indefinite diachronic process that cannot be reduced to the definite synchronic one, can a theory of developmental processes become feasible and tractable.

IX. SUMMARY

What characterizes biological developmental processes physically is the intrinsic variability exhibited by developing organisms. Although it may seem to have some commonality with the vitalistic principles envisaged by Driesch and by Bergson, intrinsic variability is purely materialistic in its origin. Equilibration as the process materializing conservedness underlying the actualization of physical conservation laws exhibits intrinsic variability through the repercussion of similar equilibrations. The materialistic carrier of equilibration, such as organisms, is infinitely degenerate with the variability to follow. Inherent to the process of equilibration is the field of internal disequilibrium in energy, in forces, and in material flow. Internal agents that identify those internal disequilibria and then implement the process of equilibration are ubiquitous in the developmental process. This by no means implies the resurrection of the vitalistic viewpoint. The emergence of such internal agents is a direct consequence of implementing physical conservation laws in reality. The interplay among those internal agents is symmetry-breaking in its operation because the variability of each internal agent is successively conditioned by what the others have done for the sake of fulfilling physical conservation laws.

One instance of equilibration in development is the embryonic cleavage process. Force equilibration is involved in the cleavage process, because forms and structures that emerge in the morphological space assume the forces to maintain them as such. The cleavage process in the making is subject to force equilibration, whereas the particular manner of cell division in the product has been under genetic influence. Still, genes in developing organisms are subject to force equilibration in the morphological space. Genes are equipped with their variability owing to force equilibration functioning on the level of the carrying organisms. One more instance of equilibration is cellular growth, in which material flow equilibration is involved. Material flow equilibration influences genes as much as it is influenced by them.

The intrinsic variability of development is seen in the indefiniteness of the fate map, cytoplasmic determinants, induction, polarity, and the like. Similarly, the intrinsic variability of genes during development is shown by intracellular exchanges of gene segments, such as transposable genic units and jumping genes. If intrinsic variability were eliminated, the biological developmental process would be reduced to a synchronic process as most physical processes are. In reality, however, the developmental process is diachronic in such a way that it cannot be reduced to a synchronic one. What cannot be reduced to the definite synchronic ubiquity is intrinsic variability inherent to the process of equilibration. Development in biology stands out in admitting the presence of irreducible indefiniteness and endogenously incorporating into itself the capacity of generating variations by constraining the extent of indefiniteness within what is being transformed, though not exhaustively.

REFERENCES

1. **Slack, J. M. W.,** *From Egg to Embryo: Determinative Events in Early Development,* Cambridge University Press, Cambridge, 1983.
2. **Gerhart, J. C.,** Mechanisms regulating pattern formation in the amphibian egg and early embryo, in *Biological Regulation and Development,* Vol. 2, Goldberger, R., Ed., Plenum Press, New York, 1981, 133.
3. **Nieuwkoop, P, P.,** Origin and establishment of embryonic polar axes in amphibian development, *Curr. Top. Dev. Biol.,* 11, 115, 1977.
4. **Brachet, J.,** An old enigma: the grey crescent of amphibian eggs, *Curr. Top. Dev. Biol.,* 11, 133, 1977.
5. **Keller, R. E.,** Vital dye mapping of the gastrula and neurula of Xenopus laevis. I. Prospective areas and morphogenetic movements of the superficial layer, *Dev. Biol.,* 42, 222, 1975.
6. **Keller, R. E.,** Vital dye mapping of the gastrula and neurula of Xenopus laevis. II. Prospective areas and morphogenetic movements of the deep layer, *Dev. Biol.,* 51, 118, 1976.
7. **Jacobson, M. and Hirose, G.,** Clonal organization of the central nervous system of the frog, *J. Neurosci.,* 1, 271, 1981.

8. **Freeman, G.,** The role of cleavage in the localization of development potential in the ctenophore Mnemiopsis leidyi, *Dev. Biol.*, 49, 143, 1976.
9. **Freeman, G.,** The effect of altering the position of cleavage planes on the process of localization of developmental potential in ctenophores, *Dev. Biol.*, 51, 332, 1976.
10. **Sander, K.,** Specification of the basic body pattern in insect embryogenesis, *Adv. Insect Physiol.*, 12, 125, 1976.
11. **Cooke, J.,** Scale of body pattern adjusts to available cell number in amphibian embryos, *Nature (London)*, 290, 775, 1981.
12. **Slack, C. and Warner, A. E.,** Intracellular and intercellular potentials in the early amphibian embryo, *J. Physiol.*, 232, 313, 1973.
13. **Edelman, G. M.,** Cell adhesion and the molecular processes of morphogenesis, *Ann. Rev. Biochem.*, 54, 135, 1985.
14. **Kirscher, M., Gerhart, J. C., Hara, K., and Ubbels, G. A.,** Initiation of the cell cycle and establishment of bilateral symmetry in Xenopus eggs, *Symp. Soc. Dev. Biol.*, 38, 187, 1980.
15. **Gerhart, J., Ubbels, G., Black, S., Hara, K., and Kirschner, M.,** A reinvestigation of the role of the grey crescent in axis formation in Xenopus laevis, *Nature (London)*, 292, 511, 1981.
16. **Crick, F. H. C. and Lawrence, P. A.,** Compartments and polyclones in insect development, *Science*, 189, 340, 1975.
17. **Kimble, J. E.,** Strategies for control of pattern formation in Caenorhabditis elegans, *Philos. Trans. R. Soc. London Ser. B*, 295, 539, 1981.
18. **Matsuno, K.,** How can quantum mechanics of material evolution be possible? symmetry and symmetry-breaking in protobiological evolution, *BioSystems*, 17, 179, 1985.
19. **Maynard Smith, J., Burian, R., Kauffman, S., Alberch, P., Campbell, J., Goodwin, B., Lande, R., Raup, D., and Wolpert, L.,** Developmental constraints and evolution, *Q. Rev. Biol.*, 60, 265, 1985.
20. **Polanyi, M.,** Life's irreducible structure, *Science*, 160, 1308, 1968.
21. **Sonneborn, T. M.,** Gene action in development, *Proc. R. Soc. London Ser. B*, 176, 347, 1970.
22. **Dover, G.,** Molecular drive: a cohesive mode of species evolution, *Nature (London)*, 299, 111, 1982.
23. **Ho, M. W. and Saunders, P. T.,** Beyond neo-Darwinism — an epigenetic approach to evolution, *J. Theor. Biol.*, 78, 573, 1979.
24. **Charles, R.,** Morphology and biochemistry, *Acta Morphol. Neerl. Scand.*, 21, 15, 1983.
25. **Meinhardt, H. and Gierer, A.,** Applications of a theory of biological pattern formation based on lateral inhibition, *J. Cell Sci.*, 15, 321, 1974.
26. **Gierer, A.,** Some physical, mathematical and evolutionary aspects of biological pattern formation, *Philos. Trans. R. Soc. London Ser. B*, 295, 429, 1981.
27. **Matsuno, K.,** Physical basis of evolutionary morphogenesis, in *Evolution and Morphogenesis*, Mlikovsky, J. and Novak, V. J. A., Eds., Academia, Prague, Czeckoslovakia, 1985, 19.
28. **Goodwin, B. C. and Trainor, L. E. H.,** A field description of the cleavage process in embryogenesis, *J. Theor. Biol.*, 85, 757, 1980.
29. **Oster, G., Odell, G. and Alberch, P.,** Mechanics, morphogenesis and evolution, *Lect. Math. Life Sci.*, 13, 165, 1980.
30. **Yumura, S. and Fukui, Y.,** Reversible cyclic AMP-dependent change in distribution of myosintick filaments in Dictyostelium, *Nature (London)*, 314, 194, 1985.
31. **Goodwin, B. C.,** A relational or field theory of reproduction and its evolutionary implications, in *Beyond Neo-Darwinism*, Ho, M. W. and Saunders, P. T., Eds., Academic Press, London, 1984, 219.
32. **Webster, G.,** The relations of natural forms, in *Beyond Neo-Darwinism*, Ho, M. W. and Saunders, P. T., Eds., Academic Press, London, 1984, 193.
33. **Bryant, P. J. and Simpson, P.,** Intrinsic and extrinsic control of growth in developing organs, *Q. Rev. Biol.*, 59, 387, 1984.

Chapter 9

EQUILIBRATION IN EVOLUTION

I. INTRODUCTION

Evolution of reproductive units or organisms assumes two processes. One is the generation of variations, and the other is the fixation of variations thus generated. If the generation of variations were completely separable from their fixation, natural selection of random variants would remain legitimate as an evolutionary mechanism by letting natural selection be an agent of fixing a limited member of those generated randomly. In reality, however, the generation of variations is inseparable from their fixation. Reproduction does not proceed without generating variations. Reproductive fidelity is always in the process of enhancing itself during evolution. The underlying dynamics of enhancing reproductive fidelity is connected with material flow equilibration in view of the fact that organisms are material units processing material flows entering and leaving them. Material flow equilibration as the process materializing flow-continuity in organisms serves as an agent of generating variations as well as enhancing their reproductive fidelity through successive generations. Natural selection is viewed as an *a posteriori* consequence of material flow equilibration. Fitness to the environment is an overall derivative character of surviving organisms as agents of material flow equilibration. Adaptability of an evolving organism to the environment is an *a posteriori* characteristic of variability inherent in organisms.

Variability on the part of organisms is originated not only in external agents but also in the organisms themselves. Organisms can initiate variations during material flow equilibration so that they are not totally controlled externally. Organisms are active in holding their intrinsic variability while being subject to the physical process of material flow equilibration. The intrinsic variability of organisms is exclusively materialistic due to the fact that the underlying law of motion is inseparable from its boundary conditions. If one understands evolution within the physical scheme of the law of motion of the one-to-one mapping type supplemented by completely identifiable boundary conditions, the origin of variability of organisms would have to be sought exclusively within external agents and their adaptability would have to be solely passive toward the environment. Fitness to the environment would refer to the best among the possibilities it provides. On the other hand, if one duly understands evolution within the scheme of the law of motion of the one-to-many mapping type supplemented by intrinsically indefinite boundary conditions, the adaptability of organisms will confirm their active participation vis à vis the environment while their variability is maintained strictly within permissible physical processes. The fitness of organisms is under their own active control.

II. REPRODUCTIVE FIDELITY

Tracing the evolutionary sequence of biologically reproductive units first requires an identification of those basic characteristics that are applicable to any stage of evolution. One strategy for this identification is to pay attention to the rate of reproduction and its variation. The separation between reproduction and its variation inevitably forces us to accept that reproduction that is free from its variation assumes perfect reproductive fidelity. This supposition of perfect reproductive fidelity has its basis solely in the theoretical commitment that enables one to define the rate of reproduction unambiguously.

Insofar as the rate of reproduction is definable as the most fundamental entity characterizing the process of protobiological and biological reproduction, the mechanism responsible for producing variations in reproduction would have to be detached from the very process of reproduction. The process of generating evolutionary variations would have to be deduced in a manner totally independent of the process of reproduction. This forced separation is a consequence of employing the rate of reproduction as the most fundamental entity in protobiological and biological evolution. The mechanism of generating variations which is compatible with the forced separation would be that of random variation, for it is supposedly free from any interference from the process of reproduction.

Random variation is however not definable as such unless a frame of reference against which randomness is defined is available. Such a frame of reference is a very special one selected from many conceivable alternatives and cannot be variable in a random manner. One weak definition of random variations is made against the reference of natural selection in such a manner that the generation of variations does not presuppose the antecedent consequence of natural selection. The utility of this definition of random variations still depends entirely on how one understands natural selection. If natural selection is understood as an *a posteriori* result of evolutionary variations generated thus far, any variation would fulfill the present definition of randomness.

On the contrary, if natural selection is taken to be involved in the very generation of variations, they cannot be random because of the former definition of randomness that asks them to be independent of natural selection. Random variation separated from the process of natural selection, if properly understood, is no more than a tautology, otherwise randomness would have to be totally rejected. This separation manifests a difficulty in letting the generation of variations be isolated from the process of reproduction. In spite of its practical usefulness, especially, in the mathematical theory of evolutionary processes, the dichotomy between reproduction and variations on the former faces an intractable problem of how to deduce reproduction being free from variations and of where to find the origin of variations other than in the process of reproduction.

One way out of the present impasse is to abandon the superficial dichotomy between reproduction and variations on it. An alternative is to call attention to the reproductive difference or fidelity between successive generations. Reproductive fidelity refers to the difference, if any, between reproductive progenitors and their progenies without making any special effort to distinguish between the process of reproduction and variations imposed on it while viewing both indiscriminately. The dynamics of reproductive fidelity is inherent to reproductive units or organisms processing material flows entering and leaving them. One predominant property of the dynamics of reproductive fidelity, which has been confirmed on the basis of empirical observation, is its evolutionary enhancement.

Nonenzymatic replication of RNA on its complementary template that would have been thought as an evolutionary precursor of its enzymatic replication exhibits the reproductive fidelity such that each residue is likely to make an error in every 10^2 replications. In contrast, the reproductive fidelity of the biologically full-blown replication of DNA is about one error in every 10^9 replications per residue.[2] The decrease of replication errors from once in every 10^2 to once in every 10^9 replications per residue undoubtedly attests to an evolutionary enhancement of reproductive fidelity. In fact, the evolutionary enhancement of reproductive fidelity is a derivative characteristic of reproductive units processing material inflow and outflow. Material flow equilibration as the process materializing flow continuity is self-constraining in its operation because of limiting the actualized interaction configurations only to those that fulfill material flow continuity *a posteriori*. Prerequisite to the evolutionary enhancement of reproductive fidelity is the presence of material units of organisms open to material flow.

Material flow equilibration that is indispensable in organisms assumes the law of motion of the one-to-many mapping type supplemented by intrinsically partially fixed boundary conditions, because of the inseparability of the law of motion and its boundary conditions.

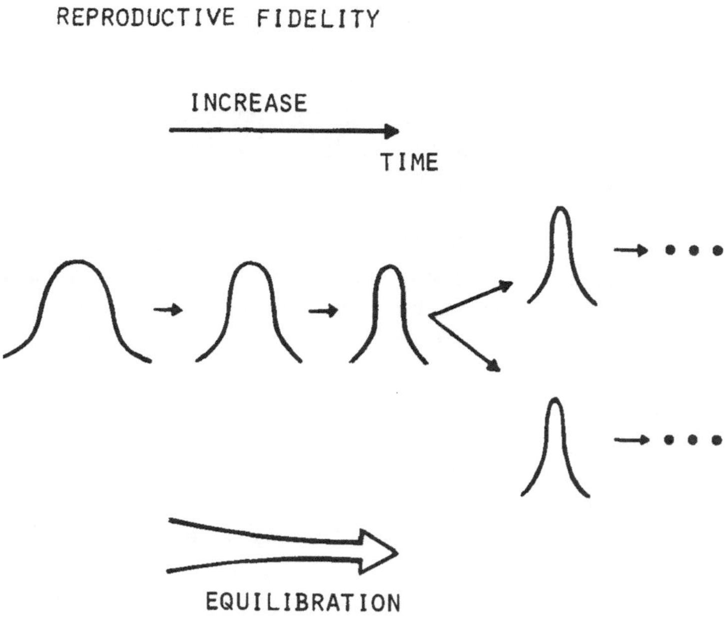

FIGURE 1. Time evolution of reproductive fidelity. The process of equilibration naturally enhances reproductive fidelity. Speciation is also one of its consequences.

Material flow equilibration lets material flow continuity serving as an infallible boundary condition also be involved in the process materializing flow continuity. Evolutionary enhancement of reproductive fidelity rests upon the law of motion of the one-to-many mapping type supplemented by indefinite boundary conditions. If the law of motion of the one-to-one mapping type that is necessarily accompanied by completely identifiable boundary conditions were invoked instead, the evolutionary enhancement of reproductive fidelity would have to be preprogrammed within the boundary conditions. In other words, if exogenous boundary conditions that were completely separable from the law of motion are taken literally, evolving material units would simply have to be passive in following the one-to-one transformation of what the boundary conditions have specified.

However, organisms cannot be totally passive toward their environment in the respect that the formation of organisms is a spatial embodiment of material flow equilibration that has to obey the law of motion of the one-to-many mapping type. Organisms are also active in initiating the change in interactions with their exterior so as to fulfill material flow continuity there *a posteriori*. Room for initiating such a change is found in the indefiniteness in separation between the law of motion and its boundary condition. Although the impetus for initiating the change in interaction so as to fulfill material flow continuity comes from a preceding similar change, the way to initiate the change is not fully preprogrammed thanks to the law of motion of the one-to-many mapping type. Material flow equilibration is reactive in its succession due to the fact that it points to the successive repercussion for material flow continuity. But, it is active in the sense that the details of reaction are not fully preprogrammed externally.

Evolutionary enhancement of reproductive fidelity as shown in Figure 1 is due to the reactively active process of material flow equilibration in organisms. The reactively active process differs from the spontaneously active one in assuming a preceding stimulus, but also differs from the reactively passive one in maintaining the corresponding response as not uniquely definable. The reactively active capacity of material flow equilibration is unfailingly materialistic in its origin, being different from the one expected from the spontaneously active character of immaterialistic origin. In particular, the reactively active character is seen in the enhancement of the reproductive fidelity of organisms owing to the genesis of successive constraints during the transformation from the possible to the actual by eliminating the impossible. The reactively active process of reproduction naturally enhances reproductive fidelity by eliminating the impossible latent in the formerly possible when transformed into the actual, which in turn successively serves as the source of the later possible.

Reproduction and variation upon it are inseparable in their genesis. What is more, a consequence of the inseparability between the two is the evolutionary enhancement of reproductive fidelity. This character exhibits a marked difference compared with the case of the reactively passive reproduction in which any change in reproduction must be sought outside the reproduction process. Although it assumes perfect reproductive fidelity unless perturbed externally, reactively passive reproduction could be conceivable only in theory, not in reality. The dichotomy between reproduction and variation upon it necessarily assumes that reproduction would be reactively passive. In contrast, the primacy of evolutionary changes in reproductive fidelity over the superficial dichotomy between reproduction and variation upon it is firmly based upon the reactively active process that is materialistically justifiable.

Reactively active reproduction seeks its materialistic basis in the law of motion of the one-to-many mapping type supplemented by intrinsically partially fixed, or indefinite, boundary conditions. The law of motion of the one-to-many mapping type renders itself inseparable from its boundary condition. In other words, the movement following the law of motion of the one-to-many mapping type can vary the boundary condition to which it has formerly been subject. Evolutionarily unique in this regard is the emergence of open material aggregates or protocells that can vary their spatial boundaries by processing material flows entering and leaving them. Those open material aggregates are reactively active in controlling material inflow and outflow. Reactively active formation of open material aggregates depends not only upon the amount of material inflow and outflow controlled externally, but also upon the presence of possible alternatives they internally allow in implementing the material flows. If the extent of possible alternatives that open material aggregates can maintain in processing material inflow and outflow is quite limited, the process of reproduction could be almost reactively passive.

Take, for instance, layer replication of montmorillonite through adsorption.[3] The material inflow for adsorption can be uniquely determined once the ingredient of solution mixture and the layer structure are identified. The matrix layer cannot be flexible enough to allow a wide variety of adsorption flows because of its rigid crystalline structure. The similar reactively passive process applies to nonenzymatic residue-by-residue replication of RNA.[4]

An evolutionarily significant epoch is the time when the transition from reactively passive to reactively active reproduction took place. Potentially, reactively passive reproduction can be alternated with reactively active reproduction if it is possible for both to compete for the same material resources. The reactively active reproduction that takes advantage of having many alternatives in fulfilling material flow continuity can accomplish greater material accumulation in time than its reactively passive counterpart does. Reactively active reproduction would be dominated by those reproductive units with the greater material accumulation rate among many possible alternatives available internally. This transition is equivalent to the one from the situation in which the law of motion of the one-to-one mapping type remains as a good approximation of the real dynamics to the situation in which the law of motion of the one-to-many mapping type has to be properly appreciated. Emergence of reproductive protocells would just coincide with the transition from the reactively passive to the reactively active reproduction from an evolutionary perspective, for reproductive protocells vary their spatial boundaries and exhibit the inseparability of the law of motion and its boundary conditions. The inseparability is a principal characteristic of the law of motion of the one-to-many mapping type.

Reproductive fidelity turns out to be more fundamental than the dichotomy between reproduction and variation upon it once the reactively active reproduction sets in. The material carrier transmitted from generation to generation to guarantee the reproductive fidelity is genotypic. The genotype is at the same time under the influence of enhancing the reproductive fidelity through the reactively active production that is phenotypic. The genotype can be variable in the phenotypic development, while the latter has followed the instruction from the former genotype. This convoluted coupling between genotype and phenotype is common to the reactively active process of reproduction. The progenitor reproductive unit or organism moves up, with time, the time-extensive hierarchy of protobiological and biological organisms.[5]

Evolutionary enhancement of reproductive fidelity is seen in the branching of the phylogenetic tree. The progenitor reproductive units are more degenerate with those to come later, and the progeny units are less degenerate comparatively. A branching of the phylogenetic tree is just one example that shows how reproductive fidelity comes to attain finer tuning by way of reactively active reproduction. If one tries to understand the branching of phylogenetic trees in terms of the dichotomy between reproduction and variation upon it, the origin of extinction and speciation would have to be sought in the external environment, no matter how small the likelihood may be. In contrast, the process enhancing reproductive fidelity that is associated with the law of motion of the one-to-many mapping type supplemented by indefinite boundary conditions is already latent within reproductive species. Speciation is a form of enhancing reproductive fidelity, though its very origin is within the level of each individual organism that imposes self-limitation upon itself because of material flow equilibration. The progenitor individual organism is degenerate with species and speciation to come. The process dissolving such degeneracy is associated with the fine tuning of reproductive fidelity.

The material flow equilibration that underlies evolutionary enhancement of reproductive fidelity at individual organisms is microscopic in its operation in the sense that those adjustments of materialistic interaction for material flow continuity and their repercussions are operative over microscopic time intervals. The extent of each interaction adjustment is also microscopic because it is imputed to the preceding microscopic adjustment of the similar nature. On the other hand, the branching of phylogenetic trees or speciation is macroscopic in the sense that such an event is intermittent or sporadic in time. This does not, however, imply an intervention of macroscopic events of exogenous origin. Speciation in process is assumed by each individual organism, but its amplification takes place at the population level. Although a population of reproductive units can sometimes exhibit a macroscopic variation as being subject to microscopic fluctuations of exogenous origin if the attenuation of those fluctuations is not uniform over their available range,[6] the macroscopic variation in the population induced by those microscopic fluctuations[7] is still within the context of the dichotomy between reproduction and variation upon it. The population needs to be examined in the light of reproductive fidelity, instead of the dichotomy assuming both reproduction having perfect fidelity and variations upon it, in order to explicate the nature of macroscopic variations observed on the level of population.

Evolutionary enhancement of reproductive fidelity proceeds at each individual organism. In contrast, the carrier of reproductive fidelity is rather a species-specific entity, because reproductive fidelity refers to the operational difference between reproductive progenitors and their progenies. The relationship between a species and those individuals belonging to it is such that the species is situated at a higher level in the time-extensive hierarchy of protobiologically and biologically reproductive units. The species has an emergent character as such not found in each individual comprising the species, while it serves as a glue to hold those constituent individuals together.[8] The real dynamics of reproductive fidelity refers to cross-hierarchical causations[8] between each reproductive individual and the covering species. What is unique to reproductive fidelity is its intrinsic variability, which is reactively active, due to the facts that each organism as a carrier capable of varying reproductive fidelity is open to material flow and that material flow equilibration acting there constantly generates variations in the carrying organisms in a manner not fully controlled externally.

Enhancement of the specificity in reproductive fidelity that takes place at the individual level can induce a speciation at the level of species. The actualization process of reproductive fidelity is cross-hierarchical in both upward and downward causations along the time-extensive hierarchy,[9] because the progenitor individuals move up the hierarchy with time and influence their progeny while variations in the reproductive fidelity operate at the level of individual organisms that is at the lowest in the time-extensive hierarchy. At the same time, those reproductive individuals at the lowest level of the time-extensive hierarchy also appear at almost all levels of the spatiotemporally localized hierarchy of biological entities, as with nested ones such as populations in the ecological hierarchy. The spatiotemporally localized hierarchy is, on the other hand, only a time slice of the time-extensive hierarchy. Those variations upon the spatiotemporally localized hierarchy originate in the internal dynamics of the time-extensive hierarchy. Variations upon reproduction originate in the internal dynamics of reproductive fidelity.

III. NATURAL SELECTION

The common understanding of natural selection refers to differential reproduction of biological species. Natural selection can be equated with differential reproduction and vice versa. This equality remains legitimate at least logically because differential reproduction is an expression of natural selection as much as natural selection is actualized in differential reproduction. But the relationship between natural selection and differential reproduction is more than just a logical one. Some form of causality may be involved between the two. A question to be raised in this regard is how natural selection and differential reproduction come to operate in reality. Differential reproduction does necessitate its own mechanism. Likewise, natural selection assumes its own mechanism. Insofar as differential reproduction and natural selection are logically identical in their implication, they would have to share the same mechanism that has characterized without invoking either differential reproduction or natural reproduction.

A typical law of motion that is physically admissible is of the one-to-one mapping type supplemented by completely identifiable boundary conditions. If one tries to understand differential reproduction within the scheme of the law of motion of the one-to-one mapping type, the origin of variations in reproduction has to be sought in the boundary condition that is exogenous to the law of motion. The generation of variations and the process preserving some of those variations are separated. It is in this context of preserving what has been generated and applied externally that natural selection can be seen as a selective sieve. Natural selection within the scheme of the law of motion of the one-to-one mapping type remains passive toward generating variations. The law of motion of the one-to-one mapping type does no more than preserve what the boundary condition of exogenous origin has provided.

The selective behavior of the law of motion of the one-to-one mapping type originates in the nonnegativity of the population of each reproductive unit. A reproductive species that cannot maintain its population above zero any more due to a certain variation in the boundary condition is destined to be wiped out. This process of being phased out unfailingly follows the law of motion of the one-to-one mapping type, but still can be seen as selective *a posteriori*. The law of motion lets natural selection be a mere passive transformation of what the boundary condition has prescribed. Exogenous variations in the rate of reproduction can be directly equated with the cause of natural selection. The law of motion of the one-to-one mapping type is however by no means selective by itself. What makes it superficially selective under the guise of natural selection is a preparation of boundary conditions that cannot guarantee the nonnegativity of population of some species any more. Natural selection as the law of motion of the one-to-one mapping type supplemented by the nonnegativity of population is thus completely controlled by its boundary condition that is identifiable in principle in its every detail. The consequence of natural selection is already evident when the boundary condition is provided.

Variations of exogenous origin determine the outcome of natural selection. Indeed, there have been many attempts to view the process of natural selection as the law of motion of the one-to-one mapping type supplemented by the nonnegativity of population, since this kind of theorizing is quite ubiquitous in physics. Nevertheless, this scheme of one-to-one mapping is intrinsically incompetent in the control of boundary conditions. The law of motion of the one-to-one mapping type has to be supplemented by completely identifiable boundary conditions, but the agent controlling the boundary condition is foreign to the law of motion. It is prohibited as a matter of principle to examine the mechanism of generating variations in the boundary condition within the framework of the law of motion of the one-to-one mapping type in which only the external agent controls those variations. If the origin of variations in the boundary condition is really to be examined physically, one must get rid of the fetter imputed to the theoretical imposition of the law of motion of the one-to-one mapping type even if this scheme may seem to approximate the reality of natural process to a limited extent. As a matter of fact, the law of motion of the one-to-one mapping type is an approximation to the law of motion of the one-to-many mapping type that is real in view of the fact that no interaction change can be propagated at a velocity exceeding light velocity.

Once the law of motion of the one-to-many mapping type is taken seriously, natural selection will assume a completely different perspective and implication. An arbitrary evolutionary precursor is latent with what will be actualized later, and the actual is always limited compared with the possible because of the involvement of a physical constraint such that conservation laws have to be fulfilled *a posteriori* in any case. The transformation from the possible to the actual is both self-constraining and selective in eliminating the impossible that had been previously latent. Furthermore, the actual subsequently serves as a carrier of the possible to come. The materialistic origin of the difference between the possible and the actual is in the process materializing conservedness. The process communicating how conservation laws are fulfilled on the part of interacting material aggregates is undoubtedly materialistic and physical in view of the fact that materialistic interaction is responsible for the communication.

Indefinite repercussions for conservedness make each interacting unit degenerate with endogenous variations to come. The latent variability in the process communicating conservedness is a dominant characteristic of the possible, and it cannot be exhausted insofar as the propagation velocity of interaction changes remains finite, as it should. Unless such a theoretical artifact is imposed that those repercussions of materialistic interaction for conservedness may be terminated at a certain arbitrary stage, the self-constraining transformation from the possible to the actual can survive indefinitely. Materialistic interaction within the scheme of the law of motion of the one-to-many mapping type can thus yield what natural selection is all about.

The nonnegativity of population takes on a new look when the law of motion of the one-to-many mapping type is rightly appreciated. Natural selection within the theoretical framework of the law of motion of the one-to-one mapping type serves only as a selective sieve on what external agents supply. Its consequence is elimination of those species that fail to fulfill the nonnegativity of population. On the other hand, the law of motion of the one-to-many mapping type lets natural selection be operative even if the nonnegativity of population is not violated. Natural selection is certainly visualized in the nonnegativity of reproductive population. But, natural selection is more than what such nonnegativity would imply.

The transformation from the possible into the actual necessarily eliminates the impossible that would fail in the communication of conservedness internally even if the nonnegativity of population is guaranteed. Elimination of such an impossibility is a form of generating variations that are somewhat biased in the sense of referring to the process materializing conservedness. This character, however, does not lead to the observation that variations would be generated in anticipation of how natural selection would work. Quite to the contrary, natural selection is a consequence of the process generating those constrained variations. What is anticipated is infallible conservation laws, and not natural selection.

Although the standard understanding of natural selection within the law of motion of the one-to-one mapping type is logically transparent in letting the origin of variations be totally exogenous to the law of motion, one major practical problem with this scheme is the claim that extremely unlikely events would have had to occur in order to give rise to and to maintain observed biological evolution.[10] Such an unlikelihood alone,[11] however, cannot disprove the utility of the law of motion of the one-to-one mapping type in the evolutionary process. The real difficulty with the present one-to-one mapping scheme is its failure to recognize the fact that the law of motion of the one-to-one mapping type is no more than an approximation to that of one-to-many. Biology surely shows that the law of motion of the one-to-many mapping type is in action, as demonstrated in the case of biological motility. One needs to appreciate the operation of natural selection in such a manner that the law of motion and its boundary condition are inseparable.

The real variations generated in the process of natural selection have to anticipate conservation laws that are infallible. Likewise, the operation of natural selection anticipates the similar conservation laws. These two do not, however, mean that the variation would be generated in anticipation of how natural selection will proceed. The process materializing, or communicating, conservedness underlies the materialistic anticipation of conservedness. Both natural selection and the genesis of those variations are in fact derived from the process materializing conservedness. Natural selection as a law of motion remains inseparable from those variations imposed upon it, and those variations imposed as a boundary condition also remain inseparable from natural selection. It is the process materializing conservedness that makes both inseparable.

Natural selection as the process communicating conservedness applies to the spatiotemporally localized class of individual organisms in which the internal communication for conservedness actually materializes. What is unique to the process communicating conservedness is that each material unit involved is plastic enough to allow many alternatives in implementing conservation laws, especially in the manner of interaction with other material units. If one equates the spatiotemporally localized class of individual organisms that are closely related with a reproductive species, its speciation or extinction can be associated with a failure of the process communicating conservedness in the former class of reproductive individuals as a coherent unit of the communication of conservedness. Such a speciation cannot be a passive outcome from resource limitation. Conversely, it originates in the reactively active capacity of each species as a spatiotemporally locally homogeneous class of material units to make choices from many possible alternatives for the sake of infallible conservation laws, resulting in more than two distinctively different classes.

The materialistic capacity for making choices is inseparable from the operation of natural selection. The present materialistic capacity simply points to the fact that a firm frame of reference against which any amount of indefiniteness could be mitigated even to a slightest degree is available only *a posteriori* in the form of fulfilling conservation laws. Unless

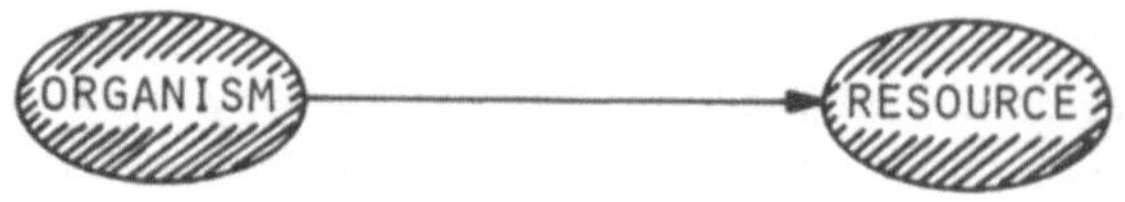

RESOURCE LIMITATION : MECHANISTIC CAUSALITY
NO CAPACITY OF ENDOGENOUS SELECTION

(A)

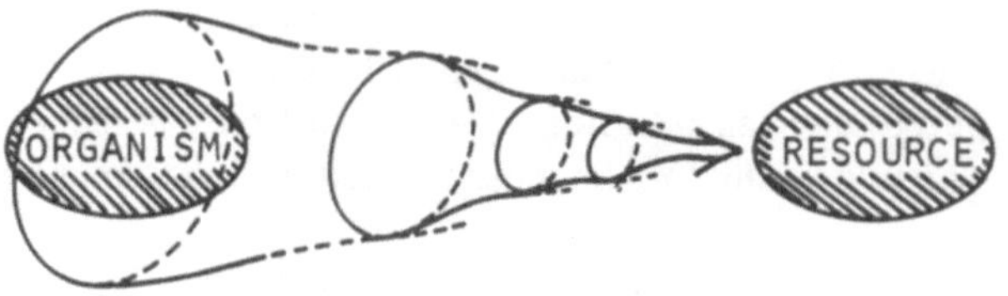

RESOURCE LIMITATION : FINAL CAUSALITY
ENDOGENOUS CAPACITY OF SELECTION

(B)

FIGURE 2. Resource limitation in (A) the mode of mechanistic causality and in (B) the mode of final causality. A distinctive character of resource limitation in the mode of final causality, or resource exploitation, is within the possession of materialistic capacity of making choices on the part of organisms.

articulated artificially, there cannot be any boundary conditions that could determine how the process materializing conservedness would proceed. Natural selection is thus unique in defying its operation under an articulation of fixed boundary conditions. It persistently undermines what looks like boundary conditions. Natural selection under fixed boundary conditions, such as a resource limitation, necessarily makes itself subordinate to the separation between the law of motion and its boundary condition. Resource limitation as a theoretical imposition is however different from the similar one in reality.

Resource limitation as an imposed boundary condition assumes a certain external agent that allows such an articulated imposition to be observed, whereas a similar limitation in reality is communicated internally in an endogenous manner. The endogenous communication necessitates an involvement of such internal agents that can recognize and then can act for resource limitation. Resource limitation serves as a final objective of the endogenous communication. The intermediate process for the communication can be visualized as the one enhancing the extent of self-constraint. Although resource limitation might seem a mechanistic cause of natural selection if it is taken as an attribute of boundary conditions to be imposed, the reality is just the opposite. Resource limitation is a final cause toward which natural selection drives itself as pictorially contrasted in Figure 2. Final cause can be allowed only in the case that the boundary condition upon the law of motion remains intrinsically indefinite, otherwise the definite boundary condition that can be identified in advance makes itself a mechanistic cause driving the law of motion. Natural selection in the framework of indefinite boundary conditions is thus a materialistic process operating in the mode of final causality. Only when the law of motion can be separated from its boundary condition will the mechanistic counterpart of causality surface. Mechanistic causality is more stringent than the final causality in that the former claims definite boundary conditions in advance.

We thus recognize that natural selection is admittedly a physical process with the reservation that it is not necessarily subject to mechanistic causality. Although it is possible to approximate natural selection by a process obeying a mechanistic causality supplemented by definite boundary conditions, such an approximation is admissible only at the expense of introducing an artificial agent that would transform the final into the mechanistic causality. The forced transformation from final to mechanistic causality would require the process communicating conservedness to proceed at an infinite velocity, because the cause driving the law of motion has to be defined as such in advance and because the same cause has to fulfill conservation laws at the very moment it is claimed to be definable. Final causality is more fundamental as a physical process than mechanistic causality because in the latter the process communicating conservedness is forcibly eliminated. Natural selection in the mode of final causality is accordingly more fundamental than that in the mode of mechanistic causality, because in the latter an intrusion of an articulated theoretical device to make a mechanistic cause or its equivalent of boundary conditions unambiguously definite is inevitable.

In particular, resource limitation is a boundary condition to be recognized and to be maintained as such by every participating material unit, instead of being a mere authoritative reinforcer upon its subordinates. The genesis of evolutionary variations can thus find their legitimate field in the multitude of inexhaustive possible alternatives in implementing resource limitation in the mode of final causality, whereas its operation in the mode of mechanistic causality exhausts the inventory of accumulated variations because resource limitation serves there as a mechanistic cause of the underlying selective process. Resource limitation in the natural setting is a characteristic experienced by the insiders, and the way of experiencing the limitation is through the process communicating conservedness. If the same resource limitation is taken as a mere theoretical imposition, a formidable problem would arise with regard to who would be the reinforcer and how such a reinforcement could be possible at all. The similar charge applies to the operation of mechanistic causality, especially as to how the initial cause comes into existence. A modest way to cope with the present problem is to leave it unanswered. Instead, it is desirable to identify the goal of the process whatever its initial condition may be. The process communicating conservedness is thus called for. Natural selection, when understood properly in biology, is a physical process in the mode of final causality. Mechanistic causality is too much to ask for.

Resource limitation in the mode of final causality also implies resource exploitation on the part of participating material units because the way the limitation is experienced cannot be definitely specified in advance due to the intrinsic indefiniteness of boundary conditions. It exhibits a distinct contrast to the one in the mode of mechanistic causality where the dynamics of resource limitation do necessitate their own mechanistic cause that can be explicitly identified in advance. Resource limitation in the mode of mechanistic causality presumes that the inventory of means for resource exploitation has already been available unless external agents further intervene, in which each individual organism is supported to carry some of these means. This would lead to the observation that the unit of natural selection may be each individual organism. Insofar as mechanistic causality is taken seriously, the present conclusion as to the significance of individual organisms as units of natural selection would remain irrefutable. However, the real difficulty in this scheme is with the unsupported claim for mechanistic causality that requires more than final causality does. Natural selection understood correctly within the framework of final causality allows the endogenous process of resource exploitation to continue indefinitely.

Each individual organism as a carrier of resource exploitation is succeeded by its progeny. This successive alternation disproves each individual organism as a unit of natural selection. The problem of which serves as a unit of natural selection, either the individual organism or the species, thus becomes illusory.[12-16] The carrier of natural selection operating as resource exploitation in an endogenous manner turns out to be the time-extensive hierarchy of organisms in which the progenitor organism moves up indefinitely with time by influencing its progenies. Differential reproduction is simply a consequence of the operation of natural selection along the time-extensive hierarchy. The reproduction rate of individual organism can be measured *a posteriori*,[17] but it cannot be identified as such in advance. The reproduction rate is a summary expression of how the individual organism under examination and its progenitors have exploited material resources and survived up to the moment it is expressed.

The summary expression, by definition, does not and cannot influence how the reproduction rate will develop from the moment it has been identified. If the inventory of means for resource exploitation were definite and remained invariant, it would be appropriate at least theoretically to claim an exhaustive list of reproduction rates of all of the possible individual organisms in advance. But, a simple fact of both the prevalence of indefinite boundary conditions and the law of motion in the mode of final causality negates such a theoretical possibility of an exhaustive identification of how reproduction will vary and evolve. Although external agents, if any, can manipulate the manner of reproduction and vary its rate, those influences of exogenous origin are admittedly beyond any theoretical control. One way to overcome the present difficulty is to leave the inexhaustible capacity of resource exploitation unspecified without imposing upon it any articulated constraint. Its candidate is natural selection in the mode of final causality operating along the time-extensive hierarchy of individual organisms. Speciation and extinction are just an expression of how the time-extensive hierarchy has come to be branched.

Natural selection in the mode of final causality derives from the operation of final causality per se. The origin of natural selection is in final causality, and not vice versa. Final causality is confirmed by the process materializing conservedness, in which the conservedness serves as a final cause. The unique character of final causality is that the impetus to dynamics is endogenous, whereas mechanistic causality assumes a dynamics whose impetus has to be of exogenous origin. Insofar as we stick to mechanistic causality, the origin of natural selection would have to be an exogenous event no matter how unlikely it may have been. However, mechanistic causality cannot be contrasted with final causality on an equal basis. Mechanistic causality is simply a derivative from the causality of final counterpart supplemented by a theoretical articulation claiming the presence of definite boundary conditions that are supposed to be clearly separated from the underlying law of motion. What must be examined is the appropriateness of such a theoretical articulation. Unless the present articulation is guaranteed against a certain referential standard, mechanistic causality cannot survive. Instead, the more fundamental final causality has to be invoked. Natural selection invokes the materialistic process in the mode of final causality that is more fundamental than the mechanistically materialistic process that is ubiquitous in physical theorizing.

IV. VARIABILITY

Resource limitation in the mode of mechanistic causality draws on those variations in reproduction that have been available, whereas the same limitation in the mode of final causality can generate variations in the way reproduction proceeds. The difference depends on whether the law of motion is of the one-to-one mapping type or of the one-to-many. We have already observed that the law of motion underlying any materialistic process is intrinsically of the one-to-many mapping type because of the involvement of the process materializing conservedness. The law of motion of the one-to-many mapping type is in fact a form of generating variations because of the fact that the one-to-many temporal mapping leaves the predecessor intrinsically incompetent to determine what the successor will be. The successor can be identified *a posteriori*, but the predecessor alone cannot determine every detail of the successor in advance. The incompetence of the predecessor holds at any moment. The intrinsic impossibility of externalizing or explicating the process materializing conservedness makes the generation of *de novo* variations tenable. The observation that the capacity of materializing conservedness is exclusively within atoms, molecules, and their aggregates defies the imposition of such an articulated boundary condition that lets them follow an unambiguously predetermined track without leaving anything indefinite.

Nevertheless, approximation of the one-to-many temporal mapping by the one-to-one mapping is ubiquitous, especially in physics. A direct consequence of the present approximation is that the generation of *de novo* variations is not allowed any more. An invariant ensemble of variants is claimed to exist unless external agents intervene. In particular, statistical mechanics is competent in identifying the most likely variants in the invariant ensemble of those variants. The minimization of thermodynamic potential or thermodynamic free energy is a common form of conceiving the invariant ensemble of variants. However, the price we have to pay for adopting the present postulation of one-to-one temporal mapping is in the lack of an explanation of how the most likely variants evolve with time.

Insofar as the law of motion of the one-to-one mapping type is assumed, irrespective of its own problem of plausibility, such a theory necessarily renders its boundary condition totally exogenous. The evolutionary process within the scheme of the law of motion of one-to-one mapping type consists in how its boundary condition evolves. The law of motion of the hierarchy one step higher thus has to be sought.[18] The logical consequence is, however, an infinite regression because of the imposed separability between the law of motion and its boundary condition in the sense that the dynamics of the boundary condition on a given law of motion does necessitate the law of motion one step higher.[19] The law of motion of the one-to-one mapping type no longer serves as a good approximation if the ensemble of variants does not remain stationary. The law of motion of the one-to-many mapping type has to be taken seriously.

Each reproductive unit, whether protobiological or biological, can vary its boundary and the interaction with its exterior through the process materilizing conservedness or material flow equilibration. Interaction changes imputed to individual organisms for the fulfillment of material flow continuity there are a form of resource exploitation. It could happen that some individual organisms might not be able to change their interaction with the exterior as responding to those variations originating elsewhere. Those individual organisms or species that cannot initiate their own variations turn out to be subject to resource limitation in the mode of mechanistic causality. In fact, resource limitation in the mode of mechanistic causality is not capable of raising evolutionary variations. It is resource limitation in the mode of final causality or resource exploitation that can give rise to them.[20] The law of motion of the one-to-many mapping type that is justifiable on a materialistic basis makes natural selection comprehensible on the basis of resource limitation in the mode of final causality. Thus, the genesis of variations is necessarily followed.

Natural selection that is properly understood refers to the process of resource exploitation, and natural selection as resource limitation in the mode of mechanistic causality is an extremely special and limiting case of this. For instance, an *in vitro* experimentation of replicating RNA molecules can undoubtedly simulate the operation of natural selection as resource limitation in which the experimental boundary condition that is forcibly separated from the law of motion acting there serves as a mechanistic cause. But, it is less competent compared with similar replication *in vivo* because in the latter the boundary condition is allowed to vary with the dynamics so as to internally aim at the conservedness serving as a final cause as in the case of material flow equilibration.

Resource limitation in the mode of mechanistic causality is no more than a hypothetically limiting case of a similar limitation in the mode of final causality, which is real. Only those individual organisms or species that are relatively incompetent in resource exploitation can be seen approximately as subject to the mechanistic causes set by those with the greater competence. Resource limitation has two fundamental characteristics, though they are closely interconnected. Each individual organism is a carrier embodying how resource exploitation has been practiced. At the same time, each organism is also equipped with the potential capacity for exploiting resources in a *de novo* manner, not completely controlled by the past. The evolutionary origin of a materialistic carrier embodying and expressing the manner of resource exploitation in the past is due to the fact that those material aggregates with the materialistic memory of how resource exploitation has been practiced are better prepared compared with those lacking such a memory in the exploitation to come. For the material aggregate carrying the memory of how resource exploitation has been practiced can utilize it in the subsequent material aggregation. But, the extent to which the past memory instructs future development is not rigid. The past does not determine every detail of the future, though the future is under the influence of the past. Both materialistic memory and its utilization can originate in resource exploitation on the part of open material aggregates, since their emergence is a particular form of resource exploitation. Resource exploitation is however always more than simply what is has accomplished because it maintains in itself intrinsic variability due to the very nature of the one-to-many temporal mapping.

Resource exploitation in the record can be genotypic in the sense that it describes how resource exploitation has been practiced. Similarly, resource exploitation in progress is phenotypic in that it refers to material accumulation in the making. Phenotypic development is under the influence of the genotype, but the latter does not fully control the former. A phenotypic development that is not found in the past record can be possible. This points to the fact that the phenotypic process can give rise to variations that have not yet been fixed in the materialistic record. If one claims that every detail of those phenotypic variations can be transformed into a memorized form in a materialistic carrier, this viewpoint would reduce to Lamarckianism. However, such an inheritance of acquired traits has long been refuted on the basis of Weismannian separation between soma and germ cells. If one claims, on the other hand, that no part of those phenotypic variations can be transformed into a memorized form, the neo-Darwinian separation between genotype and phenotype could be saved. The origin of genotypic variations has to be sought outside phenotypic development, and the latter can be variable only within the extent that the genotype permits.

The present neo-Darwinian interpretation renders the genotype a fixed boundary condition on the phenotypic dynamic process unless perturbed externally. The genotype can be preserved during development, implying that the phenotypic process is of one-to-one temporal mapping with respect to preserving the genotype. Nevertheless, the law of motion of the one-to-one mapping type has been refuted on the materialistic basis unless it is conceived as merely an approximation of reality. We are thus asked to seek an alternative that can avoid both the Lamarckian insistence on the inheritance of acquired traits and the neo-Darwinian clear-cut separation between genotype and phenotype.[21,22]

One clue in coping with the relationship between genotype and phenotype is the recognition that the denial of Lamarckian inheritance of acquired traits is not logically equivalent to affirming the neo-Darwinian separation between genotype and phenotype, and vice versa. Resource exploitation can bring in *de novo* variations that have been neither found in nor specified by the materialistic carrier of the past record. The point is that either a complete transformation of the generated variation into a memorized form or a lack of transformation of any part of those variations into memory would be a hypothetically limiting case of reality.

What remains real instead is the situation that some part of *de novo* variations can be transformed into a memorized form, which again influences the subsequent developmental process. If one considers the relationship between genotype and phenotype within the framework of the law of motion of the one-to-one mapping type, the genotype could constitute a fixed part of the boundary condition comes from the environment exerted by external agents. Consequently, phenotypic variability in the scheme of neo-Darwinian one-to-one temporal mapping originates totally in the environment, and leaves the genotype intact. The present scheme proceeds at the expense of admitting the environment whose scrutiny goes beyond any fathomable theory.

The phenotypic process following the law of motion of the one-to-many mapping type, on the other hand, can yield a variation of the materialistic carrier memorizing how resource exploitation has been practiced to the extent, of course, that it has been recorded. The phenotypic variability can induce variations in the genotype. The genotype as a boundary condition upon the phenotypic dynamic process is inseparable from the law of motion presiding over the very phenotypic process as far as the law is of the one-to-many mapping type. We thus note that genotypic variability is inseparable from phenotypic variability in the real process of one-to-many temporal mapping, whereas phenotypic variability in the neo-Darwinian process of one-to-one temporal mapping, by definition, leaves the genotype intact and invariant.

Strangely enough, however, both Lamarckian and neo-Darwinian schemes seek their basis upon the law of motion of the one-to-one mapping type by insisting upon the one-to-one correspondence between genotype and phenotype. The origin of variation is either in genotype or in phenotype, or in both. The difference between the two schemes has to do with how the one-to-one temporal mapping is implemented. Lamarckian inheritance of acquired traits lets the phenotype be the boundary condition upon how the genotype develops because variations are supposed to originate first in the phenotype. Moreover, the genotype also serves as a boundary condition on phenotype because of its definition. In fact, Lamarckian one-to-one correspondence between genotype and phenotype has been appreciated by dispensing with the distinction between the two.

In contrast, neo-Darwinian one-to-one correspondence lets the genotype be the boundary condition on the phenotype due to the fact that variations first occur in the genotype, while the phenotype is made with what is being formed instead of with what is forming. The present consideration urges us to admit that insofar as genotype is uniquely defined as a boundary condition upon the phenotypic process that proceeds in real physical time-space coordinate, the neo-Darwinian scheme has to be selected.[23] The Lamarckian framework would necessarily turn out to be false because of its insistence upon the phenotype serving as the boundary condition on the genotype. This lack of distinction between genotype and phenotype now makes the one-to-one correspondence between the two superfluous. Consequently, the law of motion of the one-to-one mapping type that maintains a clear separation from its boundary condition logically necessitates the adoption of the neo-Darwinian scheme.

However, once the law of motion is rightly recognized as being of the one-to-many mapping type, the situation will be drastically changed. The inseparability between the law of motion and its boundary condition makes it possible that phenotype can serve as a boundary condition upon genotype as much as the latter serves as the boundary condition upon the former. The Lamarckian claim of phenotype serving as a boundary condition on genotype, that would necessarily end up with a logical fallacy within the one-to-one temporal mapping framework, can now claim its legitimate currency in the scheme of one-to-many temporal mapping. Of course, the genotype also serves as the boundary condition upon the phenotype. This is not simply a matter of juxtaposing the Lamarckian and neo-Darwinian viewpoints. Far from it: the point of demarcation is whether the law of motion is of the one-to-one mapping type or of the one-to-many.

A neo-Lamarckian view of genotypic variability[24] that goes along with phenotypic one remains legitimate only when the law of motion of the one-to-many mapping type is properly appreciated. In particular, neo-Lamarckian phenotypic variability differs from its neo-Darwinian counterpart in that in the latter, genotypic variability does not follow from phenotypic variability by definition. Their difference is admittedly manifested in the manner of imple-

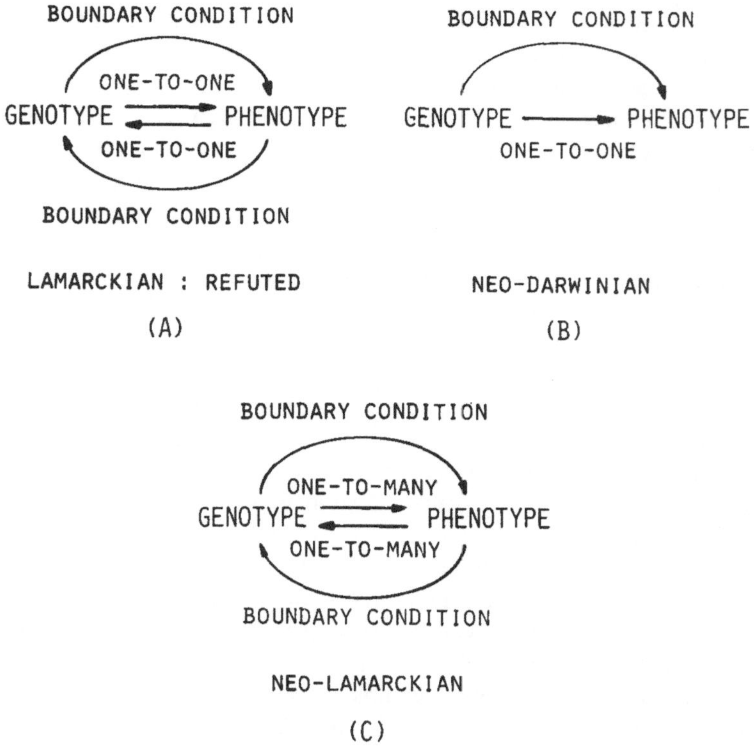

FIGURE 3. Evolutionary dynamics in terms of the law of motion and its boundary condition: (A) Lamarckian, (B) Neo-Darwinian, and (C) Neo-Lamarckian.

menting resource limitation in the phenotypic space. Resource limitation in the mode of mechanistic causality makes phenotypic variability strictly obedient to the impetus from the environment, leaving the genotype intact vis à vis variations upon the phenotype. This is what neo-Darwinian separation between genotype and phenotype requires.

On the other hand, resource limitation in the mode of final causality, or resource exploitation, allows the genotype that serves as the boundary condition on the dynamic phenotypic process to be variable during the process, for mechanistic causality requiring an invariant genotype is not invoked there. Resource exploitation is neo-Lamarckian, of course, within the realistic phenotypic process following the law of motion of the one-to-many mapping type. Phenotypic plasticity thus acquires totally different implications depending upon whether one sticks to a neo-Darwinian or to a neo-Lamarckian viewpoint. The neo-Darwinian interpretation is in fact logically consistent in claiming both resource limitation in the mode of mechanistic causality and the genotype, along with the environment serving as a mechanistic cause for the phenotypic process. Neo-Darwinian phenotypic plasticity strictly refers to the capacity of the genotype to allow an intervention from the environment into the phenotypic process, while the genotype is left intact and invariant. Insofar as one can define the environment that is clearly separated from the genotype, such a theoretical articulation of neo-Darwinian type would remain invulnerable. The question to be raised at this point is, however, whether the claimed clear separation between the genotype and the environment is valid.[25,26]

In contrast, neo-Lamarckian phenotypic plasticity within the framework of one-to-many temporal mapping implies genotypic variability or plasticity during the phenotypic process as depicted in Figure 3. What is unique to neo-Lamarckian genotypic-phenotypic plasticity is that it would keep evolutionary variability running in an inexhaustible manner indefinitely, as pictorally epitomized in the Red Queen hypothesis,[27,28] though the latter allows a wide variety of interpretations depending on the scenes on the stage.[29] We understand that unless one employs an anthropomorphic separation of individual organisms from their environment, evolutionary variability has to be sought within the evolutionary process. Evolutionary variability of the neo-Lamarckian type is connected with the fact that the phenotype serves as a boundary condition on the dynamics of genotype as much as the reverse. Preceding variation in the genotype serves as the boundary condition that induces subsequent variation in the phenotype, which in turn serves as the renewed boundary condition to induce further variations in the genotype. Crucial to the present observation is that variations in the genotype and those in the phenotype, both of which serve as the boundary condition for the subsequent process, do not occur in a simultaneous manner, because the effect of each process is propagated at a finite velocity not exceeding light velocity. The present sequential character of variation in the genotype and in the phenotype guarantees evolutionary variability that keeps operating indefinitely. The very origin of variability is within the law of motion of the one-to-many mapping type leaving the boundary condition variable during the motion thus bounded. Variability remains inexhaustible so long as the one-to-many temporal mapping remains legitimate.

Neo-Lamarckian genotypic-phenotypic plasticity departs from the original Lamarckian claim for the inheritance of acquired traits in that in the former there does not exist a one-to-one correspondence between genotype and phenotype. How phenotype influences genotype is left unspecified except that it obeys the law of motion of the one-to-many mapping type. The one-to-many temporal mapping exhibits an intrinsic asymmetry between *a priori* indefiniteness and *a posteriori* explicitness. This way of appreciating the intrinsic asymmetry renders evolutionary products the generator of subsequent evolutionary variations. For instance, a DNA molecule in this perspective serves as a generator inducing variations on the same DNA. A scheme allowing DNA to be the generator of variations is manifested in the presence of mobile genetic elements in maize,[30] certain bacterial phages including λ and Mu, and bacterial transposons.[31-33] Various mechanisms of genic excision and reintegration are latent in neo-Lamarckian genotypic-phenotypic plasticity in that the phenotypic process at any moment is latent with the one to appear later.

The phenotypic process continually dissolves its degeneracy through the operation of one-to-many temporal mapping. Neo-Lamarckian genotypic variability is no more than an indication of the phenotypic process dissolving its degeneracy. If one considers the possibility of mobile genetic elements within neo-Darwinian phenotypic plasticity that has nothing to do with its genotypic counterpart, on the other hand, a formidable problem would arise as to how such a mechanism for maintaining genotypic variability could come into existence in the first place. The neo-Darwinian dichotomy separating the law of motion from its boundary condition would be totally incompetent and at a loss with respect to the problem of how the genotype set aside as a boundary condition varies with the motion. A difficult part of the neo-Darwinian separation is in how to explain the origin of endogenous variability while admitting only exogenous variability. The present impasse cannot be resolved as long as one sticks to the neo-Darwinian scheme. Only when the neo-Darwinian separation is examined on a much deeper ground, can one expect a way out of the impasse. In fact, neo-Lamarckian genotypic-phenotypic plasticity offers such a possibility.

It is not evolution that brings in variability, but variability that brings in evolution. Endogenous variability assumes a key role in evolution. The emergence of sex,[34,35] for instance, installed a new form of genotypic variability by means of chromosome crossing and recombination. If one tries to deduce the emergence of sex on the basis of the neo-Darwinian separation between the law of motion and its boundary condition, it would be necessary to compare more than two different sets of neo-Darwinian separations, such as nonsexual and sexual reproduction, in the face of the common resource limitation. Insofar as the boundary condition driving the law of motion is definitely definable without leaving any amount of inseparability between the two, the present resource limitation would proceed in the mode of mechanistic causality. One can thus examine whether nonsexual reproduction could be alternated with sexual reproduction. However, the difficulty with the neo-Darwinian separation is that it requires an agent who prepares many different sets of the law of motion supplemented by its boundary condition and compares them. What serves as such an agent is persistently left unanswered even though it is invoked as a theoretical imperative. More than that, the neo-Darwinian separation would immediately exhaust evolutionary variability unless external agents to vary boundary conditions intervened.

On the other hand, neo-Lamarckian genotypic-phenotypic plasticity views the emergence of sex as an evolved form of evolutionary variability. The operation of one-to-many temporal mapping continually transforms the capacity of evolutionary variability into a more constrained form. Earlier evolutionary variability is more degenerate compared with later variability. It is the law of motion of the one-to-many mapping type that dissolves the degeneracy. Real genotypic variability in the form of chromosome crossing and recombination, when viewed from the neo-Lamarckian perspective, is evidence for evolution of evolutionary variability.

The evolution of variability also touches on the nature of variations thus generated especially with regard to whether they are microscopic or macroscopic. The neo-Darwinian separation of the law of motion and its boundary condition is however unable to cope with the present problem of whether evolutionary variations are microscopic or macroscopic, for the variability is by definition totally exogenous in its origin. In contrast, neo-Lamarckian genotypic-phenotypic plasticity continually affects the mechanism generating evolutionary variations, and accordingly specifies the nature of variations.

If the continual process of one-to-one temporal mapping is the case, the boundary condition would be continuously transformed, preserving its every detail. The extent of those variations generated as a continual transformation of boundary conditions would always be microscopic, keeping their continuity in time unless macroscopic variations of exogenous origin intervened. Although one may say that those variations could be macroscopic on a macroscopic time scale, such macroscopic variations would be superficial. An anthropomorphic arbitrary choice of time scale cannot serve as a cause of macroscopic variations.[36-38] One-to-one temporal mapping, unless exogenously disturbed macroscopically, is continual and continuous, thus rendering variations always microscopic and gradual if a sufficiently small time scale is chosen.

The real evolutionary process, on the other hand, obeys the law of motion of the one-to-many mapping type. An evolutionary predecessor is always degenerate with its evolutionary successors, and the degeneracy is constantly dissolved through the operation of one-to-many temporal mapping. The operation of the law of motion is continuous in time, but the effect can be discontinuous in time in eliminating some of those degenerate within the evolutionary predecessor. The growth of phylogenetic trees in fact is the result of a continuous evolutionary process supplemented by discontinuous effects as demonstrated in its branching. We thus observe that the operation of neo-Lamarckian genotypic-phenotypic plasticity is continuous in process, but discontinuous in effect. Macrovariation in evolution with a discontinuous character is more than just a theoretical artifact depending upon the choice of time scale. It is a fundamental indication of the operation of one-to-many temporal mapping.

Variability in evolution is based on the operation of final causality, but the final objective has been and never will be attained in the sense that the process materializing conservedness keeps running indefinitely through its successive repercussions. Mechanistic causality requiring an unambiguous boundary condition in advance for its own sake cannot cope with such an open-ended variability, because it contradicts the initial premise. Of course, mechanistic causality remains legitimate in its own right, and the neo-Darwinian separation between the law of motion and its boundary condition is irrefutable insofar as its initial premise is observed. Natural selection viewed in this way, for instance, must be understood in terms of resource limitation in the mode of mechanistic causality, leaving the mechanism of generating variations totally independent of the operation of the selection process.

There is nothing wrong logically with the operation of resource limitation in the mode of mechanistic causality. Any analysis of evolution based on an irrevocable and unequivocal premise is necessarily mechanistic in the sense that such a premise entails a mechanistic cause throughout. But, what we have observed is final causality, instead of mechanistic causality. It would be inappropriate to refer to natural selection as a mechanistic cause of the evolutionary process. On the contrary, natural selection will simply be a consequence of the operation of final causality as embodied in the process materializing conservedness.

Final causality underlying the process materializing conservedness necessarily renders the evolutionary precursor indefinite in its implication because of its intrinsic variability. If one tries a theory of evolution in which the logical antecedent may be assimilated to an evolutionary precedent, it would destroy evolutionary variability altogether. For the analytic drive toward the logical consequence is explicitly latent within the logical antecedent without allowing any amount of indefiniteness, otherwise every analytical theory would be destined to fail. As a matter of fact, evolutionary variability persistently refutes such a logical analysis. The analytic theory of evolution is self-contradictory. What is possible instead is to identify the nexus concatenating the evolutionary predecessor with its successor even if the inclusion of some sort of indefiniteness is inevitable.[39,40] If the nexus remains indefinite in its implication as it really should, it cannot serve as a logical antecedent to drive the subsequent logical analysis that has to turn out to be unambiguous. The nexus we have observed is the process materializing conservedness characterized by its own variability.

The evolutionary process viewed in terms of the process materializing conservedness is by no means its analysis, but its unfolding at best.[41] Such an evolutionary unfolding is more than just metaphorical.[42] Even if the evolutionary precedent is unfolded through the process materializing conservedness, what has not yet been unfolded persists. The presence of something that has not yet been unfolded defies complete analysis. The unfolding process admitting something that has not yet been unfolded is much weaker than the analytic process driving toward the logical consequence starting from its antecedent, in the latter everything involved has already been unfolded. When evolution remains as an incomplete unfolding process, its theoretical analysis would be too much to ask for. The alternative is to preserve what has not yet been unfolded as it is. Intrinsic variability is in fact a fundamental character of the materialistic carrier of evolution that persistently defies its complete unfolding.

V. ADAPTABILITY

Evolutionary variability constantly dissolves the degeneracy inherent in the evolutionary predecessor, and its implication is in the adjustment of interaction within each individual organism and among those individuals. The adjustment of interaction is such that at the least physical conservation laws such as those on energy and on matter have to be observed. Take, for instance, conservation of matter or its equivalent form known as material flow continuity. We recognize that conservation refers to both the state of being conserved and the process of becoming conserved, that is to say, the process materializing conservedness. However, these two cannot coexist physically. Only the process materializing conservedness is admissible in the realistic law of motion of the one-to-many mapping type. What survives during the adjustment of interaction is the process of becoming adjusted instead of the state of being adjusted. At the same time, the adjustment allows within itself such a dichotomy as what is adapting to and what is being adapted to, because the adjustment is completed when these two fit right into together. The relationship between adapting to and being

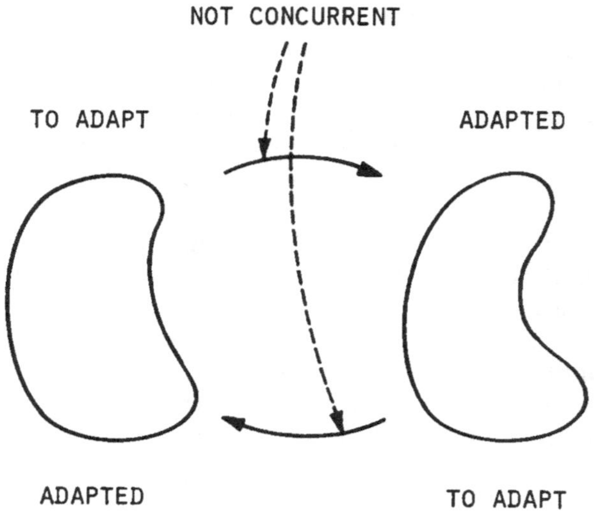

FIGURE 4. Interplay between to adapt and to be adapted. Because of the absence of concurrency between the two processes, adaptation keeps surviving without being trapped by a stalemate lacking further adaptability.

adapted to as demonstrated in Figure 4 is bilateral in the sense that what is being adapted to is also viewed as being what is adapted to by its counterpart.

It is thus found that the process of becoming adjusted internally accommodates the processes of becoming adapted. This internal mechanism of adjustment now gives rise to what we call adaptation. The adjustment of a whole body presumes the adaptation of each of its parts to all the others.[43]

Adaptation can also imply both the state of being adapted and the process of becoming adapted or the process materializing adaptedness. By only the process of becoming adapted is realizable in time.[44] Adaptation in the sense of the process of becoming adapted locally belongs to the global adjustment in the mode of final causality. A unique feature of the process of becoming adapted is that although its consequence will have materialized adaptedness, the state of being adapted is yet to come. Aptation[45] as a final cause of adaptation is not specific enough to identify what will have been the concrete context of exaptation[45] recorded as having been adapted. The trio of aptation, adaptation, and exaptation in essence refers to the process materializing adaptedness to come, in progress, and already in the record, respectively.

As a matter of fact, an advantage of employing the present triad lies in a special character of adaptation that lets the process materializing conservedness be deciphered in terms of the dichotomy between adapting to and being adapted to. This dichotomy, however, cannot and does not remain rigid. An agent of adapting is also viewed as an object to be adapted by its counterpart, and vice versa. The internal process of becoming adapted is bilateral between adapting and being adapted.[46] The process materializing conservedness is thus an integration of the internal processes of becoming adapted. Adaptation is local and internal, while conservation is wholistic and global.

The global adjustment of interaction for conservation is also explicable by identifying how each local part becomes fitted with its surrounding through their mutual interaction. The adjustment in the mode of final causality presumes the process of becoming fitted that proceeds in a bilateral manner, in which any material aggregate involved participates in the process of becoming fitted. We notice here that both adaptedness and fittedness are used interchangeably. Fitness as an attribute of the state of being fitted constantly avoids its completion as much as the state of being adapted persistently evades its full complementation.[27,28] The law of motion of the one-to-many mapping type that renders the process of becoming fitted or adapted real thus dispenses with fitness.

On the other hand, however, if the hypothetical law of motion of the one-to-one mapping type is taken instead, a totally different perspective arises. The global interaction adjustment for conservation is accomplished instantaneously at every moment, thus dispensing with the process of becoming adapted or fitted. The state of being adapted is to be actualized at any moment as manifesting its fitness. Adaptation embodied in the state of being adapted is strictly static, and it cannot induce dynamic processes by itself. What is conceivable instead is at most comparative statics that can tell which is the fittest when there are more than two states available. Adaptation within the scheme of comparative statics works only when fitness is definable and various states to be compared preexist. Fitness as a measure of reproductive efficiency or reproductive success, if definable, can determine which of these preexisting variants is selected as being adapted. Such comparative statics can yield the state of being adapted.

Nevertheless, comparative statics raise two serious, insurmountable questions. One is how the preexisting variants could be prepared and by whom. The other is who measures the fitness and when, before, after, or right in the middle of events. The source of variants is evolutionary variability that is not within the reach of the law of motion of the one-to-one mapping type, and measured fitness is about a result of becoming fitted, not about fitness per se. In spite of this, comparative statics steadfastly avoid direct confrontation with these two questions by invoking external agents whenever needed. Even if one says that adaptation would proceed in the direction along which the fitness or the inclusive fitness increases,[47,48] what is implied by the present adaptation is still comparative and static. For the motive force of the superficial dynamics is the definability of the fitness and the preexisting variant states to be examined in reference to their fitness. Insofar as fitness is claimed to be unambiguously definable even in the form of a law of motion, the consequence of comparative statics is explicitly latent right at the moment variant states are prepared. Although one might conceive that certain variants among those prepared would become dominant with time as wiping out others and that the increasing dominance would be associated with adaptation, such an adaptation was explicit when the fitness along with variant states was defined. Adaptation and fitness are made forcibly indistinguishable from each other within the dynamics of mechanistic causality.

Once one decides to retreat from rigid mechanistic causality and to appreciate the reality of final causality, fitness, as an attribute of the state of being fitted or adapted, has to be reexamined. What emerges instead is the state of being fitted as an *a posteriori* consequence of the process of becoming fitted or adapted. Adaptation is the process of becoming adapted. Even if one tries to associate fitness with a system of measuring the genetic capability of organisms in terms of net reproductive efficiency, what is actually meant by this fitness is the process of becoming fitted because it can be identified as such only after the net efficiency has been measured. Despite the *a posteriori* character of measurement, if one still insisted on both the utility of and the definability of fitness even before measurement, it would necessarily turn out to be a mechanistic cause. And again, insurmountable problems with mechanistic causality would creep in as to who defines fitness, on what ground, and how. The real operation of final causality, on the other hand, lets adaptedness or the realization of the process of becoming fitted instead of fitness, be an internal objective toward which the process known as adaptation persistently drives itself.

Adaptation as the process of becoming adapted is admittedly teleological in its operation. But it does not presume the state of being adapted as its final objective. The state of being adapted is constantly fading out during the process of becoming adapted. This is because, when the present becomes the immediate past, the process is so constrained as not to break physical conservation laws. The teleology associated with the process of becoming adapted is due to the constraint inherent to the transference of the present into the immediate past. Although it would be perfectly legitimate to say that the immediate past has developed into the present, it does not necessarily follow that the immediate past would uniquely determine the present. The teleology intrinsic to adaptation works when the immediate past cannot uniquely determine the present. It asserts that whatever the relationship between the immediate past and the present may be, physical conservation laws have to be observed when the present is transferred into the immediate past that remains irrevocable. Of course, it is legitimately admitted that there must be a certain mechanism driving the immediate past into the present. This does not, however, imply that a complete identification of the mechanism would also be required in order to claim such a mechanism in the first place. In fact, the teleological functioning we find in the process of becoming adapted is a common denominator among those possible mechanisms driving the immediate past into the present.

Adaptation as the process of becoming adapted should be distinguished from adaptation as the process of increasing fitness, even though the latter may seem teleological in its outlook. If the final state is clearly defined in advance as much as the fitness is unambiguously determined, the drive toward the goal would already be explicit in the premise. Adaptation conceived as the process increasing fitness would thus proceed in the mode of mechanistic causality. The theory of adaptation is somewhat ironical. If one understands that any theory should be constructed on an invariant premise having an unambiguous content, such a premise would serve as the mechanistic cause to drive the whole theory. It is in this sense that adaptation as the process increasing fitness is mechanistic in its operation despite its teleological outlook. However, adaptation in reality cannot be mechanistic because of the impossibility of drawing a rigid and irrevocable demarcation between what is adapting to and what is being adapted to. The nonmechanistic character of adaptation is rooted in physical conservation laws that are fundamentally teleological without taking mechanistic causality for granted in advance. Imposition of mechanistic causality upon physical conservation laws is a theoretical artifact at best.

Final causality remains legitimate in reality irrespective of any analytic articulation on the part of theory construction, while mechanistic causality can be possible only at the expense of the more fundamental final one. Adaptation as the process of becoming adapted is a way of appreciating the final causality intrinsic to the process materializing conservedness as preventing an intrusion of mechanistic causality.

A natural exclusion of mechanistic causality from the process of becoming adapted is due to the bilaterality of adaption. An arbitrary pair of interacting material aggregates are involved in the process of adapting to each other. Adaptability to become adapted is latent in every material aggregate, such as individual organisms and their constituent elements. Adaptability is thus a local expression of evolutionary variability.[49] Local variability that can be identified as adaptability functions both with and within individual organisms. Consequently, adaptability is classified into two groups, genotypic and phenotypic. Phenotypic adaptability is in the realization of resource exploitation among individual organisms, whereas genotypic adaptability is within each organism as a descendant of those organisms that have completed resource exploitation.

Resource exploitation in individual organisms is phenotypic, whereas similar exploitation at the level of germ cells or their evolutionary equivalent is genotypic. Genotypic adaptability exhibits itself in the evolutionary enhancement of reproductive fidelity through the genealogical sequence of surviving organisms because of the involvement of material flow equilibration there. The genotype can be variable and adaptive only to the extent that its reproductive fidelity is enhanced with time. The adaptive capacity of a materialistic carrier of reproductive fidelity is thus constrained with time under the condition that the carrying organism has survived through its genealogical sequence. At the same time, the genealogical persistence has relied upon phenotypic adaptability of those intervening individual organisms in exploiting their resources, indicating that the genotype allows for phenotypic adaptability.

When one says that adaptation proceeds in the phenotypic space, this is because phenotypic adaptability dominates resource exploitation compared with the genotypic counterpart. The decoupling between genotype and phenotype lies in the contrast between the specificity in the manner of resource exploitation in the past and the multitude of possible alternatives faced by resource exploitation in progress. Phenotypic adaptability originates in the fact that the past process of becoming adapted cannot control its future development. But, the final causality intrinsic to resource exploitation is operative. Resource compensation[49] propagates in the network of bilateral processes of becoming adapted.

Any individual organism that is exploited by others compensates its depletion by exploiting resources elsewhere, otherwise its covering species could not have survived. Likewise, any organic element in an organism is in the process of compensation with others in terms of energy and material resources. The compensation process within an organism is more than being homeostatic, because what would be called a referential level to be reestablished is also involved in the process of internal compensation. The capacity of compensation as embodied in material flow equilibration at the level of individual organisms is adaptability. What is more, adaptability goes beyond the genotype. The very origin of adaptability is entailed by the law of motion of the one-to-many mapping type. It is legitimate to say that the genotype constantly constrains the extent of phenotypic adaptability, but the latter could always, at any moment, imply more than what the genotype prescribes. Phenotypic adaptability precedes the genotype and its adaptability. Evolutionary separation between genotype and phenotype is just a consequence of phenotypic adaptability.

VI. SUMMARY

Evolution viewed from the legitimate law of motion of the one-to-many mapping type lets the endogenous enhancement of reproductive fidelity be the most fundamental process. Although it might be conceived that reproductive units with perfect fidelity, perturbed intermittently by disturbances of exogenous origin, could trace the evolutionary process, this viewpoint based on the hypothetical law of motion of the one-to-one mapping type could be saved only at the expense of the legitimate one-to-many temporal mapping. The uniqueness of one-to-many temporal mapping is in its final causality, allowing at the least physical conservation laws to be observed *a posteriori*, whereas the law of motion of the one-to-one mapping type necessarily operates in the mode of mechanistic causality. An approximation of one-to-many temporal mapping in reality by one-to-one mapping is not admissible in the sense that final causality cannot be approximated by mechanistic causality as a matter of principle.

Natural selection properly viewed from the underlying one-to-many temporal mapping is also final in its operation. The resource limitation that natural selection faces proceeds in the mode of final causality. Resource limitation in the mode of final causality lets individual organisms be involved in resource exploitation. Differential reproduction is simply a consequence of resource exploitation on the part of participating individual organisms and species. Resource exploitation among individual organisms yields evolutionary variations that are phenotypic. The one-to-many temporal mapping that makes a clear separation between the law of motion and its boundary condition impossible turns the genotype into the boundary condition on the phenotype and vice versa. If the law of motion were of the one-to-one mapping type, only the neo-Darwinian perspective that renders the genotype to be a boundary condition on the phenotype would remain legitimate and the reversed influence from phenotype to genotype would admittedly be prohibited. The Lamarckian claim that the phenotype serves as a boundary condition on genotype would be illegitimate. On the other hand, however, if the law of motion of the one-to-many mapping type is rightly appreciated as such, a neo-Lamarckian insistence on genotypic variability that can be associated with phenotypic variability will resume its currency. The primary difference between neo-Darwinian and neo-Lamarckian viewpoints rests on whether one takes the law of motion of the one-to-one mapping type or that of the one-to-many.

One-to-many temporal mapping lets adaptation as the process of becoming adapted proceed in the mode of final causality. The process of becoming adapted is bilateral in the sense that an object of being adapted to turns out to be an agent of adapting to and vice versa. Indefinite repercussions among the bilateral processes of becoming adapted prevents an establishment of the state of being adapted. Fitness as an attribute of the state of being adapted cannot be definable in reality because its full realization is never achieved. Adaptability as the capacity for becoming adapted depends on evolutionary variability. Phenotypic adaptability serves as a premise for the evolution of the genotype. Evolution is an example of a physical process that operates in the mode of final causality.

REFERENCES

1. **Bernstein, H., Byerly, H. C., Hopf, F., Micod, R. A., and Vemulapalli, G. K.,** The Darwinian dynamic, *Q. Rev. Biol.*, 58, 185, 1983.
2. **Englisch, U., Gauss, D., Freist, W., Englisch, S., Sternbach, H., and von der Hoar, F.,** Error rates of the replication and expression of genetic information, *Angew. Chem. Int. Ed. Engl.*, 24, 1015, 1985.
3. **Weiss, A.,** Replication and evolution in organic systems, *Angew. Chem. Int. Ed. Engl.*, 20, 850, 1981.
4. **Orgel, L. E.,** Selection in vitro, *Proc. R. Soc. London Ser. B*, 205, 435, 1979.
5. **Vrba, E. S. and Eldredge, N.,** Individuals, hierarchies and processes: towards a more complete evolutionary theory, *Paleobiology*, 10, 146, 1984.
6. **Newman, C. M., Cohen, J. E., and Kipnis, C.,** Neo-Darwinian evolution implies punctuated equilibria, *Nature (London)*, 315, 400, 1985.
7. **Matsuno, K.,** Nonsteady stochastic properties of Markovian kinetics and macroscopic fluctuations, *Phys. Rev. A*, 12, 639, 1975.
8. **Pattee, H. H.,** The problem of biological hierarchy, in *Towards a Theoretical Biology*, Waddington, C. H., Ed., Edinburgh University Press, Edinburgh, 1970, 117.
9. **Campbell, T.,** "Darwinian causation" in hierarchically organized biological systems, in *Studies in the Philosophy of Biology*, Ayala, F. J. and Dobzhansky, T., Eds., University of California Press, San Francisco, 1974, 179.
10. **Noda, H.,** Probability of life: rareness of realization in evolution, *J. Theor. Biol.*, 95, 145, 1982.
11. **Raup, D. M. and Valentine, J. W.,** Multiple origins of life, *Proc. Natl. Acad. Sci. U.S.A.*, 80, 2981, 1983.
12. **Lewontin, R. C.,** The units of selection, *Annu. Rev. Ecol. Syst.*, 1, 1, 1970.
13. **Stanley, S. M.,** A theory of evolution above the species level, *Proc. Natl. Acad. Sci. U.S.A.*, 72, 646, 1975.
14. **Hull, D.,** Individuality and selection, *Annu. Rev. Ecol. Syst.*, 11, 311, 1980.
15. **Byerly, H. C.,** Natural selection as a law — principles and processes, *Am. Nat.*, 121, 739, 1983.
16. **Vrba, D. S.,** What is species selection?, *Syst. Zool.*, 33, 318, 1984.
17. **Arnold, S. J.,** Morphology, performance and fitness, *Am. Zool.*, 23, 347, 1983.
18. **Rosen, D. E.,** Do current theories of evolution satisfy the basic requirements of explanation?, *Syst. Zool.*, 31, 76, 1982.
19. **Cracraft, J.,** Species selection, macroevolutionary analysis, and the hierarchical theory of evolution, *Syst. Zool.*, 34, 222, 1985.
20. **Lewontin, R. C.,** The organism as the subject and object of evolution, *Scientia*, 118, 65, 1983.
21. **Eden, W.,** Inadequacies of neo-Darwinian evolution as a scientific theory, in *Mathematical Challenges to the Neo-Darwinian Interpretation of Evolution*, Moorhead, P. S. and Kaplan, M. M., Eds., Wistar Institute Press, Philadelphia, 1967, 5.
22. **Gould, S. J.,** Darwinism and the expansion of evolutionary theory, *Science*, 216, 380, 1982.
23. **Orzack, S. H.,** The modern synthesis is partly Wright, *Paleobiology*, 7, 128, 1981.

24. **Buscaglia, M.,** Conceptions of biological evolution and an approach to phylogeny-ontogeny, *Hum. Dev.*, 27, 240, 1984.
25. **Kaplan, R. H. and Cooper, W. S.,** The evolution of developmental plasticity in reproductive characteristics: an application of the ''adaptive coin-flipping'' principle, *Am. Nat.*, 123, 393, 1984.
26. **Osse, J. W. M.,** Morphology and evolution, *Acta Morphol. Neerl. Scand.*, 21, 49, 1983.
27. **Van Valen, L.,** A new evolutionary law, *Evol. Theor.*, 1, 1, 1973.
28. **Van Valen, L.,** Two modes of evolution, *Nature (London)*, 252, 298, 1974.
29. **Maynard Smith, J.,** A comment on the Red Queen, *Am. Nat.*, 110, 325, 1976.
30. **McClintock, B.,** Mechanisms that rapidly reorganise the genome, *Stadler Genet. Symp.*, 10, 25, 1978.
31. **Shapiro, J. A.,** *Mobile Genetic Elements*, Academic Press, New York, 1983.
32. **Middaugh, C. R., Brandau, D. T., Micanovic, R., and Katzeustein, G. E.,** Entropic DNA, *Med. Hypoth.*, 17, 387, 1985.
33. **Hartl, D. L., Medhora, M., Green, L., and Dykhuizen, D. E.,** The evolution of DNA sequences in escherichia coli, *Philos. Trans. R. Soc. London Ser. B*, 312, 191, 1986.
34. **Maynard Smith, J.,** *The Evolution of Sex*, Cambridge University Press, Cambridge, 1978.
35. **Bernstein, H., Byers, G. S., and Michod, R. A.,** Evolution of sexual reproduction: importance of DNA repair, complementation, and variation, *Am. Nat.*, 117, 537, 1981.
36. **Gingerich, P. D.,** Punctuated equilibria-where is the evidence? *Syst. Zool.*, 33, 335, 1984.
37. **Penny, D.,** Two hypotheses on Darwin's gradualism, *Syst. Zool.*, 34, 201, 1985.
38. **Scudo, E. M.,** Darwin, Darwinian theories and punctuated equilibria, *Syst. Zool.*, 34, 239, 1985.
39. **Riedl, R.,** *Order in Living Organisms*, John Wiley & Sons, London, 1979.
40. **Bohm, D.,** *Wholeness and the Implicate Order*, Routledge and Kegan Paul, London, 1980.
41. **van der Hammen, L.,** Unfoldment and manifestation: the natural philosophy of evolution, *Acta Biotheor.*, 32, 179, 1983.
42. **Schwabe, C. and Warr, G. W.,** A polyphyletic view of evolution: the genetic potential hypothesis, *Perspect. Biol. Med.*, 27, 465, 1984.
43. **McDonald, J. F.,** The molecular basis of adaptation: a critical review of relevant ideas and observations, *Annu. Rev. Ecol. Syst.*, 14, 77, 1983.
44. **Krimbas, C. B.,** On adaptation, neo-Darwinian tautology, and population fitness, in *Evolutionary Biology*, Vol. 17, Hechet, M. K., Wallace, B., and Prance, G. T., Eds., Plenum Press, New York, 1984, 1.
45. **Gould, S. J. and Vrba, E. S.,** Exaptation — a missing term in the science of form, *Paleobiol.*, 8, 4, 1982.
46. **Gould, S. J. and Lewontin, R. C.,** The spandrels of San Marco and the panglossian paradigm: a critic of the adaptationist program, *Proc. R. Soc. London Ser. B*, 205, 581, 1979.
47. **Hamilton, W. D.,** The genetical evolution of social behavior. I and II, *J. Theor. Biol.*, 7, 1, 1964.
48. **Wilson, E. O.,** *Sociobiology: The New Synthesis*, Harvard University Press, Cambridge, Mass., 1975.
49. **Conrad, M.,** *Adaptability: The Significance of Variability from Molecule to Ecosystem*, Plenum Press, New York, 1983.

Chapter 10

EQUILIBRATION IN ECOLOGY

I. INTRODUCTION

Adaptation as the bilateral processes of becoming adapted operates not only in the genotype and in the phenotype, but also on the population level. The ecological niche as a spatio-temporal locale for a reproductive species depends upon how each individual organism belonging to the species recognizes its environment. The way of recognition as a particular mode of materialistic interaction is intrinsically degenerate with those possible alternatives that can appear later. Dissolution of the degeneracy proceeds with material flow equilibration in a material aggregate as the sum total of individual organisms that constitute a species belonging to a particular ecological niche of its own. Ecological adaptability as the capacity of a species occupying an ecological niche to adjust the interaction with its environment is inherent in the process of each species coming to establish and to recognize its own niche. Material flow equilibration at the ecological level induces ecological variability and adaptability, and dissolves the degeneracy latent in evolutionarily earlier niches. If ecological adaptation proceeds without affecting both the genotype and the phenotype of the individual organisms involved, the process would reduce to ecological succession, leading to its climax. Ecological climax can be perturbed by disturbances of exogenous origin, and accordingly, succession could be resumed. However, ecological climax cannot remain stationary indefinitely because ecological variation cannot be completely isolated from phenotypic variations at the individual level, which in principle do not operate in a manner completely separated from genotypic variations. Ecological evolution is rooted in ecological adaptability that could yield variations in the genotype in the long run. Ecological interaction sets a boundary condition upon genotype as much as the latter does to the former, with the phenotype of individual organisms serving as an intermediary. Evolution from prokaryotes to eukaryotes could have been a consequence of ecological adaptability that can latently induce variations in the genotype.

II. ECOLOGICAL NICHE

Evolutionary variability and adaptability can be identified within each individual organism. But, individual organisms do not exist in isolation. They interact among themselves at least on the population level. Insofar as one recognizes that the process of becoming adapted is bilateral in the sense of influencing as much as being influenced, the evolutionary process cannot be adequately understood without paying due attention to the dynamics among individual organisms.

Individual organisms as reproductive open material aggregates have their own unique way of exploiting material resources. When a reproductive species is defined in such a manner that both the progenitor and the progeny individuals belonging to the species share the same recognition of their material resources and share the same pattern of their exploitation, each ecological niche defined as a spatio-temporal locale for resource exploitation for a species can maintain one and only one species. This is in essence what Gause's competitive exclusion principle[1] tells us. If there were more than two different reproductive species sharing the same ecological niche in the sense of employing the same pattern of resource exploitation, only one species would actually survive. The one with the greater net reproductive rate would survive. This is, however, not a consequence of competing with each other in the natural setting. Comparative statics is quite theoretical in origin. Although it might be possible for an experimenter to artificially prepare more than two different species in an environment that would allow only one niche, this is not what nature does. Each of the reproductive species comes to occupy its own ecological niche without overlapping any other.

The real dynamics of each reproductive species has the corresponding niche as its boundary condition. In fact, the law of motion of the one-to-many mapping type renders the law of motion inseparable from its boundary condition. This inseparability underlies the conformity between a reproductive species and its niche. Consequently, a way of looking at evolutionary variability and adaptability on the global level comes to refer to the reproductive dynamics of each species and to its boundary condition as a niche.

If one tries to deduce the emergence of and the presence of ecological niches within the scheme of the hypothetical law of motion of the one-to-one mapping type, a formidable problem would arise with regard to how to justify its boundary condition. One-to-one temporal mapping would leave no alternatives other than having recourse to the external agents to erect and to keep maintaining its boundary condition. Although it might be conceivable that the external environment could be solely responsible for establishing a particular niche for a particular reproductive species, this would simply be an admission of our inability to decipher how the external agents work. The more fundamental problem, however, is whether we would be actually incompetent as alleged. In particular, the law of motion of the one-to-one mapping type leaves the dynamics of how its boundary condition develops exclusively intact because of its basic characteristic of being isolated from the law of motion. If the problem of how the external environment works in establishing and maintaining ecological niches becomes of primary concern, the realistic one-to-many temporal mapping actually has to be invoked.

A reproductive species occupying a certain ecological niche maintains, as a material aggregate, the sum total of those individual organisms belonging to it. The material aggregate on the species level takes in material resources through its boundary identified as its ecological niche. Material flow equilibration on the species level functions much as the similar equilibration operates on the level of individual organism, where the physical boundary of each organism serves as its endogenous boundary condition.

A carrier of material flow equilibration maintains a coherent coordination of interactions with its exterior, otherwise the carrier would disintegrate into smaller ones maintaining their locally coherent coordinations instead. An individual organism certainly serves as a carrier of material flow equilibration because the whole body is physiologically well coordinated while taking in and then letting out material resources. Similarly, the material aggregate on the species level can also maintain a coherent coordination of interactions with its niche by establishing communication among those individual organisms involved. A representative method of communication for establishing coherent coordination within the species is reproductive interaction. The coherent coordination of interactions on the species level is embodied in the manner of recognizing and exploiting material resources.

When the law of motion of the one-to-many mapping type is deciphered on the species level, the law of motion is manifested in how each individual organism, as a representative of the covering species, recognizes and exploits its material resources and how the niche to be exploited serves as its boundary condition. However, the intrinsic inseparability between the law of motion and its boundary condition does not leave the relationship between exploiting and being exploited invariant, as a matter of principle. Each reproductive species is involved in both activities, exploiting material resources and being exploited by others in one way or another. Of course, reproductive dynamics on the species level and the corresponding ecological niche are an expression of the underlying fundamental law of motion of the one-to-many mapping type, but both of them are constantly in the process of influencing each other. Speciation could occur when a certain material aggregate on the species level ceases to be a carrier of material flow equilibration that can maintain a coherent coordination

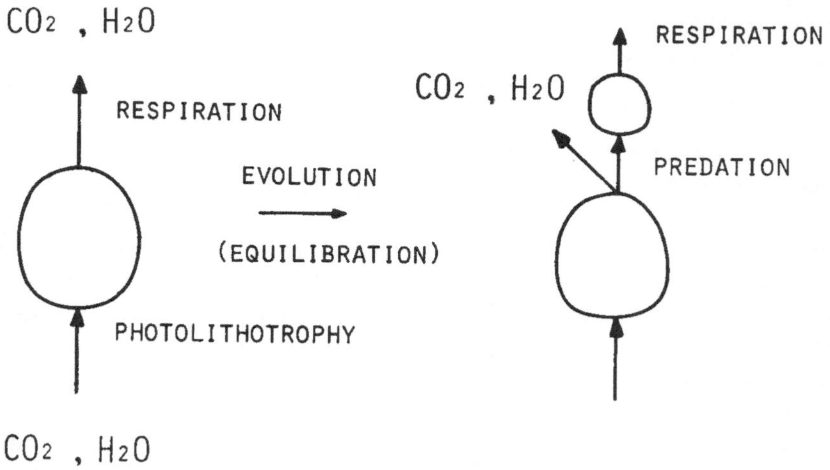

FIGURE 1. Evolution from prokaryotes or prokaryote-eukaryote complexes to eukaryotes. The lowering of the turnover rate of organic carbon is associated with the evolutionary transition.

of interactions with its niche. Reproductive isolation is just one example of speciation, indicating that a former species has lost a spatially coherent coordination of resource exploitation. The dichotomy between the reproductive species as a unit of resource exploitation and the corresponding niche it exploits is no more than an incomplete expression of material flow equilibration on the species level.

Material flow equilibration on the species level constantly dissolves the degeneracy in earlier ecological niches through its process of self-constraining. In particular, a most significant dissolution of the degeneracy is seen in the emergence of the niche for eukaryotes from the preceding one where only prokaryotes were found or both coexisted in a degenerate manner.[2] The prokaryotic way of recognizing and exploiting material resources was through the synthesis of organic matter by photolithotrophic bacteria, and the accumulated organic matter was exploited by oxidative phosphorylation.[3,4] In fact, a major consequence of the Precambrian prokaryotic evolution was the establishment of approximately 1.5-V difference in the oxidation-reduction potential.[5] Material flow equilibration at the ecological niche for prokaryotes was such that some of the organic matter produced by photosynthesis was left in the environment without being exploited by oxidative phosphorylation. One of its consequences was the oxidation of the atmosphere. This could help establish a *de novo* manner of exploiting organic matter by eukaryotic predation (as shown in Figure 1) in addition to prokaryotic oxidative phosphorylation. The emergence of a new ecological niche for eukaryotes could follow. The prokaryotic ecological niche was degenerate with the eukaryotic one, and the dynamics internal to prokaryotes were responsible for dissolution of the degeneracy.[6,7]

Ecological niches are maintained between the two processes, exploiting material resources and being exploited by others. The Precambrian evolution of prokaryotes kept the ecological niche for prokaryotes between the two processes, photolithotrophy by autotrophs and oxidative phosphorylation by heterotrophs. The excess food which heterotrophs could not oxidize entered into the lithosphere as permanent sediments. The flow of organic matter into permanent sediments, while oxidizing the atmosphere, was a form of the prokaryotic ecological niche to be exploited by its external environment. The prokaryotic niche, however, did not remain stationary. When the degeneracy with the niche was dissolved due to its internal dynamics and accordingly the ecological niche for eukaryotes emerged, the bilateral process of resource exploitation became more complex because of the dissolution of the former degeneracy.

Eukaryotes exploited material resources by predation, while they were exploited in the form of detritus by prokaryotes.[8] Dissolution of the degenerated ecological niches into the actual ones presupposes that the turnover rate of organic carbon round the photosynthesis-respiration cycle in the whole ecosystem would decrease as a consequence of the emergence of the new resource exploitations. The reducing carbon compounds as the material resources to be exploited must be utilized before they are consumed by respiration. However, the decrease in the turnover rate of organic carbon in the whole ecosystem is not what dissolves the degeneracy within earlier ecological niches. Resource exploitation at material aggregates, whether macroscopic or microscopic, is a primary cause of the decrease in the turnover rate. More fundamentally, resource exploitation is rooted in the law of motion of the one-to-many mapping type. The actual emergence of new resource exploitation proceeds through material flow equilibration at the level of ecological niches.

Eukaryotic ecological niches have further dissolved their degeneracy with evolutionary time, and niches for secondary and tertiary consumers in addition to primary ones have been established. Each ecological niche has its own reproductive dynamics. Nevertheless, the relationship between the reproductive dynamics of a species and its boundary condition identified as its niche is necessarily incomplete in that the way the niche is influenced by other species is left unspecified. One way to overcome the present incompleteness is to invoke the turnover rate of organic carbon in the whole ecosystem because it serves as a thread connecting all of the niches in the ecosystem. Every ecological niche is involved in processing that reduces carbon compounds in one way or another between the two ends of photosynthesis and respiration.

The complex ecosystem of prokaryotes and eukaryotes, which emerged at the end of the Proterozoic era[9,10] and has lasted even up until the present time, is a dissipative system taking in and letting out carbon dioxide. The ecosystem is in fact a form of throughflow of organic carbon. Dissipation in the form of respiration is actually measured by the molecule number of carbon dioxide emitted from the ecosystem, in which both prokaryotes and eukaryotes are involved in respiration. Since each molecule of carbon dioxide is temporally and statistically independent of those emitted either earlier or later, the degradation process emitting carbon dioxide can be seen to be stochastic, though the ecosystem is constantly compensated by photosynthesis. Probabilistically, the most likely degradation process, while compensated by other means, is the one that minimizes the decay rate that measures the average molecule number of carbon dioxide going out per unit time per unit amount of organic carbon existing in the ecosystem.[11,12]

The minimum degradation flow of carbon dioxide is actually identical with what the least turnover rate of organic carbon in the ecosystem implies. Of course, neither the minimum respiration flow of carbon dioxide per unit amount of organic carbon nor the least turnover rate of organic carbon serves as a final cause for the evolution of ecosystem. What is final in its operation is material flow equilibration in each ecological niche. The decrease in the turnover rate of organic carbon or in the respiration flow of carbon dioxide could materialize only through the dissolution of the degeneracy within earlier ecological niches, resulting in complexity increase of the actual ecological niches.

III. ECOLOGICAL ADAPTABILITY

Protobiological and biological evolution of the ecosystem can sometimes be perturbed by disturbances of exogenous and nonbiological origin, such as the impact of asteroid, oceanic turbulence, and forest fires to name a few. The ecosystem recovers its natural course of development after such disturbances.[13] However, the recovery process, usually referred to as ecological succession, differs from the evolution of the ecosystem in that the genotype and the phenotype of involved individual organisms do not change appreciably. Ecological succession is initiated by forced changes upon the existing ecological niches and sets its own course of development without inducing variations in the genotype and in the phenotype of individual organisms. The goal as the climax of such an ecological succession is the one that has already been established prior to the disturbances applied externally during the natural evolution of ecosystem, since both the genotypes and the phenotypes are supposed to remain invariant. The actual ecological niches are constructed so as to realize the least possible turnover rate of organic carbon in the whole ecosystem. A fundamental characteristic of ecosystem dynamics is the decrease in the turnover rate with time irrespective of whether in evolution or in succession because of the involvement of material flow equilibration, especially with regard to the flow of organic carbon at each ecological niche. Ecological succession can decrease the turnover rate only by endogenously adjusting ecological niches,[14] while an evolutionary process can decrease the turnover rate by varying individual genotypes in addition to the endogenous adjustment of those niches.

The natural reconstruction of ecological niches after disturbances applied externally certainly exhibits ecological adaptability. However, ecological adaptability is more than a simple shift of the course of development of ecological niches. Dissolution of the degeneracy within earlier ecological niches can also bring in variations in the genotype of individual organisms. The emergence of eukaryotic ecological niches from the proceding prokaryotic ones manifested ecological adaptability as being capable of adjusting the web of ecological niches. What is more, the emergence of eukaryotes was accompanied by corresponding variations in the genotype. Ecological adaptability can serve as a cause of variations in the genes of individual organisms. Genotypes, phenotypes, and ecological niches are mutually causative. This is because each of the triad of genotype, phenotype, and ecological niche can serve as a boundary condition upon the dynamics of the remaining two.

The actual law of motion of the one-to-many mapping type that makes itself inseparable from its boundary condition does not allow for any stand-still boundary condition that remains indifferent to the prevailing motion. If one takes the hypothetical law of motion of the one-to-one mapping type that is invulnerable, by definition, against whatever happens on the phenotypic and on the ecological levels, variations in individual phenotypes and in the ecological niche for those individual organisms would be a consequence of the dynamics whose boundary condition is set by the genotype. Variability would be unidirectional in its influence, always from the genotype. However, once the reality of one-to-many temporal mapping is appreciated, the bilateral variability of each of genotype, phenotype, and ecological niche is found to sustain indefinite repercussions of variations among the three. The neo-Lamarckian bilaterality in variations could reduce to the neo-Darwinian unilaterality, beginning with genotype only when the underlying law of motion is approximated by that of the one-to-one mapping type. Approximation cannot, however, undermine reality. In fact, ecological variability and adaptability can induce variations in the genotype of an individual organism as much as the phenotypic variations can.

Ecological adaptability does not by any means point to the competition for the preexisting niches. Resource exploitation does not necessarily imply that individual organisms compete with each other noncooperatively in a mutually defective manner. Individual organisms belonging to a species that occupies a certain ecological niche can exploit their material resource cooperatively, also. Altruism is in fact an example of cooperative resource exploitation. The causative factor of cooperation[15] among individual organisms is resource exploitation. If natural selection were taken to affect differential reproduction at the level of individual organisms instead of to effect it, the evolutionary emergence of cooperation would have to become a very difficult obstacle to overcome. For natural selection would be forced to operate in the direction opposite to what it would have claimed when those altruistic individuals who sacrifice themselves for the benefit of others are raised. The present difficulty is, however, due to an inappropriate understanding of natural selection. Insofar as one rightly takes natural selection to be the process giving rise to differential reproduction,[16] the evolution of cooperation can be grounded upon its effect.

Natural selection which gives rise to differential reproduction is actualized in resource exploitation, and the real content of the latter always remains to be identified and cannot be deciphered in an exclusive manner in advance. Only those species that have survived exhibit what kinds of exploitation strategy have really materialized. Natural selection, when properly understood, is a process of dissolving the degeneracy of resource exploitation strategies existing within earlier individual organisms. This is no more than expression of material flow equilibration in material aggregates that keep maintaining themselves by varying the interactions within and with their exterior. Resource exploitation strategies in individual organisms come to materialize only through resource exploitation.

The evolutionary emergence of a cooperative strategy for resource exploitation is not a consequence of comparing it with the noncooperative one. It is of course possible for a theoretician to examine how the cooperative strategy could remain stable against the invasion of those individuals with the noncooperative one. One can understand the stability of the cooperative strategy on a comparative basis. Nevertheless, it is not allowed to expect that the same comparative statics could also work in real nature. The difficult part has to do with who could prepare those alternative strategies of resource exploitation to be compared and how. It is one thing to examine evolutionarily stable strategies[17] for resource exploitation on the basis of comparative statics but quite another to seek how those strategies have developed in nature. If natural selection were taken to affect resource exploitation in advance and to be made explicit, say, in terms of the reproductive rate of individual organisms, the comparison among many alternatives would seem to gain some plausibility, though only at the expense of discarding evolution as a process of dissolving the preceding degeneracy. The evolution of cooperation among individual organisms clearly shows that natural selection is effected by resource exploitation.

The surviving species consists of those individual organisms of the similar genotype that share the same ecological niche. The spreading of the similar genotype in the population has its own mechanism. If one follows the neo-Darwinian modern synthesis, the ultimate source of variability would be within point mutations in genes and evolution would be gradual allelic substitution within the population. Natural selection thereof would serve as an *a posteriori* sieve to prevent most of the preexisting mutant genes from spreading in the population, thus conferring upon itself the role of editor, and not that of author. The common underlying assumption is that chromosomal crossing-over of equal length, as the mechanism of allelic substitution eventually establishes an equilibrium distribution of genes in the population and does not induce any further genotypic variations. If allelic substitution finally leads to a stationary equilibrium distribution of genes, the substitution mechanism involved would not yield genotypic variability on its own. However, there is accumulated evidence of other mechanisms of allelic substitution, such as chromosomal unequal crossing-over, gene conversion, and transposition. The problem thus raised is whether allelic substitution in reality could not induce any further genotypic variability, being in consonance with the neo-Darwinian claim.

Allelic substitution is a form of rearranging genes. Unless it is theoretically articulated so as to lead to an equilibrium distribution of genes in the population, allelic substitution of any type can bring in variations in gene. In fact, such a genic rearrangement exhibiting variability is a particular form of the generator of evolutionary variations, since any material aggregate appearing on the evolutionary stage is an open material aggregate processing material flow equilibration that constantly turns its product into the generator of variations to come. Genic rearrangement is a form of the generator of genic variations that is intrinsically degenerate with those generators to emerge later. Unless it is viewed from the hypothetical perspective of the law of motion of the one-to-one mapping type, evolution is the process of dissolving the degeneracy within the generator of variations as generating variations. Genic rearrangement is no exception. Material flow equilibration at the genic level induces variations in the gene and further drives them toward a more constrained form with time. Genotypic variability due to material flow equilibration at the genic level is a mechanism of molecular drive[18] that manifests the organismal process of genic exchange within and among chromosomes. What is more, genotypic variability due to material flow equilibration is an attribute of natural selection in the legitimate framework of the law of motion of the one-to-many mapping type.

The organismal process of genic exchange can survive indefinitely. The indefinite longevity of genotypic variability may seem to contradict the outcome from a commonly held theoretical practice suggesting that the genic distribution in the population would eventually approach its stationary equilibrium state if allelic substitution rates are definable as a logical antecedent.[19] This is a special case of the Markovian postulate of stochastic process claiming the establishment of stationary probability distribution upon the premise of definite transition probability rates. If the present logical antecedent were equated with the causal precedent, the stationary equilibrium distribution could actually be realized in time. But, a serious difficulty is unavoidable with respect to how to convert the logical antecedent into the causal precedent.

The Markovian stochastic process is in fact a law of motion of the one-to-one mapping type on the probability distribution which is subject to its own set of transition probability rates prepared as a separable boundary condition. Markovian dynamics will come to claim its legitimacy only when the set of transition probability rates as the boundary condition can be defined independently of the dynamics. If the separation between the dynamics of the genic distribution and those allelic substitution rates were satisfied, the logical antecedent about those rates could also be converted into the causal precedent by claiming to be able to *identify* those values separately and independently. The equilibrium genic distribution not accompanied by any further genic variability could finally be established.

One could of course determine the values of those allelic substitution rates upon the premise of the separability. If one dares to interpret empirical data on the basis of the separability and to ascribe any deviation from the premise to disturbances of exogenous origin, this theoretical commitment would be in its essence incontrovertible because those exogenous disturbances could work as an omnipotent arbitrator. But if the mechanism of producing variations is really examined, the *a priori* theoretical commitment to the separability would have to be reexamined. Mere empirical measurement does not guarantee the separability between the dynamics of the genic distribution and its allelic substitution rates. How allelic substitution rates develop and vary has to be examined, as does the genic distribution in the population. A resolution in this matter is rather modest in abandoning the *a priori* separability between the law of motion and its boundary condition. In particular, material flow equilibration at the genic level is a mechanism varying both the genic distribution and the allelic substitution rates involved in an inseparable manner indefinitely.

Allelic substitution rates (as a factor conditioning the population of a species) are intrinsically interdependent with the ecological niche of the species because of the bilaterality internal to the process of becoming adapted. Figure 2 represents intrinsic interdependence. Adaptability at the three different levels, ecological, phenotypic, and genotypic, is actually indivisible.

IV. ECOLOGICAL EVOLUTION

If the law of motion were of the one-to-one mapping type, then both the genotype and the ecological niche would serve as the imposed boundary conditions upon the population dynamics. However, these two boundary conditions cannot be set arbitrarily with each other.

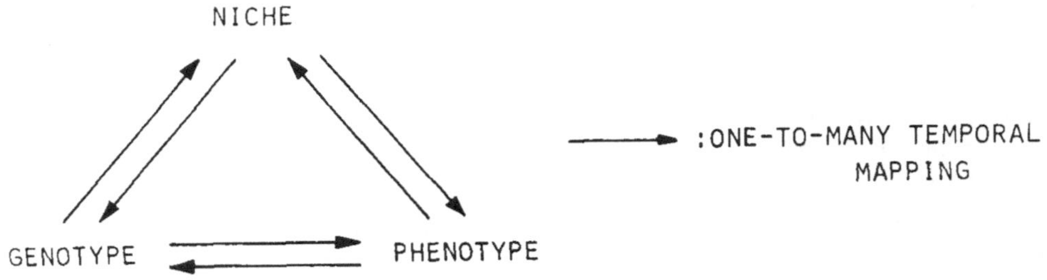

FIGURE 2. Tripartite relationship among ecological niche, genotype, and phenotype. The dynamics relating any two of them is one-to-many temporal mapping.

The resulting conceptual difficulty with the one-to-one temporal mapping is that there is no definitive way of fixing the two boundary conditions in a mutually consistent manner. Both the genotype and the ecological niche, even including their variations and fluctuations, are the effect of the operation of a common mechanism. Material flow equilibration as the common mechanism proceeds at least at the two seemingly different levels, ecological and genotypic.

Resource exploitation by the species as a whole is seen in material flow equilibration at the level of ecological niche, whereas genotypic variability is seen in a similar equilibration at the genic level. The interrelatedness of the two equilibrations lets the genotype of individual organisms be influenced by the variation in their covering ecological niche, and vice versa. Variations in the ecological niche are the aftereffects of how those members that have been exploited come to vary their own ways of resource exploitation. Consequently, changes in the pattern of resource exploitation by other species eventually yield variations in the genotype through the resulting modification of the ecological niche.

If one calls variations of the ecological niche of a species caused by interaction with other species fluctuations of the environment, the origin of genotypic variability can be traced back to the environmental fluctuations. The origin of those environmental fluctuations is within how each species comes to recognize and to establish its own ecological niche for resource exploitation while interacting with others. Repercussions of readjustment among ecological niches last indefinitely since both the ecological niche and the genotype serve as the boundary condition on the dynamics of the other.[20] Genotypic variations of one species are involved in the genesis of environmental fluctuations upon the other species.

The interplay between genotype and ecological niche proceeds along the time-extensive genealogical hierarchy, in which the genotype of progenitor individuals moves up the hierarchy with time influencing their progeny. The genotype serves as a temporal boundary condition, whereas the ecological niche serves as a spatial one on the time-extensive hierarchy. The inseparability between the law of motion and its boundary condition yields a similar inseparability between temporal and spatial boundary conditions. The mutual influencing between temporal and spatial boundary conditions underlies ecological evolution.

V. SUMMARY

Protobiological and biological evolution is a physical process in which boundary conditions are endogenously variable. Although the boundary condition upon the law of motion of the one-to-one mapping type is clearly definable in its own right because of its hypothetical separation from the law of motion, the evolutionary process is unique in its inseparability between the law of motion and its boundary condition. Boundary conditions in evolution are by no means intrinsic, but rather relative in that they are supposed to remain approximately invariant only over a limited interval. The interaction dynamics among boundary conditions is inevitable. Boundary conditions of evolutionary significance involve ecological niches, not to mention the genotype of individual organisms. Ecological niches and individual genotypes influence each other. The individual genotype, phenotype, and ecological niche, are mutually adaptive through the bilateral processes of becoming adapted.

The manner of resource exploitation on the part of individual organisms is an expression of how each individual comes to recognize its own ecological niche. The resource exploitation strategy that connects the individual genotype and the covering ecological niche is intrinsically degenerate, and its degeneracy is dissolved with the progression of resource exploitation. The emergence of eukaryotic predation as a *de novo* strategy of resource exploitation was a consequence of dissolving the degeneracy within the earlier prokaryotic strategy of oxidative phosphorylation. The population of individual organisms of a species occupying its ecological niche can keep generating its genic variations. The genic distribution in the population constantly avoids being trapped in a hypothetical equilibrium distribution. Such an indefinite genesis of genic variations in the population is due to the process of mutual influence between the genotype and the ecological niche, where the genotype serves as the temporal boundary condition while the latter serves as the spatial one. The earlier ecological niche is degenerate with the ones to emerge later and its degeneracy is dissolved with material flow equilibration occurring in the ecological niche.

REFERENCES

1. **Gause, G. F.,** *The Struggle for Existence,* Hafner, New York, 1934.
2. **Hunt, L. T., George, D. G., Yeh, L.-S., and Dayhoff, M. O.,** Evolution and prokaryote and eukaryote lines inferred from sequence evidence, *Origins Life,* 14, 657, 1984.
3. **Cloud, P.,** Evolution of ecosystems, *Am. Sci.,* 62, 54, 1974.
4. **Lovelock, J. E.,** *Gaia,* Oxford University Press, Oxford, 1979.
5. **Broda, E.,** *The Evolution of the Bioenergetic Processes,* Pergamon Press, Oxford, 1975.
6. **Rich, P. H.,** Trophic-detrital interactions: vestiges of ecosystem evolution, *Am. Nat.,* 123, 20, 1984.
7. **Rich, P. H.,** Trophic vs. detrital energetics: is detritus productive?, *Bull. Mar. Sci.,* 35, 312, 1984.
8. **Slobodkin, L. B.,** Energy in animal ecology, in *Advances in Ecological Research,* Vol. 1, Cragg, J. B., Ed., Academic Press, New York, 1962, 69.
9. **Margulis, L.,** *Symbiosis in Cell Evolution: Life and Its Environment on the Early Earth,* Freeman, San Francisco, 1981.
10. **Skyring, G. W. and Donnelly, T. H.,** Precambrian sulfur isotopes and a possible role for sulfite in the evolution of biological sulfate reduction, *Precamb. Res.,* 17, 41, 1982.
11. **Matsuno, K.,** Prebiological molecular evolution as physico-chemical processes, *J. Phys. Soc. Jpn.,* 42, 1691, 1977.
12. **Matsuno, K.,** Evolution of dissipative system: a theoretical basis of Margalef's principle on ecosystem, *J. Theor. Biol.,* 70, 23, 1978.
13. **Odum, E. P.,** The strategy of ecosystem development, *Science,* 164, 262, 1969.
14. **Margalef, R.,** *Perspectives in Ecological Theory,* University of Chicago Press, Chicago, 1968.
15. **Axelrod, R. and Hamilton, W. D.,** The evolution of cooperation, *Science,* 211, 1390, 1981.
16. **Vrba, E. S.,** Macroevolutionary trends: new perspectives on the roles of adaptation and incidental effect, *Science,* 221, 387, 1983.
17. **Maynard Smith, J. and Price, G. R.,** The logic of animal conflict, *Nature (London),* 246, 15, 1973.
18. **Dover, G.,** Molecular drive: a cohesive mode of species evolution, *Nature (London),* 299, 111, 1982.
19. **Ohta, T. and Dover, G. A.,** Population genetics of multigene families that are dispersed into two or more chromosomes, *Proc. Natl. Acad. Sci. U.S.A.,* 80, 4079, 1983.
20. **Whitham, T. G. and Slobodchikoff, C. N.,** Evolution by individuals, plant-herbivore interactions, and mosaics of genetic variability: the adaptive significance of somatic mutations in plants, *Oecologia (Berlin),* 49, 287, 1981.

Chapter 11

EPILOGUE

Final causality, that is to say, *causa finalis* has long since been discarded as the physical mechanism of producing and transforming variations. Instead, mechanistic causality originating in *causa efficiens* has exclusively been appreciated for this role. The dominance of mechanistic causality has been vindicated by the law of motion of the one-to-one mapping type supplemented by its completely identifiable boundary conditions in classical mechanics, in quantum mechanics, and in stochastic process. However, a serious problem we have faced is whether the law of motion is exclusively of the one-to-one mapping type or whether causality acting in materialistic processes is unquestionably destined to be mechanistic. The frame of reference against which the present problem is examined should not be a hypothetico-deductive theory in one form or another, but has to be incontrovertible, empirical facts. Physics has boasted of possession of such a solid frame of reference established on the empirical principle of conservation of energy, matter, and the like. We thus can hope that the physical principle of conservation may provide a basis on which the actual role of causality is legitimately examined.

The principle of conservation has at least three different faces; the state of being conserved, the process preserving conservedness, and the process materializing conservedness. The weakest form among the three is the process materializing conservedness, because the empirical observation is limited only to the measurement that has been done and because the consequence of the process materializing conservedness is simply the observation of conservation in the record. The state of being conserved implies more than what empirical observation can tell *a posteriori*. It stands alone without allowing any interference from actual measurements. The process preserving conservedness refers to the temporal transference of the state of being conserved.

The temporal transference identified as a law of motion is separable from the state of being conserved set as the boundary condition. Both the state of being conserved and the process preserving conservedness are compatible within the scheme that the law of motion is completely separable from its boundary condition. The separability is in fact witnessed by the law of motion of the one-to-one mapping type on the trajectory in classical mechanics and on the wavefunction in quantum mechanics, since the boundary condition defined independently of the law of motion is transferable through the one-to-one temporal mapping. Conservedness is one attribute of the boundary condition. The transference of the state of being conserved manifests mechanistic causality, because the causal precedent can determine what will appear later with the help of the law of motion that can be identifiable irrespective of whatever the precedent may be.

On the other hand, however, the process materializing conservedness cannot and does not imply the state of being conserved in view of the fact that no physical processes are allowed to be propagated at a velocity exceeding light velocity. Only when the process communicating conservedness would be hypothesized to proceed in the medium at an infinite velocity, could the state of being conserved be deemed to be established at every moment. The process materializing conservedness does not distinguish between the law of motion and its boundary conditions, because the principle of conservation cannot be set aside as a separable boundary condition on the motion. The inseparability is evident in the law of motion of the one-to-many mapping type in which the preceding boundary condition is necessarily degenerate with those to appear later. Biological motility in fact manifests the process communicating conservedness being propagated at a velocity over ten orders of magnitude less than light velocity.

The process materializing conservedness is final in its operation because the impetus of the process is to meet the principle of conservation. At this point, we find legitimate room for final causality in the realm of materialistic processes. Although physics has traditionally appreciated the principle of conservation only from the perspective of the state of being conserved and the process preserving conservedness, the biological side of materialistic processes comes to pay due attention to the process materializing conservedness in place of the former two.

Biology, being different from the physics of being, has kept the appreciation of the difference among the state of being conserved, the process preserving conservedness, and the process materializing conservedness, though implicitly. Instead of referring to conservation directly, biology has adaptation at its core. Adaptation implies at least the three different characteristics: the state of being adapted, the process preserving adaptedness, and the process materializing adaptedness. Both the state of being adapted and the process preserving adaptedness are compatible with each other, accordingly dispensing with adaptability as the further capacity of adaptation. The dynamic impetus to temporally transfer the state of being adapted is not adaptation itself. However, empirical observation certifies inexhaustible adaptability. What is implied by inexhaustive adaptability is the process materializing adaptedness as a special case of the process materializing conservedness, because biological adaptation, whatever it may mean, has to obey the physical principle of conservation.

Adaptation as the process materializing adaptedness is final in its operation as much as the process materializing conservedness is. Final causality, we have been emphasizing, is no more than a mere derivative from the materialistic process entailed by and leading to physical conservation laws.

Final causality makes natural selection comprehensible as effecting, instead of affecting, differential reproduction. Natural selection expressed as a selective coefficient measuring the degree of differential reproduction remains legitimate only *a posteriori*. Similarly, biological fitness is a characteristic that final causality makes evident only as an effect. Adaptation, natural selection, and fitness, when properly understood, thus point to the fundamental attributes of a materialistic process that proceeds in the mode of final causality.

The origin of final causality is thoroughly materialistic and physical in that the process communicating conservedness does not proceed at a velocity exceeding light velocity. Nevertheless, physics has been practiced exclusively in the mode of mechanistic causality. The dismissal of final causality has been accomplished by presuming a more or less universal theoretical artifact that would have the process communicating conservedness proceed in the medium at an infinite velocity. One form of such a theoretical artifact is the separability of the law of motion and its boundary conditions. Another form is the so-called adiabatic approximation, that is ubiquitous in physical theory, claiming that when a physical process of any type is divided into the two subgroups, faster and slower, the faster process could be deemed to proceed at an infinite velocity. Physics has underestimated or even negated the role of final causality. In contrast, biology has appreciated the operation of final causality in its own way. The physical underpinning of biology is thus in the appreciation of the role of final causality, though this has been depreciated in physics per se.

The process communicating conservedness in reality staunchly rejects the presence of an omnipotent observer that could claim an instantaneous and simultaneous bird's eye view on the state of conservation on a global level, though conceivable in theory. The state of being conserved would be evident to such an external observer, but there is no external inspector able to control and to supervise how the principle of conservation materializes. Instead, the process communicating conservedness necessitates the involvement of many internal watchdogs whose role is to recognize their manner of interaction with others and then to adjust it so as to fulfill the principle of conservation altogether *a posteriori*. An important difference from the hypothetical external inspector is that each of these endogenous watchdogs recognizes how others adjust their interaction only after they do so, not simultaneously, though all of them participate in upholding the principle of conservation.

Each interacting material aggregate assumes the role of endogenous watchdog. As a result, final causality comes to prepare the ground upon which the system of many interacting internal observers becomes real. Observation necessarily dissolves the degeneracy latent in the observer by decreasing the extent of indefiniteness as a consequence of measurement. The dynamics of internal observers obeys the law of motion of the one-to-many mapping type because of the nature intrinsic to the process of measurement that constantly constrains with time the capacity of measurement yet to come. The process of internal measurement is in fact the subject matter of quantum mechanics, and the quantum mechanics of internal measurement furnishes the law of motion of the one-to-many type. Quantum mechanics that admits final causality underlies biology, whereas physics has traditionally practiced quantum mechanics in the mode of mechanistic causality, especially with regard to the wave function. A summary expression is found in Figure 1.

The physical underpinning of biology thus faces a tricky dilemma. If one tries to theorize protobiological and biological evolution in its extreme form by identifying its logical antecedent, the resulting analysis would inevitably render the evolution a mechanistic process by equating the antecedent with the causal precedent. Only when mechanistic causality is guaranteed on other grounds, would such a theory that does not leave anything indefinite be saved. But, protobiology and biology are simply not the case of mechanistic causality. The involvement of internal observers is inevitable. One cannot tell in advance what these internal observers will measure. A physical appraisal of final causality is necessarily accompanied by internal observers who unfailingly maintain the capacity of measurement in the future.

One should not deprive those internal observers of the capability of measurement in the name of a biological theory. At the same time, the effort to provide a physical underpinning to biology cannot be stopped. Physics has previously limited itself to the mechanistic mode of causality and enjoyed its theoretical definiteness and clarity. In contrast, biology is unique in addressing itself to how the indefiniteness latent in final causality is constrained with time.

The conception of biology in physics is already implicit in mechanics, especially with the materialistic capacity of making choices. If the force field is uniquely related to the displacement of the medium, the law of motion of the one-to-one mapping type would survive as in celestial mechanics, which admits a unique relationship between moving bodies and the potential force field. This would imply that the force field simultaneously follows the displacement of these moving bodies in an unambiguous manner. However, such a simultaneous adjustment of the force field in the entire medium would require the propagation of changes in force at an infinite velocity. A more realistic picture of mechanics is sought with due consideration to the fact that no changes in physical action and reaction are allowed to be propagated over light velocity. Even if the displacement of the medium remains the same, the force acting upon it would come to differ after it has suffered from a change in the propagating force field originating elsewhere. No unique relationship exists between the displacement of the medium and the propagating force field acting there. Still, one constraint remains. This is the conservation of energy.

When a portion of the medium is subject to a change in the force field originating elsewhere, the aftereffect of experiencing the newly arrived change would be either in the mechanical acceleration of the local portion or in a change in the force field induced therefrom, or in a juxtaposition of the two, while observing the conservation of energy. In particular, there exists some intrinsic arbitrariness in how to impart the incoming energy of the propagating force field to the mechanical acceleration and the induced force generation. One thus observes an *a priori* indefiniteness about the relationship between the displacement of the medium

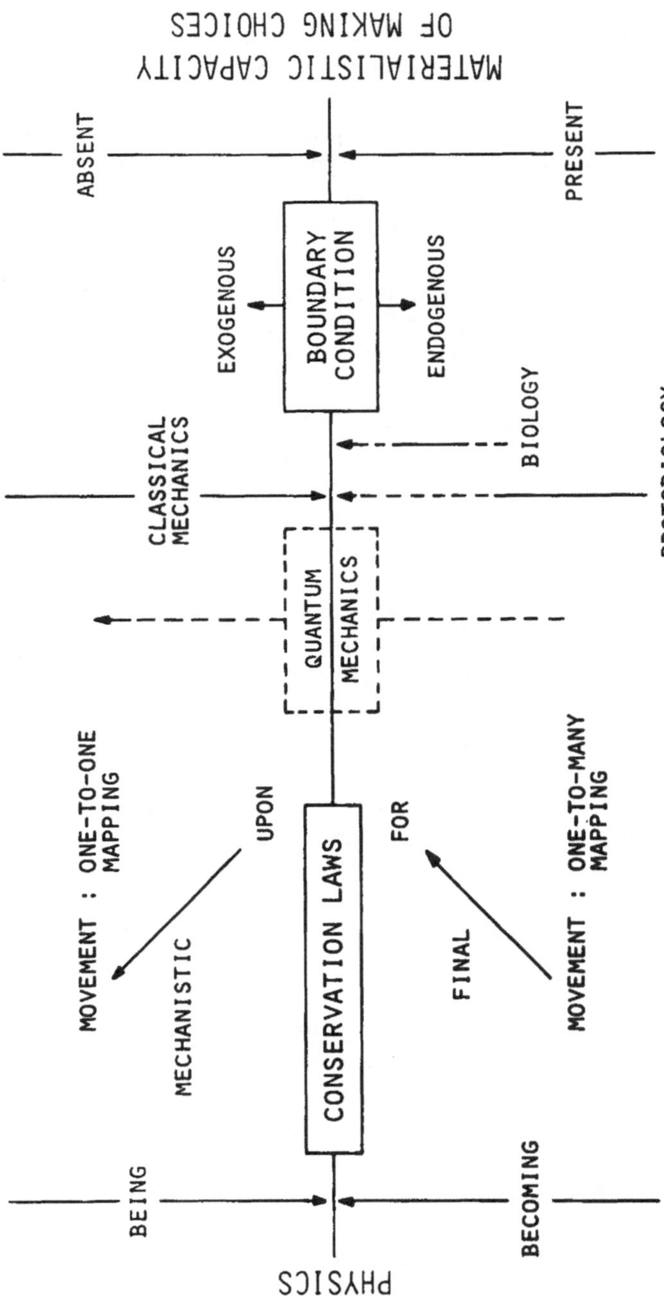

FIGURE 1. Protobiology as a physics of becoming that addresses the endogenous dynamics for physical conservation laws. Material capacity for making choices originates in the physical movement for conservation laws.

and the propagating force field, but the relationship is found to be definite *a posteriori* in the sense of being irrevocable in the record. The materialistic capacity of making choices is latent in the process of transforming what is indefinite *a priori* into what has been definite *a posteriori*.

Protobiology and biology are a form of physics founded on the materialistic capacity of making choices.

INDEX

A

Abiotic molecules, 136
Abiotic peptide synthesis cycle, 149—150
Acquired phenotypic variation, 172
Actin, 20, 170—171
Actin contraction, 171
Actin filaments, 20—24, 31, 60, 67, 101
Action-at-a-distance, 12—13, 58—59, 61—62
Action-through-medium, 13, 59—62
Active agents, 38
Active transport, 136—137, 139—140
Acto-myosin complex, 60, 171
Actualization, 95, 111, 113, 139, 182
Adaptability, 177, 194—199, 201, 204—207, 212
Adaptation, 27, 195—198, 201, 212
Adaptedness, 195, 196, 212
Adapting to, 194—195, 197
Adaptive capacity, 197
Adenine, 102, 143, 145
Adenine nucleotides, 143
Adenine-thymine rich DNA, 145—146
Adenosine diphosphate (ADP), 25, 136, 143
Adenosine moiety, 143
Adenosine monophosphate (AMP), 143
Adenosine triphosphate (ATP), 20, 24—25, 60—61, 136, 143, 171
Adiabatic approximation, 71—72, 212
ADP, see Adenosine diphosphate
Adsorption, 145, 180
Aggregation, 62
Aldehydes, 144
Allelic substitution, 206
Allelic substitution rates, 207
Altruism, 205
Amino acids, 63, 93, 98—101, 105—106, 109—111, 114—116, 120—126, 136, 144—145, 148, 150—154
Ammonia, 100
AMP, see Adenosine monophosphate
Amphibian embryo, 168
Anthropomorphic information, 109—116, 122—123, 125, 126
Antibiotics, 151
Approximation, 188
Aptation, 195
Arginine, 115
Association, 98, 136
Asteroid, impact of, 204
Atoms, 136
ATP, see Adenosine triphosphate
ATPase, 20
ATP process, 140—143
Autocatalysis, order of, 116, 118—121, 123, 152—153
Autocatalytic cycle, 119, 123
Autocatalytic polymerization, 117—119, 122—123, 148
Autocatalytic polypeptide, 119, 123
Autocatalytic proteins, 118
Autocatalytic reactions, 97, 116, 118
Autotrophic protoreproduction, 151
Autotrophic reaction cycle, 151
Autotrophic reproductive units, 150
Autotrophs, 203
Avogadro's number, 1, 9
Axoneme, 16—17

B

Base pairing, 129, 131
Beat cycle, 25
Benard cells, 135
Bending moment, 17
Bending movement, 17—18
Bending wave, 25
Bifurcation, 71
Big Bang cosmology, 15, 85—86, 88
Bilayer, 140
Bilayer membrane, 136—137, 140, 142
Bilayer vesicles, 136—137, 139—142
Biological evolution, 3, 178, 184
Biological information, 27
Biological motility, 1, 15, 20—21, 23—25, 28, 31, 57, 62, 184, 211
Biomonomers, 145
Boltzmann's entropy, see Boltzmann's statistical entropy
Boltzmann's H-theorem, 5, 79
Boltzmann's statistical entropy, 81—84, 90
Boundary conditions, 1—6, 39—41, 51, 64, 67—68, 96—97, 125—126, 129, 158, 163, 177, 194, 202
  additiveness of, 42
  completely fixed, 1—3, 5—15, 20, 27, 31—32, 36—37, 47, 51, 54, 57, 59
  conservedness, 211
  endogeneity, 146—150, 152—154
  endogenous, 101, 106, 130, 146
  evolutionary dynamics, 191
  exogenous, 146
  experimental controllability, 33, 35, 41, 70
  externalization of, 106
  fixed, 161—162
  flow-controlled, 69—70
  force-controlled, 69—70
  force equilibration, 61
  indefiniteness, 1, 3, 9
  limited controllability of, 35, 51
  limited experimental controllability, 36, 46, 78
  local, 66, 72—74, 80—84, 88, 90, 93, 98
  macroscopic, 68
  microscopic, 68
  molecular replication, 134
  morphological space, 105
  nonequilibrium thermodynamics, 72—73

partially fixed, 2—3, 5, 9—15, 20, 24—25, 28, 31, 59, 94
production process, 144
resource limitation, 186
role of internal measurement, 57
spatial, 146—147
spatially localized, 135
theoretical controllability of, 34, 48
unlimited controllability, 132
unlimited theoretical controllability of, 6, 34—37, 41, 46, 48—52, 78, 80—81
Branching, 193
Brownian motion, 21
Budding, 138

C

Calcium, 171
Calcium regulation mechanism, 171
Carbon dioxide, 144, 204
Carboxyl, 140
Catalytic cycle, 145
Catalytic polymerization, 148
Celestial mechanics, 1, 12
Cell differentiation, 161, 163
Cell divisions, 158, 169, 174
Cell growth, 169
Cell mixing, 158—159
Cells, 52, 119, 129—130, 171
Cell surface, 169—171
Cellular growth, 172—174
Cellular membrane, 137
Cellular multiplication, 134—135
Cellular protoreproduction, 150
Cellular reproduction, 138
Cellular reproductive fidelity, 138, 149—150
Chaotic trajectory, 34, 68
Charge-separated bilayer vesicles, 141—143, 146—147, 154
Charge separation, 137, 139—141
Chelating capability, 143
Chelation, 143
Chemical affinity, 151—154
Chromosomal crossing-over, 167, 192—193, 206
Chromosomal unequal crossing-over, 206
Chrososome recombination, 192—193
Chromosomes, 207
Chromosomic disjunctions, 167
Circularity, 11
Classical mechanics, 1—2, 5—6, 8, 11—12, 15, 26, 31—36, 41, 57—58, 61—62, 67—68, 80, 86, 105, 211
Classical trajectory, 14
Clay-mediated oligomerization, 145
Clays, 145
Cleavage process, 157, 169—172
Coenzyme A, 143
Comparative statistics, 201
Compensation process, 198
Competence, 159—161
Competent tissue, 160
Complexity, 105—159
Complexity increase, 105—106, 109, 204
Conservation, principle of, 6—11, 37—38, 48, 52, 64, 78—79, 82, 211—212
Conservation laws, 7, 57, 157, 164
Conservation of energy, 7—8, 36, 61, 84, 90, 141, 157, 211, 213
Conservative force, 12
Conservedness, 7—8, 31, 57, 164, 183
boundary condition, 211
communication velocity of, 8—11
process communicating, 8, 10—11, 13, 15, 32, 36, 38, 52, 183—184, 186, 211—212
process materializing, 8, 10, 32, 37—40, 47—54, 61—63, 70—78, 82, 90, 93, 98, 101—105, 118, 124, 158—169, 174, 183, 187—188, 194, 211—212
process preserving, 8, 10, 37, 39, 50, 163, 211—212
Constraint-fluctuation complex, 69—70, 73—74
Contemporary proteins, 109, 115—116, 118, 120
Continuum mechanics, 13—14, 61
Contour lines, 171
Contractile force, 20
Controllability, 14, see also Boundary conditions
Convection flow, 135
Cooperation, 205—206
Copenhagen interpretation, 42
Coulomb force, 12, 61
Covalent interaction, 136
Creatinine, 140
Crossbridges, 171
Cross-hierarchical causations, 182
Crosslinking, 171
C-terminal, 115
Cyclic polymerization, 123, 148—149
Cyclic polypeptide synthesis, 150
Cyclic reaction, 119
Cytoplasm, 171
Cytoplasmic determinants, 159—163, 165—167, 174
Cytosine, 102, 145
Cytosine moiety, 143
Cytoskeleton, 171

D

Decision making, 51
Definiteness, 173
Deformation motion, 169
Degeneracy, 10—11, 34, 65, 84, 94—105, 138—139, 149—150, 158—159, 166, 173, 181, 193, 201—209
Degrees of freedom, 2, 4, 6, 9, 65—66, 95—96, 100, 118, 148, 152
Dehydration, 140
Detritus, 204
Developing embryos, 172
Developing organisms, 173
Development, 101—106, 124, 157—175, see also Equilibration
Developmental process, 89, 157—158, 160—163, 165, 167, 172, 174
Developmental program, 103—106

Developmental symmetry breaking, 168—169
Developmental variability, 165
Diachronic process, 173—174
Dichotomy between reproduction and variation, 180—181
Differential reproduction, 182, 187, 198, 205—206
Differentiating cells, 168
Differentiating embryo, 163
Diffusive propagation, 160
Dimension analysis, 121
Dimers, 96—97
Disequilibrium, 39—40, 48, 57—58, 136, 157, 163
Dissipation, 204
Dissipative system, 204
Dissociation, 62, 98, 136
Dissolution of degeneracy, see Degeneracy
Divalent cation, 143
DNA, 50, 102—104, 112—113, 131, 153, 167, 192
Dorsoventral axis, 168
Dry weights, 169
Dynamic variables, 6—7, 10
Dynein arm, 25
Dynein-tubulin complex, 60, 62, 67

E

Earth, 129, 134, 136, 144—145, 147—148
Ecological climax, 201
Ecological niche, 201—205, 208—209
Ecological succession, 201, 204—205
Ecology, equilibration in, 201—209, see also Equilibration
Ecosystem, 204
Electrical excitability, 136
Electric discharge, 144
Electrogenic carriers, 140—141, 143
Electrolytes, 130
Electromagnetics, 87
Electron donors, 144
Embryo, 158—160, 166, 168
Embryogenesis, 159
Embryonic cleavage process, 174
Embryonic development, 169, 172
Embryonic field, 163, 165—166
Embryonic regulation, 160
Endogenous variability, 192
Endoparticles, 138
Energy, 161
Energy equilibration, 57, 61, 82—84, 141—142
Energy transduction, 129, 140—143, 154
Entelechy, 166
Entropy, 27, 77, 79—86, 90
Environmental fluctuations, 208
Enzymes, 132, 138, 145
Equilibration, 27, 31, 40—41, 57—75, 79, 82—90, 101, 103, 146, 179, see also other specific topics
  adaptability, 194—198
  cleavage process, 169—172
  development, 157—175, see also other subtopics hereunder
  developmental symmetry breaking, 168—169
  diachronic process, 173
  ecological adaptability, 204—207
  ecological evolution, 207—208
  ecological niche, 201—204
  ecology, 201—209
  evolution, 177—200, see also other subtopics hereunder
  genotype-phenotype interdependence, 167—168
  growth, 172—173
  internally regulative principle, 157—162
  natural selection, 182—187
  perpetual disequilibrium and, 163—166
  process materializing conservedness, 77—78
  process of, 27
  reproductive fidelity, 177—182
  synchronic process, 173
  time's arrow, 86—87
  variability, 187—194
Equilibration-induced fluctuations, 74
Equilibrium configuration, 39
Equilibrium state, 77—79
Eukaryote, 16, 201, 203—205
Eukaryotic predation, 209
Evolution, 52, 54, 62, 70, 100—101, 104, 124, 133, 136, 143, 148, 150, 154, 158, 168, 192, 206
  ecological, 201, 207—208
  equilibration in, 177—200, see also Equilibration
  reproductive units, 177
Evolutionarily stable strategies, 206
Evolutionary capability, 100, 135
Evolutionary enhancement of specificity, 146
Evolutionary matrix, 52
Evolutionary nexus, 115
Evolutionary potential, 100
Evolutionary process, 68, 100
Evolutionary unfolding, 194
Evolutionary variability, 192—194, 197, 199, 201
Exaptation, 195
Expanding universe, 82, 88, 90
Exponential growth, 96
External agents, 39, 177
External measurement, 32—33, 35, 38, 40—46, 53, 163
Extinction, 181, 184, 187

F

Falsification, 9
Fate map, 158—163, 165—167, 169, 174
Feedback, 104
Ferrous ion, 144
Fertilization, 161
Fertilized egg, 103, 159, 169
Field property, 163
Final causality, 185—188, 191, 193, 195—199, 204, 211—213
Final objective, 185, 193, 196
First law of thermodynamics, 77—79, 84, 90
Fitness, 147, 177, 196—197, 199, 212
Fittedness, 195

Flagella, 16, 60, 62, 67
Flagellar axoneme, 16—19
Flagellar movement, 15—16, 20, 23—24, 31, 101
Flavin adenine dinucleotides, 143
Flow disequilibrium, 64—65, 67, 172—173
Flow equilibration, 143
Fluctuation-induced instability, 69—71
Fluctuation, 14, 57, 68—69, 73—74, 181
Force, 18, 20, 139—140, 157, 161
Force equilibration, 15—17, 25—26, 57—62, 67, 74, 101, 140—142, 157, 167, 169—174
  force equilibrium differentiated, 58
  propagation velocity of, 18—25
Force equilibrium, 13, 15—17, 58, 59
Forest fires, 204
Formaldehyde, 100

G

Gap junctions, 160
Gastrulation, 166
Gause's competitive exclusion principle, 201
Gene conversion, 206
General relativity, 87
Genes, 167, 171—173, 206
Gene segments, 167
Genetic code, origin of, 129, 150—154
Genetic influence, 171, 173—174
Gene transposition, 206
Genic exchange, 207
Genic excision, 192
Genic rearrangement, 206—207
Genic reintegration, 192
Genic variability, 207
Genic variations, 206, 209
Genomes, 167, 173
Genotype, 181, 189—191, 197—201, 205—209
Genotype-phenotype interdependence, 167—168
Genotypic adaptability, 197
Genotypic variability, 190—192, 206—209
Genotypic variations, 189, 201, 208
Germ cell, 172, 189
Glutamic acid, 98—99
Glutamine, 151
Glycine, 98—99
Goat pseudogenes, 102
Growth, 172—173
Guanidine, 140
Guanine, 102
Guanine-cytosine rich DNA, 146
Guanosine moiety, 143

H

Half-cystine, 115
Heat flow, 135
Heterotrophic protoreproduction, 150
Heterotrophic reproductive units, 150—151
Heterotrophs, 203
Holoblastic cleavage formation, 103, 169
Holoblastic cleavage process, 170—171
Homogeneous universe, 86
Homopolynucleotides, 151, 154
Hydrodynamics, 8—9
Hydrogen, 144
Hydrogen cyanide, 100, 144—145
Hydrolysis, 21, 25, 60—61,143
Hydrophilic environment, 137
Hydrophilic membrane, 140
Hydrophilicity, 151
Hydrophilic properties, 146
Hydrophobic membrane, 140
Hydrophobicity, 151
Hydrophobic properties, 137, 146
Hypercycle, 120—123, 150, 153
Hypercyclic catalytic reaction, 119

I

Inclusive fitness, 196
Indefiniteness, 3, 6, 9, 12—14, 17—18, 37, 40—41, 48, 50—51, 59, 61, 65, 67, 68, 124—126, 132—133, 161—162, 166, 174, 184, 186, 192, 194, 213
Induction, 160—167, 174
Inductive signal, 159—161
Inertia, 11
Information, 27, 109—127
  carrier, 122—123, 153
  classification, 126
  content, 109—111, 114—117, 120—126, 152—153
  translation, 122—123
Inheritance of acquired traits, 189, 192
Inorganic chemistry, 145
Inorganic phosphate, 136, 140
Inorganic replication, 130—132
Inseparability, 178, 181, 190, 208, 211, see also One-to-many mapping
Interaction configuration, 39
Interaction-dominant irreversibility, 83
Interaction-free reference, 84—86, 90
Interaction-recessive irreversibility, 83, 85
Intercalating synthesis, 129—131
Intercalation, 131
Internal agents, 165—166
Internal disequilibrium, 163—164
  field of, 164—174
Internally regulative principle, 157—162
Internal measurement, 26—28, 31—32, 36—41, 46—53, 57—59, 80, 163—164, 213
Internal objective, 196
Internal regulation, 158, 162
Internal structural identification, 38—40, 51
Internal structural realization, 38—40, 51
Internal teleology, 105
Internal tensile force, 16—24, 169—171
Internucleotide bond synthesis, 118
Intervening adaptors, 154
Intracellular variability, 167
Intrinsic chemical affinity, see Chemical affinity
Intrinsic irreversibility, 77—91

entropy, 77, 79—85
first law of thermodynamics, 77—79
quantum mechanics, 77
second law of thermodynamics, 77—79
self-assembly, 93
time's arrow, 77, 85—90
Intrinsic potency, 159
Intrinsic reversibility, 80
Intrinsic variability, 158, 167, 172—174, 177, 182, 189, 194
Invagination, 137—138, 140, 142
Iron, 144
Irreversibility, 4—6, 15, 51, 82, 87—88, 90, 129, 133
Boltzmann's statistical entropy, 82
classical mechanics, 80
classification, 89
energy equilibration, 83
thermodynamic time's arrow, 86
time's arrow, 85

J

Jumping genes, 167, 174

K

Key-lock relationship, 151
Kinetic energy, 21, 25
K-meson, 88—89

L

Lamarckian inheritance of acquired traits, 189—190
Lamarckianism, 189
Lamarckian scheme, 190
Law of motion, 1—41, 51, 64, 67, 105, 118, 125—126, 158, 177, 194; see also One-to-many mapping; One-to-one mapping
classification, 37
evolutionary dynamics, 191
function, 163
internal measurement, 59
Layer-replication, 154, 180
Layer-silicates, 129
Lead, 131
Le Chatelier's principle, 77
Limb, 166
Linear decomposability, 45—47, 54, 72
Linear superposition, 43—45
Lipid-like material, 139—140, 142
Lipid-like properties, 136
Local variability, 197
Lyaponov functions, 79
Lysine, 118—119

M

Macromolecular multiplication, 150
Macroscopic boundary conditions, 68
Macroscopic conservation, 7
Macroscopic constraints, 70—71, 74
Macroscopic variations, 193
Macrovariation, 193
Many-to-one mapping, 37, 42
Markovian postulate, 207
Markovian stochastic process, 207
Mars, 147—148
Material accretion, 136
Material accumulation, 66—67, 69, 101, 118, 135, 137, 146, 148, 152, 172, 189
Material evolution, 70, 72, 145
Material flow continuity, 62—64, 118, 146—147
Material flow equilibration, 62—70, 74, 78, 98—100, 118, 120, 129, 133—154, 157, 172—181, 188, 197—208
actualization of material flow continuity, 142
fidelity of cellular photoreproduction, 138
genic level, 207
organisms, 179
Materialistic interactions, 8
Materialistic irreversibility, 85
Materialistic processes, 212
Material self-assembly, see Self-assembly
Maxwell's demon, 9, 165
Measurement, 26, 28, 31, 196, 211
apparatus, 32, 43—44, 50—51, 54, 163—164
classical mechanics, 31—36
classification, 32
quantum mechanics, 41—53
Mechanical displacement, 17—20, 58—59, 139
Mechanical force, 11
Mechanical motion in biology, 15—25
Mechanics and biology, 11—15
Mechanistic causality, 185—189, 191, 193, 194, 196—198, 211, 213
Mechanistic model, 25
Mechanochemical couplings, 25
Mechanochemical energy transformation, 24
Membrane, 136
Meroblastic cleavage process, 171
Mesodermal mantle, 166
Messenger RNA, 153, 159
Metabolism, 139, 149
Metal ions, 131, 145
Methane, 100
Microfilaments, 171
Microparticles, 70
Microscopic boundary conditions, 68
Microscopic entropy, 85, 87
Microscopic fluctuations, 69—71, 181
Microscopic interaction, 7
Microscopic irreversibility, 87—89
Microscopic states, 83—84
Microspheres, 136
Microsystems, 63
Microtubules, 170—171
Miller-Urey experiment, 100, 144
Mixed states, 48, 73
Mobile genetic elements, 192
Molecular chaos, 5
Molecular drive, 207

Molecular evolution, 52
Molecular multiplication, 141—143, 148—149
Molecular replication, 129, 134—135, 138—140, 144—150, 154
Molecular reproductive fidelity, 138, 149—150
Molecules, 129—130
Momentum, conserving of, 13—14
Monomers, 93, 96—98, 117, 138—139, 141
Montmorillonite, 129—130, 154, 180
Morphogenesis, 52, 54, 167
Morphological space, 104—106, 109, 163, 164, 167, 169, 171—172, 174
Morphology, 103
Muscle, 20
Mutation, 27, 102, 167
Myosin, 20, 22—23, 31, 62, 170—171

N

Natural selection, 27, 111, 132, 127—178, 182—187, 194, 198, 205—206, 212
Neo-Darwinian dichotomy, 167—168, 192
Neo-Darwinian interpretation, 189
Neo-Darwinian one-to-one temporal mapping, 190
Neo-Darwinian phenotypic plasticity, 191
Neo-Darwinian random variation, 132
Neo-Darwinian scheme, 168, 190, 192
Neo-Darwinian separation, 189, 191—193
Neo-Darwinian unilaterality, 205
Neo-Lamarackian bilaterality, 205
Neo-Lamarackian genotypic-phenotypic plasticity, 191—193
Neo-Lamarackian genotypic variability, 192
Neo-Lamarackian phenotypic plasticity, 191
Neo-Lamarackian phenotypic variability, 190
Neutral random variation, 132
Neutrino's mass, 86
Newton's equations of motion, 1—2, 11—12, 14—15, 26—27, 32—33, 41, 68
Niche, 201—205, 209
Nitrogen, 144
Nitrogen basses, 100
Nonanthropomorphic information, 109, 112—124, 126
  distinctive character of, 113
  reproduction, 129
Noncovalent bondings, 101
Nonequilibrium stationary states, 79
Nonequilibrium thermodynamics, 14, 68—73, 79
Nonrandomness, 93, 102, 104, 110—111, 115, 133
Nonrandom production, 104
Nonrandom products, 104, 106
Nonrandom variation, 66, 68, 98, 106, 133—134
Nonrelativistic quantum mechanics, 1—2, 12, 41, 62
Nonsexual reproduction, 192—193
Nucleases, 134
Nucleic acid first hypothesis, 145—146
Nucleic acids, 144—145
Nucleotides, 102, 144, 151
Nucleotide substitutions, 102—104, 106
Null-conservation of forces, 11—12
Null-conservedness, 12—13, 15—18, 20, 24, 26, 57—61, 140, 167, 169

O

Objective finality, 157—158, 162
Oceanic turbulence, 204
Oligomerization, 145
Oligonucleotides, 145
One-to-many mapping, 11, 13, 18—19, 23, 25—28, 31, 68, 78, 84, 86—89, 94, 96—97, 101—105, 125, 132—133, 147, 159—170, 177—198, 211
One-to-many temporal mapping, 4—5, 10, 18, 28, 31—37, 58—60, 87—97, 113, 157—162, 189—193, 198—199, 205
One-to-one mapping, 1—37, 54—68, 78, 83, 87—105, 113, 125, 132, 147, 162—170, 177—183, 189, 196
One-to-one temporal mapping, 3—5, 10, 13, 15, 20, 25, 32—41, 50, 57, 87, 94, 96, 100, 105, 163, 167, 202, 208, 211
Ontogeny, 89
Open material aggregates, 63—71, 74, 97, 98, 100, 118—119, 129, 132—138, 147—150, 180
Organic carbon, 203—205
Organic molecules, 145
Organic replication, 130
Osmotic pressure, 137
Oxidant, 137
Oxidative phosphorylation, 203, 209

P

Particle-antiparticle decay, 88—89
Peptide antibiotics, 145
Peptide bond synthesis, 118
Permutation symmetry, 99
Perpetual disequilibrium, 163—166
Phenotype, 167, 189, 191, 198—199, 201, 205, 208—209
Phenotypic adaptability, 197—199
Phenotypic development, 189
Phenotypic plasticity, 191
Phenotypic space, 191, 198
Phenotypic variability, 190—191
Phenotypic variations, 189
Phosphate, 140
Phosphate bonds, 137, 140—142, 154
Phosphate transport, 141
Phospholipid-like material, 136—137, 139—140, 142
Phospholipid-like molecules, 129
Phosphorus, 144
Phosphorylation, 141, 143
Phosphoryl donor, 143
Photochemical reducing, 144
Photolithotrophic bacteria, 203
Photons, 83
Photosynthesis, 145, 204
Photosynthetic reaction, 147

Phylogenetic tree, 181, 193
Phylogeny, 89
Physical conservation laws, 158, 161—166, 169, 174, 197, 212, 214
Physical irreversibility, 4—5, 11
Physics of becoming 1, 25, 214
Physics of being, 1
Plasma membranes, 160
Plasticity of environment, 147
Point mutation, 93, 102—103, 206
Polarity, 163, 165, 168, 174
Polarized particle, 42—43
Polyamino acid, 105, 116
Polyarginine, 146
Polylysine, 145
Polymerase, 120
Polymeric information translation, 123
Polymerization, 96—97, 105, 116, 122—123, 131, 145
Polymers, 93, 122, 126, 150
Polynucleotide, 119—120, 122—123, 126, 131—132, 145, 149—150, 152, 154
Polynucleotide replication, 139, 150
Polypeptide, 119—123, 145, 149—150, 152, 154
Polypeptide replication, 139, 149—150
Polypeptide synthesis, 126
Polypeptide synthesis cycle, 152
Population, 181—184, 201, 206, 207, 209
Positive feedback, 96
Potassium, 130
Potency, 162, 166
Potential force field, 213
Prebiotic conditions, 143
Precambrian prokaryotic evolution, 203
Precells, 129, 135
Predation, 204
Prediction, 3—4
Prediciton/retrodiction asymmetry, 3—4
Preformationism, 161
Primitive Earth, see Earth
Primordial soup, 52, 62, 66, 68—69, 72, 98, 100, 136
Prions, 151
Probability amplitude, 41—42, 59
Probability distribution, 110, 114, 207
Process-description, 90
Process of, see specific type
Process of becoming adapted, 195—196, 198—199, 201, 207, 209
Process of becoming adjusted, 194
Process of becoming conserved, 194
Process of becoming fitted, 195—196
Process-simulation, 168
Production process, 93—98, 122, 124, 144
Production space, 106
Prokaryotes, 201, 203
Prokaryotic niche, 203
Prokaryotic oxidative phosphorylation, 203
Protein first hlypothesis, 145—146
Protein-nucleic acid-simultaneity hypothesis, 146
Proteinoid microspheres, 70—71, 136—138, 148
Proteinoids, 70, 98, 136, 148
Proteinous enzymes, 145
Proteins, 20, 100, 105, 110—111, 113—114, 118, 122, 145
Protein synthesis, 120—121, 123
Proterozoic era, 204
Protobiogenesis, 138
Protobiological evolution, 178
Protobiological information, 27
Protobiology, 1—29, see also specific topics
Protocells, 119, 129, 134—141, 148, 180
Protocellular reproduction, 139
Protocellular units, 139
Protohypercycle, 119—123, 150, 153
Proton flow, 136
Protonmotive force, 137, 140
Protons, 136, 140
Proton translocation, 129
Protoreproduction, 129—158
  ATP process, 140—143
  energy transduction, 140—143
  genetic code, origin of, 150—154
  molecular replication, 144—150
  multiplication of protocells, 134—140
  origins of, 129—134
Protoreproductive cellular units, 150
Protoreproductive process, 129, 149
Protoreproductive units, 129—130, 132—134, 138
Pseudogenes, 93, 102—103, 106
Pure states, 48, 73
Puridine nucleotides, 143
Purine bases, 143
Purines, 144—145
Purposiveness, 162
Pyrimidine bases, 143
Pyrimidines, 144—145
Pyroglutamyl-glycyltyrosine, 99
Pyroglutamyl-tyrosyl-glycine, 99

Q

Quantum coherence, 47—50, 53, 54, 73
Quantum-mechanical coherence, 42
Quantum-mechanical wave function, 14, 27, 31, 41—42, 45, 47, 59, 124
Quantum mechanics, 1—2, 5—6, 8, 11, 31, 57—58, 67—68, 72—73, 80—81, 87, 89, 94, 124, 211, 213
  classification, 41
  internal measurement, 49
  intrinsic irreversibility, 77
  measurement in, 41—53
  microscopic irreversibility, 89
  time's arrow, 88—90
Quantum mechanics of interaction, see Quantum mechanics
Quantum statistical mechanics, 47—49, 52, 65—66, 72—74, 77, 81
Quantum-statistical mixed states, 73

R

Radiation field, 82
Radiative atom 45

Random mutation, 102—103
Randomness, 110—111, 132, 178
Random polymerization, 109—111, 116
Random sampling, 109
Random variation, 98, 132—134, 178
Rate of reproduction, 177—178, 183, 187
Ratio of heat energy to temperature, 80
Reaction cycle, 122, 138—139, 149—150, 152—153
Reaction-diffusion equations, 168
Reaction kinetic equations of motion, 68
Reaction kinetics, 120
Reaction rate coefficients, 68, 119
Reactively active process, 179
Reactively active reproduction, 180
Reactively passive process, 179
Reactively passive reproduction, 180
Reading machine, 105
Recapitulation process, 123—124
Red Queen hypothesis, 192
Reducing gas, 144
Reductant, 137
Regional difference, 163, 168
Regionalization, 159
Regional specification, 157, 159—160, 163
Regulative principle, 28, 66—67, 69—71
Relativistic mechanics, 15, 41
Relaxation processes, 93—94, 96—98, 100
Relaxation time, 93—94, 96, 100—101
Replica layers, 130
Replicating molecules, 135
Replicating polymers, 129, 149
Replicating RNA, 188
Replication, 134, 139
Replication cycle, 145
Replication inhibitors, 134
Replication of DNA, 178
Replication of RNA, 178
Replication process, 146
Replication rate, 134
Replication systems, 146
Reproduction, 129, 138, 180
Reproductive dynamics, 204
Reproductive fidelity, 130—134, 145, 154, 177—182, 197—198
Reproductive interaction, 202
Reproductive isolation, 203
Reproductive molecules, 129
Reproductive protocells, 180
Reproductive species, 184, 201—203
Reproductive unit, 132, 134, 138, 157, 177, 178, 181, 188—198
Residue-by-residue replication, 131, 150, 154, 180
Resistive force, 16
Resource compensation, 198
Resource exploitation, 186—189, 191, 197—198, 201, 203, 205—206, 209
Resource limitation, 185—189, 191—194, 198
Respiration, 145, 204
Retrodiction, 3—4
Ribosomal proteins, 153
Ribosomal RNA, 153
RNA, 134
RNA-polymerases, 153
RNA Qβ-replicase, 134

S

Sarcomeric structure, 20
Schrödinger's cat, 45, 49
Schrödinger's equation of motion, 1—2, 11—12, 14, 27, 31, 41—42, 46, 50—52, 54, 68, 106
Sea urchin, 25
Second law of mechanics, 12, 17, 58—60, 139, 169
Second law of thermodynamics, 4—5, 74, 77—79, 83, 90
Selection, 67, 110—111
Selection process, 31
Selective neutrality, 102
Selective sieve, 182—183
Self-assembly, 93, 103—104, 106
Self-constraining actualization, 10—11
Self-constraint, 113, 118, 129, 138
Self-limiting constraints, 133
Self-sequencing, 150
Sense strand, 102
Separability, 212, see also Boundary conditions; Law of motion
Sequence from product to production, 94, 97—98, 100—101, 106, 113, 116, 125, 133—134
Sex, emergence of, 192—193
Sexual reproduction, 192—193
Shannon's information, 109, 113
Side-chain specificity, 150
Silicate layers, 129—131
Sliding force, 16
Sodium, 130
Solation, 171
Some, 172, 189
Somatic cell divisions, 167
Sommerfeld's condition, 87
Sound waves, 8
Source matrix of information, 110—114, 116, 121—122, 126
Spatiotemporally localized hierarchy, 182
Speciation, 179, 181, 184, 187, 202—203
Specificity, 149, 157
Spermatozoa, 16, 19, 23—25, 31
Spillover, 18
Spin, 42—44
Spin-wave function, 43
Spontaneously active process, 179
Stability, 79
Starfish, 16, 19, 23—24, 31
State-description, 90
State functions, 79, 90
State of being adapted, 195—196, 212
State of being adjusted, 194
State of being conserved, 194, 211—212
State of being fitted, 196
Stationary ensemble, 81—82
Statistical ensemble, 48, 65—66
Statistical entropy, 81, 83—87, 90

Statistical fluctuations, 74
Statistical irreversibility, 83
Statistical-mechanical entropy, 27
Statistical mechanics, 5—6, 14—15, 27, 35, 132
Statistical variance, 2
Stem cells, 172
Stochastic process, 207, 211
Successive feedback, 114, 129, 138, 144
Sugars, 144
Sun, 147
Symmetry-breaking, 88, 93—107, 109, 118, 138—139, 146, 149—154, 160, 168—174
  biological, 101—105
  complexity and complexity increase, 105—106
  developmental, 168—169
  perturbations, 169
  points, 66
  production process, 93—98
  protobiological, 98—101
Symmetry of time, 11
Symmetry-preserving, 106, 168
Synchronic process, 173—174
Synchronic ubiquity, 173

T

Teleology, 105, 157, 162, 197
Template, 97, 117, 131, 138, 178
Template degraders, 134
Temporally asymmetric law of motion, 89—90
Terrestrial biological evolution, 147
Thermal agitation, 62
Thermal copolymers, 115
Thermal equilibrium, 132
Thermal motion, 100
Thermal polyamino acids, 63, 136
Thermal polymerization, 93, 98—100, 106, 114, 122—123, 126, 136
Thermal polymers, 101
Thermal proteins, 70, 98, 114—121, 148, 150
Thermodynamic constraints, 68—69
Thermodynamic entropy, 5, 85, 89
Thermodynamic equilibrium, 77, 85
Thermodynamic fluctuation, 93
Thermodynamic potentials, 93, 100
Thermodynamic relaxation processes, 93—94
Thermodynamics, 4—5, 77, 80, 143
Thermodynamic stability, 93, 100
Thermodynamic time's arrow, 85—87, 90
Thermodynamic variables, 68—69, 71, 79—80
Third law of mechancis, 11—12, 15—18, 25—28, 57—61, 67, 139, 141, 168, 170
Thymine, 102
Time-extensive genealogical hierarchy, 208
Time-extensive hierarchy, 182, 187
Time-reversal law of motion, 87
Time-reversal symmetry, 88, 101
Time's arrow, 11, 77, 85—90
Tissue, 159—161
Trajectories, 1, 5, 10—11, 27, 31—36, 68, 80, 211
Transcription, 117, 122—124
Transfer RNA, 153
Transformation process, 167
Transitional mutation, 102—103
Transition probability rates, 207
Transmembrane electrogenic carrier, 143
Transposable genic units, 167, 174
Transposons, 192
Transversal mutations, 102—103
Trimers, 96—97
Tripeptide, 98
Tripeptide isomers, 99
tRNA-synthetases, 153
Tryptophan, 115, 151—152
Tubulin, 25
Tyrosine, 98—99, 151—152

U

Unfolding, 194

V

Variability, 40, 124, 167—168, 173, 177, 187—194, 201, 205
Variations, 98, 101—102, 104, 106, 157, 180
Venus, 147—148
Vesicles, 136—137, 142

W

Wave function, 2, 41—42, 45, 47, 49, 51, 59, 72
Wave mechanics, 87—88
Weismannian separation between soma and germ cells, 189

X

X-particle, 88

Y

Yolk mass, 169

Z

Zero-point motion, 2
Zinc, 131